QUANTITATIVE REASONING

QUANTITATIVE REASONING

TOOLS FOR TODAY'S INFORMED CITIZEN

ALICIA SEVILLA & KAY SOMERS

Copyright © 2007 by John Wiley & Sons, Inc.

All rights reserved.

No part of this publication may be reproduced, stored in a retrieval system or transmitted in any form or by any means, electronic, mechanical, photocopying, recording, scanning or otherwise, except as permitted under Sections 107 or 108 of the 1976 United States Copyright Act, without either the prior written permission of the Publisher, or authorization through payment of the appropriate per-copy fee to the Copyright Clearance Center, 222 Rosewood Drive, Danvers, MA 01923, (978) 750-8400, fax (978) 750-4470 or on the web at www.copyright.com. Requests to the Publisher for permission should be addressed to the Permissions Department, John Wiley & Sons, Inc., 111 River Street, Hoboken, NJ 07030-5774, (201)748-6011, fax (201)748-6008, or online at http://www.wiley.com/go/permissions.

To order books or for customer service, please call 1(800)-CALL-WILEY (225-5945).

Printed in the United States of America.

ISBN-13 978- 0-470-41232-9

10 9 8 7 6 5 4 3 2

Contents

Section III: Statistical Reasoning

Activities

Resources on the Student CD

Annotated Contents

Section I: Numerical Reasoning

This section of the text explores topics related to pictorial representations of data and relationships and using data, numbers, and functions to help solve problems.

Graphs allow us to absorb quantitative information quickly and accurately and have become an important part of daily communication. Today's informed citizen must have not only the ability to read and understand graphs, but also the ability to create them. In this topic, we analyze different types of graphs routinely used to convey information and tell a story. We discuss examples of bar graphs, pie charts, histograms, and stemplots (stem-and-leaf graphs), and which graphs are appropriate to use when working with categorical variables and which are best to represent quantitative variables.

This topic investigates scatterplots of data sets that involve two quantitative variables: an explanatory or independent variable and a response or dependent variable. We discuss the difference between a relation and a function and also explore the importance of representing relationships in multiple ways—using tables of numerical data, symbols, graphs or charts, and words. Finally, we look at directly proportional relationships. We use SAT grades by state, calories expended for an activity, and sports data to explore these ideas.

Topic 3: Graphs of Functions 45

In this topic, we discuss characteristics of graphs of functions and how to interpret these different characteristics in the context of the function variables. This topic includes the vertical line test for graphs of functions and discussion of increasing and decreasing functions, concavity, and absolute and relative maximum and minimum. In addition, we introduce the concept of average rate of change and the relationship between concavity and variations in the rate of change of a function. We explore these ideas using federal budgets, number of students enrolled in private and public schools, and salaries of baseball players.

Topic 4: Multiple Variable Functions 69

In this topic, we examine situations where the response (dependent) variable depends on several explanatory (independent) variables. We continue to emphasize different modes of communicating relationships between variables—words, tables, symbols, and graphs. We also investigate how a response variable behaves when all but one independent variable is held constant. Contexts for these investigations include basal metabolic rate, body mass index, wind chill equivalent temperature, and blood alcohol level.

Topic 5: Proportional, Linear, and Piecewise Linear Functions 85

We review the concept of proportional relationships and discuss linear and piecewise linear functions and their equations. Examples include basal metabolic rate, medicine dosage, income tax, and the price of letter postage. We explore the significance of the y-intercept and the slope of a linear function using physical exercise–related data. We also discuss the concept of inversely proportional functions using data on the average speed of winners of the Daytona 500 auto race for several years.

Topic 6: Modeling with Linear and Exponential Functions 103

This topic begins with a discussion of how to recognize a linear relationship between two quantitative variables. We explore how and when to fit a least-squares regression line to data that are not exactly linear. Exponential relationships and exponential growth are compared to linear growth, and we look at an example in which growth is even more rapid than exponential growth. We study these ideas using population data from around the world and financial data such as salaries and federal debt over time.

Topic 7: Logarithms and Scientific Notation 119

This topic introduces the common logarithm and the exponential function, base 10, using the decibel scale for sound and the Richter scale for earthquake intensity. We explore the relationship between the graph of a function and the graph of its inverse function and review basic properties of logarithms. We also discuss the use of scientific notation to make estimates when very large or very small quantities are involved.

Topic 8: Indexes and Ratings 133

In this topic, we look at indexes, such as the Consumer Price Index and the Consumer Confidence Index, to investigate trends over time. We also examine the Fog Index, which assesses the reading difficulty of a passage of text. Rating systems that are set up to compare people or places are examined. These indexes are used to help understand trends in minimum wage data, the rise in sporting-event costs, and the readability of newspaper editorials, while rating systems are applied to cities and colleges.

Topic 9: Personal Finances 151

In this topic, we introduce terminology associated with personal financial management and discuss simple and compound interest. We analyze and use formulas for computing interest received on savings and calculate costs and payments associated with various types of loans and annuities. In particular, we consider examples that relate to saving for a down payment on a home and paying credit card debts.

Topic 10: Introduction to Problem Solving 169

In this topic, we analyze basic techniques for problem solving and identify where we have used them in the previous topics. We apply these techniques to revisit the problem of finding a person's body mass index. We also use these techniques to analyze two heating oil purchase plans and several car loan options. We consider examples in which multiple problem-solving techniques are used in tandem and investigate a credit card example.

Section II: Logical Reasoning

This section investigates inductive and deductive reasoning approaches and applies these ideas to decision making, apportionment issues, and problem-solving techniques.

Topic 11: Decision Making **185**

This topic focuses on methods for helping us make decisions in which information related to the decision is known. (These types of decisions are called decisions under certainty.) We discuss criteria that impact various decisions as well as two methods for making decisions: the cutoff screening and weighted sum methods. These methods are applied to decisions such as purchasing a digital camera and computer or deciding on which job to accept.

Topic 12: Inductive Reasoning **201**

This topic starts with a discussion of the difference between deductive and inductive reasoning. We then explore various forms of inductive reasoning—prediction, generalization, causal inference, and analogy—using a variety of examples taken from newspaper articles and other readings. We present a variety of scenarios to help the student distinguish between correct and incorrect inductive reasoning.

Topic 13: Deductive Reasoning **215**

In this topic, we focus on the basic elements of deductive reasoning. We discuss statements and their negation as well as how to formulate and analyze compound statements and their negations. The contrapositive and converse of conditional statements and quantified statements are examined, leading up to deductive arguments. We investigate these ideas using published speeches, newspaper articles, and advertising statements.

Topic 14: Apportionment **239**

In this topic, we discuss different methods of apportionment that the House of Representatives has used or that were once proposed. We investigate two quota methods: Hamilton's method and Lowndes' method. We also discuss the following divisor methods: Jefferson's method, Adam's method, Webster's method, and the Huntington-Hill method, which is the method used currently to determine the number of representatives for each state based on the population numbers determined by the most recent census. We explore all these methods using data from the 2000 census.

Topic 15: More on Problem Solving 255

This topic discusses more problem-solving techniques to add to those introduced in Topic 10 and then links these techniques with deductive and inductive reasoning. We identify how these techniques were used in previous topics and apply them to solve problems such as a salary negotiation and the best payment option to select after winning the lottery.

Section III: Statistical Reasoning

This section introduces basic concepts of probability and statistics and applies them to the study of sampling and surveys and making decisions that involve uncertain data.

Topic 16: Averages and Five-Number Summary 265

In this topic, we investigate several measures of center and spread. We discuss the concepts of mean, median, mode, quartiles, range, and interquartile range. We also calculate the five-number summary and graph the boxplot for data sets. In exploring these concepts we use data sets on number of waste sites by state, state governors' salaries, and calorie content in popular brands of brownies and ice cream bars, among others.

Topic 17: Standard Deviation, z-Score, and Normal Distributions 287

In this topic, we investigate the standard deviation as a measure of variability within a data set. We look at normal curves and compare normal curves with different means and standard deviations. We consider standardized z-scores as a way to compare values obtained from data sets with different units of measure. We apply these ideas by looking at various data sets, such as calorie content of desserts, SAT scores, and number of home runs.

Topic 18: Basics of Probability 309

We introduce basic probability concepts—random process, sample space, outcomes and events, relative frequency, probability of an event—as well as basic probability rules. We explore these ideas using coins, dice, and playing cards. We also use data on single vehicle crashes by size of vehicle and compute the probability of relevant events by computing relative frequencies.

Topic 19: Conditional Probability and Tables 325

We explore how to use a two-way table to represent data in which each individual in the data set is characterized in two different ways. We discuss how to analyze this data and look for relationships using conditional probabilities. We examine how identifying independent events helps us discover additional relationships. These ideas are investigated using Olympic gold medal data, vehicle crash data, and congressional voting records.

Topic 20: Sampling and Surveys 343

Basic components of observational studies and experiments are investigated in this topic. We examine how to understand the results of a study or experiment by identifying the sample, the population, and the relevant variables. We discuss various sampling methods and explore sampling variability and biases. We also discuss one method to elicit honest responses to sensitive survey questions. These ideas are explored through studies reported in newspapers, periodicals, and on the Internet.

Topic 21: More on Decision Making 363

In this topic, we discuss the concept of expected value (or mean or average value) of a probability distribution and how to use it to make decisions that involve uncertain information. We also examine several other approaches to decision making when uncertainty is present. To explore these ideas, we analyze decision-making problems regarding accepting a magazine raffle offer, choosing a health insurance alternative, and deciding on an investment approach.

Activities

Resources on the Student CD

Activities Contents

Graphing calculator versions of the Excel Activities are included on the Student CD packaged with the text.

Preface

Empowering students to use quantitative information to make responsible financial, environmental, and health-related decisions in their daily lives is at the core of *Quantitative Reasoning: Tools for Today's Informed Citizen*. This book's main objective is to help students become better critical thinkers by engaging them as active learners of the quantitative methods of analysis discussed in the text. Through numerous examples, explorations, and activities featuring real data, students develop the skills necessary to

» Identify, analyze, and solve real-world problems that involve quantitative information

» Reason quantitatively and make numerical arguments

» Interpret and communicate the results of quantitative analyses

» Use technology and Internet resources effectively and build skills in working with data

» Develop and improve "numerical intuition" and confidence in the ability to engage in quantitative thinking

Quantitative Reasoning: Tools for Today's Informed Citizen is organized into three sections. Section I, "Numerical Reasoning," provides a foundation for quantitative reasoning and communication. It includes topics related to using numbers, functions, and graphs and an introduction to problem solving. Section II, "Logical Reasoning," addresses different types of reasoning and applications and concludes with a further discussion of problem-solving techniques. Section III, "Statistical Reasoning," includes investigations of descriptive statistics, probability, and sampling. It concludes with a discussion of decision making when outcomes are uncertain.

Throughout the text, students use a variety of methods of analysis: inductive and deductive reasoning; tabular, symbolic, verbal, and graphical forms of functions and relations; graphs and pictorial representations of data; interpretations of probabilistic data; surveys and statistical studies. Nearly all of the examples, exercises, and activities in the text use real data or draw on real-life situations to demonstrate the significance of quantitative reasoning in students' daily lives and to illustrate misapplications of mathematics and quantitative reasoning. The use of real data highlights the relevance and practicality of the material. In this way, students gain a better understanding of the concepts. This text shows over and over again how useful and relevant mathematics is for understanding the world we live in and for making informed decisions based on the proper use of quantitative information and reasoning.

Key Features

Each of the three sections of the text consists of **topics** that introduce mathematical concepts and terminology necessary to explore different approaches to solving the problems presented. These topics can be investigated typically in one or two class periods and include the following features.

Objectives are outlined at the beginning of each topic so that students can preview the ideas they will explore.

Objectives preview the ideas students will explore in each Topic.

OBJECTIVES

After completing this topic, you will be able to

» Analyze trends in several commonly used indexes

» Use and calculate indexes to understand and compare data

» Recognize rating systems as a type of indexing system and investigate what might go into setting up a rating system

Quantitative indexes and rating systems are used to give information about general trends and to allow us to make comparisons and judgments. We'll examine some frequently used indexes and rating systems and look at how to use them and what goes into setting them up.

The **Dow Jones Industrial Average (DJIA)** is a well-publicized index that reflects the value of stock prices. The DJIA includes 30 stocks that represent a variety of industries—financial, food, technology, retail, heavy equipment, oil, chemical, pharmaceutical, consumer goods, and entertainment. The DJIA is not a simple average but is adjusted to take into account the changes in price associated with stock splits in each of the included companies. The average is calculated by summing the prices of the 30 stocks and then dividing by a constant called the divisor

Examples are integral to illustrating the concepts and tools introduced and showing how they can be applied to real-life situations using actual data. Worked-out solutions to the examples explain how to solve a problem and what the solution means in the context of the problem.

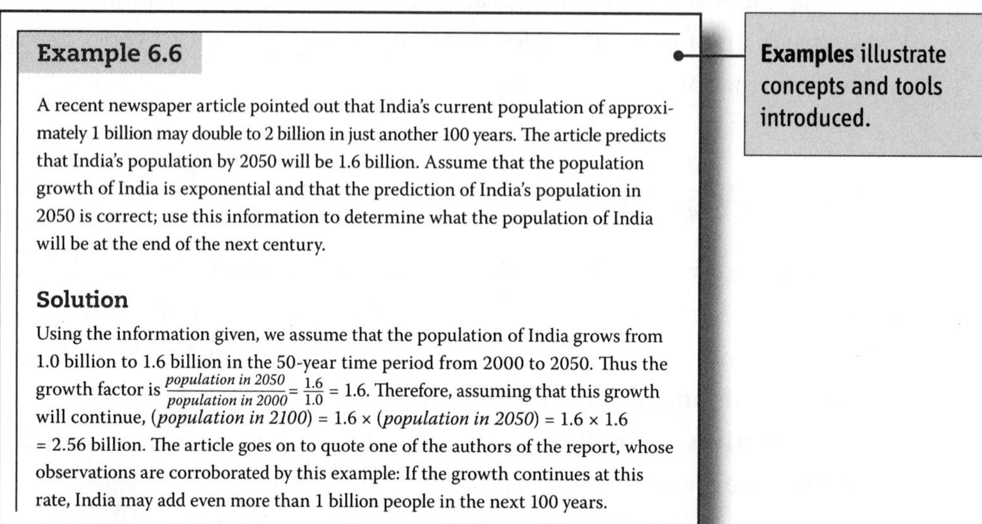

Example 6.6

A recent newspaper article pointed out that India's current population of approximately 1 billion may double to 2 billion in just another 100 years. The article predicts that India's population by 2050 will be 1.6 billion. Assume that the population growth of India is exponential and that the prediction of India's population in 2050 is correct; use this information to determine what the population of India will be at the end of the next century.

Solution

Using the information given, we assume that the population of India grows from 1.0 billion to 1.6 billion in the 50-year time period from 2000 to 2050. Thus the growth factor is $\frac{\textit{population in 2050}}{\textit{population in 2000}} = \frac{1.6}{1.0} = 1.6$. Therefore, assuming that this growth will continue, (*population in 2100*) = 1.6 × (*population in 2050*) = 1.6 × 1.6 = 2.56 billion. The article goes on to quote one of the authors of the report, whose observations are corroborated by this example: If the growth continues at this rate, India may add even more than 1 billion people in the next 100 years.

Examples illustrate concepts and tools introduced.

A **summary** of the topic reviews the concepts that are critical in helping students master the objectives introduced at the beginning of the topic.

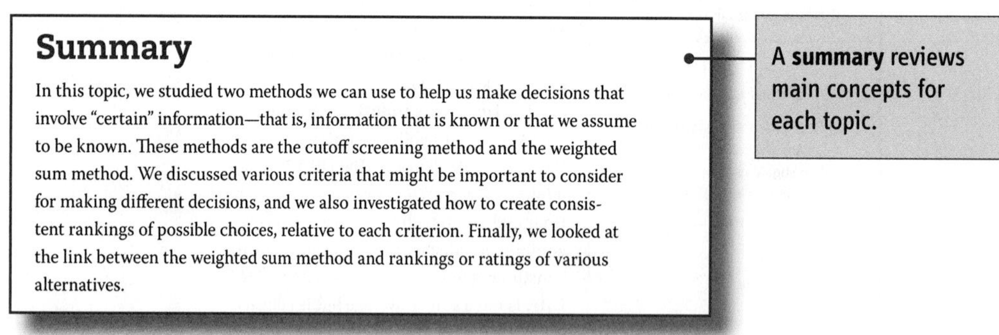

Summary

In this topic, we studied two methods we can use to help us make decisions that involve "certain" information—that is, information that is known or that we assume to be known. These methods are the cutoff screening method and the weighted sum method. We discussed various criteria that might be important to consider for making different decisions, and we also investigated how to create consistent rankings of possible choices, relative to each criterion. Finally, we looked at the link between the weighted sum method and rankings or ratings of various alternatives.

A **summary** reviews main concepts for each topic.

Explorations give students a chance to apply their understanding of the main concepts to additional real-life situations. These explorations allow students to broaden their problem-solving skills and their understanding of the mathematics involved, thus enabling them to see new contexts for the applications discussed.

Explorations

1. In each of the following situations, identify at least three explanatory variables that influence the given response variable. For each explanatory variable, identify if it is quantitative or categorical.
   a. The amount of time a student spends writing a research paper for a particular course in which he or she is enrolled
   b. A one-month electric bill for a family's residence
   c. The amount of profit made by a business during a one-year period
2. The estimated number of calories burned in one day also depends on a person's activity level. If a person is sedentary, his or her activity factor is approximately 1.3; if moderately active, the activity factor is 1.4; if very active, the activity factor is 1.7. Then,

   Total calories needed in one day = BMR * activity factor

   Find the daily calorie needs for the woman and man described in Example 4.2 if they are sedentary. By how much will their calorie needs change if they are moderately active? By how much will their calorie needs change if they are very active?
3. One way to find out whether your weight is "reasonable" for your height is to calculate your body mass index (BMI). A BMI of 20 to 25 is normal. BMI is a function of height and weight and is determined as follows: To calculate a person's BMI, divide his or her weight in kilograms by the square of his or her height in meters. (To convert into kilograms, multiply pounds by 0.4536; to convert into meters, multiply inches by 0.0254.) Thus, you can write an equation to calculate BMI:

$$\text{BMI} = \frac{0.4536 * \textit{wt. in pounds}}{(0.0254 * \textit{ht. in inches})^2}$$

   a. Describe how this equation can be used to calculate BMI for a friend who does not understand the description of BMI.
   b. Abraham Lincoln was 6 feet 4 inches in height and weighed about 180 pounds, according to one historical account. From pictures you have seen of Abraham Lincoln, do you think his BMI would fall in the normal range? Calculate Lincoln's BMI.

Explorations enhance students' understanding of the main concepts and broaden problem-solving skills.

Activities engage students as critical thinkers by investigating real-life situations using the concepts learned in greater depth. Each of the activities is designed so that the bulk of the activity can be completed by most students in a typical class period, with some time remaining for a review of background material and wrap-up. Activities can also be assigned as out-of-class work if desired.

ACTIVITY 20.1

Sampling and Surveys

Activities complement the material in topics and encourage critical thinking.

In this activity, you will investiga
also consider some issues involve

1. Here is a list of the 50 United

Alabama	Hawaii	Mas
Alaska	Idaho	Mich
Arizona	Illinois	Minn
Arkansas	Indiana	Miss
California	Iowa	Miss
Colorado	Kansas	Mon
Connecticut	Kentucky	Nebr
Delaware	Louisiana	Neva
Florida	Maine	New
Georgia	Maryland	New

a. Using your sense of the land area of each state, choose what *you think* is a representative (that is, your *subjective*) sample of six states. List them in the following table. (It might help to try to visualize a map of the U.S. to get your sample of six "representative" states.)

State	
1.	
2.	
3.	
4.	
5.	
6.	

b. Refer to the table at the end of this activity and record the land area for each of your chosen states, in square miles given to the nearest integer, in the second column of the table. (Source: *Department of Commerce, Bureau of the Census,* www.infoplease.com/ipa/A018355.html.)

c. Compute the sample mean land area for your *subjective* sample of six states and record it here: ________________.

d. Now use Excel's random number generator to generate six random integers between 00 and 49. (Recall how you generated random integers 0, 1, 2, and 3 in Activity 18.1 and adapt that technique to generate integers between 00 and 49. It is possible that you might get repeated numbers when you generate the random integers. If you do, discard the second occurrence of the repeated number and generate another integer so you

Technology

Students are encouraged to use technology in a fundamental way to help visualize data and to facilitate calculations. Because most students have access to Microsoft Excel and will use it in their work environment, the activities in the text feature Excel. Data sets to accompany the Excel Activities are included on the **Student CD** packaged with the text, as well as at the web-based **Resource Center** for the text, which can be accessed at **www.wiley.com/college.** For instructors who prefer that students use a graphing calculator, a version of the activities featuring the TI-83 or TI-84 Plus graphing calculator, instead of Excel, are available on the Student CD. Instructions for using Microsoft Excel and graphing calculator technology are integrated into the activities carefully so that students can concentrate on ideas rather than computational details when investigating problems. An index of Excel Commands by Activity and Graphing Calculator Commands by Activity are also provided in the back of the text and on the Student CD, respectively, as a quick reference tool for locating specific technology functions.

To the Student

Welcome to an experience that will help you better understand mathematical concepts and tools and the quantitative information you need to make informed decisions in our changing world. This book was developed with you in mind—to help you understand and solve problems that are relevant to your family, your community, your workplace, your country, and your world. Because all of the concepts introduced in this book are anchored in real-world situations, you will never need to ask, "What is this good for?"

The key to your success in this course is active involvement when reading and solving problems. You should read and work on examples for a particular topic before the topic is visited in class; by doing so, you will be ready with questions for your instructor and for participation in class discussion. Be an active reader, and read with pencil and paper close by to check computations, record questions, and provide alternative answers. You will find that some of the questions in this book have more than one good (or "right") answer. Just as you become proficient at a sport by practicing, you become skilled at problem solving by solving a lot of problems. You might even discover that problem solving can be just as much fun as tennis or golf or basketball!

The **examples** in this text are essential in demonstrating how mathematical concepts and tools are used in the real world. Read each example and try to come up with your own solution first. Then work through the given solution carefully, following each step in detail. Analyze the graphs and charts and think about different ways to visualize information and problems. Use technology tools, such as Microsoft Excel or a graphing calculator, to help create additional graphs. Ask yourself whether the solution makes sense; that is, ask whether or not the solution is reasonable in the context of the problem. Discuss the examples and your findings with your classmates and your instructor. Think about related questions you might ask to better understand the topic and other situations in which you might be able to apply what you have learned.

Jump in and work on the **explorations** at the end of each topic. The issues and problems covered in the explorations build on those in the topic and include additional, related concepts. In the explorations, you will bring ideas and tools together to investigate new examples and practice your problem-solving and analysis skills. Your explanations will help you better understand the concepts discussed in the topic and make connections with those covered in previous topics.

The **activities** build on the readings, examples, and problems presented in the explorations. The activities generally focus on an interesting application of the material in the topic in more depth. By completing the activities, you will better understand the topic. Because detailed instructions on how to use technology to help create graphs and perform calculations are included in the activities, you will also gain facility in using this technology.

You will gain experience and learn new techniques and different ways of looking at problems as you solve interesting and relevant problems in which quantitative information and techniques play an important role.

Student Support Materials

The **Student CD** packaged in the back of your text contains the following resources to assist you in your studies:

» Excel data bank to accompany Excel Activities

» Selected answers to explorations

» Graphing Calculator Activities

» Graphing Calculator Commands by Activity

These resources, and more, can also be accessed at the web-based **Resource Center** for this text at **www.wiley.com/college.**

To the Instructor

This textbook grew out of notes written for a course we have been teaching at Moravian College since the year 2000. This course, developed with a grant from the National Science Foundation (Grant No. 9950229), has been very successful with students who liked mathematics before coming to college and feel comfortable with quantitative analysis, as well as with students who declare on the first day of class, “I cannot do math.”

In developing this course and writing the text, we were especially conscious of the need to reach all types of students, especially this last group. Our philosophy in developing this course was to help students see the need for what they are learning; thus the emphasis of this text is to teach applications using real data. As instructors, our objective is to facilitate the learning process and support students as they learn, and with this objective in mind, we developed an activity-based course that combines class instruction with in-class student activities. This approach engages the student as an active learner and makes the classroom atmosphere much more enjoyable for both students and instructors.

To prepare students for the class in which they will be investigating a topic, we suggest assigning a section of reading from the book (sometimes a full topic, sometimes a portion of a topic) to be completed along with some **explorations** from the same topic. We use time at the beginning of the class to discuss the assigned reading material and explorations. Our students then spend the majority of the class period working on the carefully constructed **activities** for each topic, which build on and complement the assigned reading. For courses with a shorter class period, activities can also be completed during two class periods or assigned as homework. Some activities require students to work together, and we always encourage students to discuss their progress and results with one another.

Because we want students to feel comfortable with technology, and not threatened by it, we provide carefully written instructions on using Excel, or the graphing calculator. Generally, we introduce only the necessary functions in each activity so as to not overwhelm the students with the technology and cause them to lose track of the main objective of thinking critically about the activity.

We strongly recommend our activity-based approach, but this text can also be used in a more traditional setting. For example, an instructor can give an interactive lecture on a particular topic using some of the examples, explorations, and/or activities to illustrate the concepts. Additional explorations and activities can be assigned to be completed out-of-class.

The **Student CD** includes numerical answers to explorations, but it does not include solutions to explorations that require explanations or written interpretations, nor does it provide solutions to activities. It is essential that students reach their own conclusions and show their thinking processes. The **Instructor Resources**, available in both print and electronic form, contain complete solutions to explorations and activities, suggestions for teaching each topic, and information on typical student errors and misconceptions. Furthermore, suggestions for designing a course syllabus and sample questions for quizzes and exams are provided in the Instructor Resources.

Because the emphasis is on the student as an active learner and the context of each example is real, students become engaged with the course and the material quickly. Students will see that quantitative reasoning is a necessary skill in order to be an informed and productive citizen.

Instructor Support Materials

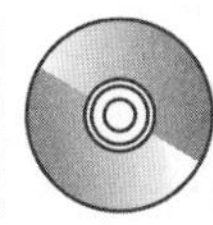

The **Instructor Resources**, available in print and on CD, contain the following resources to assist you in preparing for class quickly and effectively:

» Teaching ideas for each topic and accompanying activities

» Complete solutions to explorations, Excel Activities, and Graphing Calculator Activities

» Sample course syllabi

» Sample questions for quizzes and exams

These resources, and more, can also be accessed at the web-based **Resource Center** for this text at **www.wiley.com/college.**

Acknowledgments

The authors and publisher wish to thank the following reviewers for their outstanding and invaluable advice on guiding this text to its successful completion:

Ron Barnes, University of Houston, Downtown Campus

Linda E. Barton, Ball State University, Indiana

Pamela B. Cohen, Southern New Hampshire University

Larry Cusick, California State University, Fresno

Caren L. Diefenderfer, Hollins University, Virginia

Carrie Muir, University of Colorado, Boulder

Jessica Polito, Wellesley College, Massachusetts

Juli D'Ann Ratheal, West Texas A&M University

Mariano Rodrigues, Rhode Island College

Allan Rossman, California Polytechnic State University, San Luis Obispo

Terry Tolle, Southwestern Community College, North Carolina

And the following consultants from whose expertise we benefited:

Christian Aviles-Scott, Consultant in Educational Mathematics

Marcy Lunetta, Permissions Consultant

We gratefully acknowledge the contributions of Allan Rossman, an author of Key's *Workshop Statistics* series. His vision and preliminary work on a Workshop Quantitative Reasoning project influenced the development of our text. We also thank the National Science Foundation (Grant No. 9950229) for support that helped us create the course and materials from which this textbook grew.

We especially thank our students at Moravian College for their support and feedback as they learned from early versions of the text. Finally, we thank our families for their continuing love and encouragement, their infinite patience, and most of all, for maintaining a sense of humor.

Topics

TOPIC 1

Organizing Information Pictorially Using Charts and Graphs

OBJECTIVES

After completing this topic, you will be able to

- Distinguish between quantitative and categorical variables
- Draw bar graphs and pie charts, and interpret them in the context of the data they represent
- Compute percentage of the whole and percent change
- Interpret stemplots and histograms and group data to create them
- Decide when each type of graph is appropriate

A picture can tell a story faster than many words. Newspapers, magazines, books, and television news all use charts and graphs to present information and help the reader or viewer understand articles that contain numerical data. Because technology makes it possible for us to make graphs and charts easily, graphic representations are used frequently. In this topic, we discuss four types of graphs: bar graphs, pie charts, histograms, and stemplots (stem-and-leaf graphs).

When trying to understand some phenomena or to make sense of the relationship between two or more factors, pictures help us to see patterns, identify relationships, and describe main ideas. Graphs and charts can show patterns and trends not readily evident in the raw data. We will investigate **variables** and the various ways to represent them pictorially.

A **variable** is a characteristic of an object or a person (sometimes called an **individual** or a **case**) that can change from one object or person to the next. If the variable is assigned a number, then it is a **quantitative variable**. If it is assigned a category, then it is called a **categorical variable**. In the following example, we identify cases and variables, and decide whether each variable is quantitative or categorical.

Example 1.1

Suppose the individuals in our data set are network television specials. The following table gives the audience rating (the percentage of all U.S. households with TV sets who watch the particular show) and the network for the ten top-rated television specials in the 2002–2003 television season. (Source: *Time Almanac 2004*, page 95.)

Rank	Program Name	Network	Audience Rating (%)
1	*Academy Awards*	ABC	20.4
2	*American Idol* Special	Fox	17.2
3	*20/20* Special (2/6)	ABC	16.8
4	*Joe Millionaire* Special	Fox	16.6
5	*Everybody Loves Raymond* Special	CBS	15.3
6	*Grammy Awards*	CBS	14.7
7	*Will & Grace* Clip Show	NBC	14.5
8	*My Big Fat Greek Life* Special	CBS	14.3
9	*Primetime* Special Ed (12/4)	ABC	13.8
10	*Friends* Special (2/6)	NBC	13.6

a. Identify the cases, and whether the following variables are quantitative or categorical: rank, network, audience rating.

b. Why would we want to classify variables as quantitative or categorical?

Solution

a. The cases are the programs. Each case (or program) has the three characteristics of rank, network, and audience rating associated with it. When data are given in a table format, each row represents a case and each column represents a different variable. The variables of rank and audience rating are quantitative variables; the network is a categorical variable.

b. These classifications can help us determine an appropriate way to present the information graphically.

Bar Graphs and Pie Charts

A **bar graph** and a **pie chart** are two ways of representing categorical variables pictorially. (Note that the terms *graph* and *chart* are often used interchangeably.) These tools are also used to represent quantitative variables when the numbers fall into only a few categories. Because many people will look only at the graph in a news article (and not read the whole write-up), bar graphs and pie charts should be labeled so they are easy to understand. On the other hand, the graph should not be too cluttered with words and other symbols that mask the basic point. When examining a set of data, we will sometimes want to look at one variable at a time; other times we will want to study relationships between two or more variables. Next, we look at some examples of bar graphs and pie charts, and think about what information they give us.

Example 1.2

The following bar graph and pie charts accompanied the article "A Movable Epidemic," which appeared in *The New York Times* on September 9, 1999.

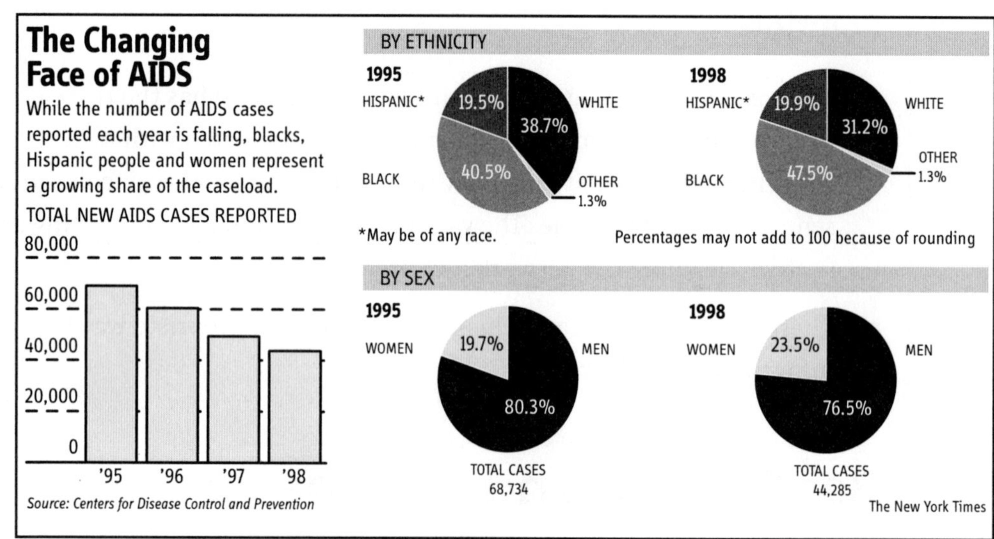

Explain what the graph and charts show about the "movable" AIDS epidemic.

Solution

The bar graph shows how the total number of new AIDS cases reported has fallen fairly steadily over the years from 1995 to 1998, from close to 70,000 to slightly more than 40,000. The pie charts tell us that the percentage of new cases among Hispanics has remained relatively stable from 1995 to 1998, while the percentage of new cases involving white patients has decreased, and those involving black patients has increased. Similarly, the percentage of new cases among women has increased while men represent a smaller percentage of the new cases.

When constructing a bar graph or pie chart for a categorical variable or a quantitative variable for which the numbers fall into one of several categories, we first need to decide what the groupings or categories will be. We then determine how many cases fall into each of the categories. If we are creating a bar graph, we can represent the height of the bars either as the total number of cases that fall into each category (as the graph in Example 1.2 does, showing total new AIDS cases reported) or as the proportion or percentage of the total number of cases that fall into that category. Which option we choose (total number of cases, proportion of total, or percentage of total) will determine how the vertical axis of the graph is labeled, but the form of the bar graph and relative heights of the bars will be the same for all three options.

Example 1.3

The following graph appeared in *The New York Times* on October 25, 1999, and accompanied the article titled "You've Got Mail, Indeed" that made the point that "despite the Internet direct-mail marketers grow even more prolific."

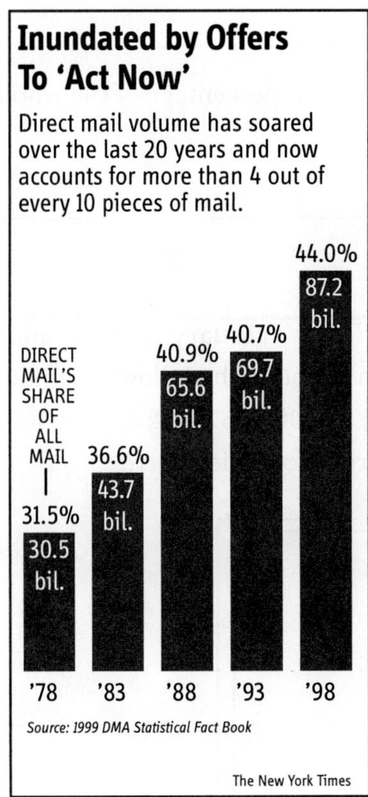

Explain what the labels on this graph represent; specifically, why is the bar for '88 labeled 65.6 bil. and 40.9% while the bar for '93 is labeled 69.7 bil. and 40.7%?

Solution

The height of the bar for each of the years 1978, 1983, 1988, 1993, and 1998 gives the number of pieces of "junk mail" for each of those years. The percentage given for each year is the percentage of all pieces of mail handled during that year by the post office that are "junk mail." In 1988, the number of direct mail pieces handled by the post office was 65.6 billion, which was 40.9 percent of the total number of pieces handled that year. In 1993, the number of direct mail pieces handled by the post office was 69.7 billion. This is over 2 billion more than in 1988 but represents only 40.7 percent of that year's total mail. So the proportion of all mail that was "junk mail" was larger in 1988 than in 1993. When giving percentages, we need to make sure we know what the percentage represents.

Pie charts are useful when we want to show what percentage of the whole each category represents.

Example 1.4

The following bar graph appeared in *USA Today*, February 17, 2000, and shows the percentage of winter travelers who plan to spend the following amounts on winter travel: less than $500; $500 to $999; $1,000 to $2,499; and $2,500 or more. Would the information as given be appropriate for a pie chart?

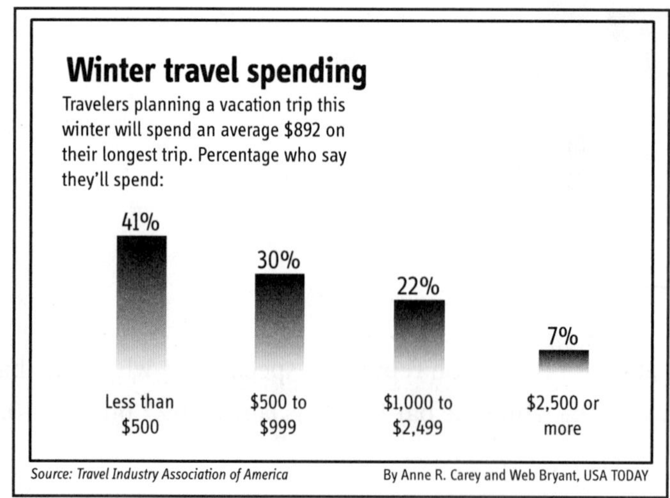

Solution

Because the information given for each category is a percentage of all winter travelers, we could construct a pie chart using the given information as shown. Since there are four categories, the pie is divided into four pieces, using the appropriate percentage of the circle to represent each category. We can quickly see what part of the whole each group occupies.

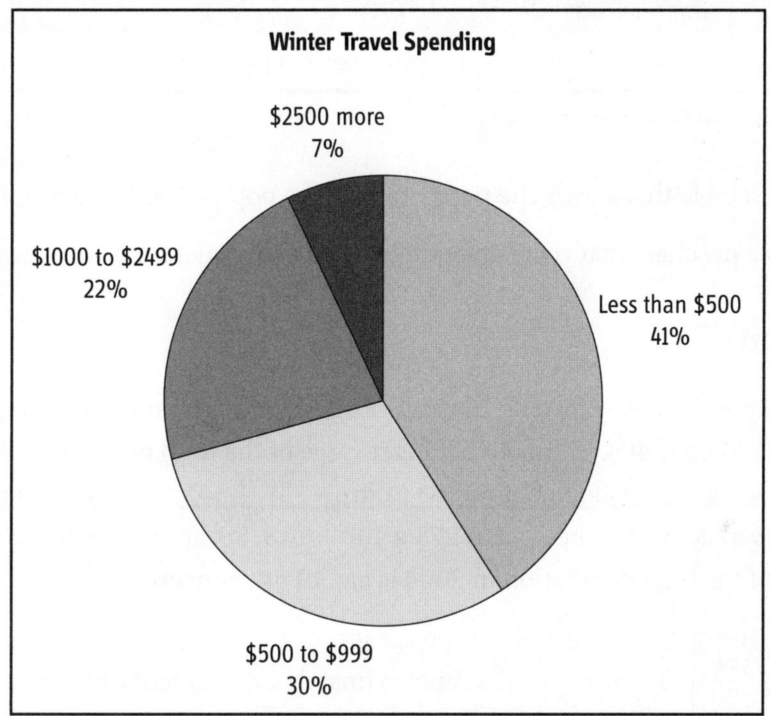

To construct a pie chart from raw data, we need to find what percentage of the whole each category represents. If we are drawing the pie chart by hand, it is helpful to represent each category not only as a percentage, but as a portion of the whole circle that will be easy to visualize as, for example, halves, quarters, eights, or sixteenths. We discuss how to do this in the following example. (Although computer programs can easily be used to draw nice pie charts, it is useful to know how to create a pie chart without the aid of a computer. This way we gain a deeper understanding of such graphs and we can better tell whether or not a chart created on a computer is correct.)

Example 1.5

The total population of the U.S. was estimated to be 293,655,404 on July 1, 2004. The number of people in each of six age groups is as follows:

Age Range	19 Yrs and Younger	20 to 39 Yrs	40 to 59 Yrs	60 to 79 Yrs	80 to 99 Yrs	100 Yrs and Older
Population	81,551,798	82,055,558	81,164,640	38,463,652	10,358,920	60,836

(Source: *U.S. Census Bureau*, www.census.gov.)

a. Make a table that shows the percentage of the population in each age group.

b. Draw a pie chart that represents the estimated population by age group.

Solution

a. There are 81,551,798 people 19 years of age or younger, and the total population is 293,655,404. To find what percentage of the total population 81,551,798 represents, we divide to obtain the ratio $\frac{81{,}551{,}798}{293{,}655{,}404} \approx 0.278$, and then multiply by 100 to convert to percent: $0.278 \times 100 = 27.8$. So, approximately 27.8 percent of the total population in 2004 is age 19 or younger.

Since the number of people in the age group 20 to 39 years is 82,055,558 and $\frac{82{,}055{,}558}{293{,}655{,}404} \approx 0.279$, we see that approximately 27.9 percent of the population in 2004 is between 20 and 39 years old. In the same manner, we compute the percentage of the total population that corresponds to each of the remaining age groups. The following table shows the percentages:

Age Range	19 Yrs and Younger	20 to 39 Yrs	40 to 59 Yrs	60 to 79 Yrs	80 to 99 Yrs	100 Yrs and Older
Percentage	27.8%	27.9%	27.6%	13.1%	3.5%	0.02%

b. A pie chart for these data will be a circle divided into six sectors, each representing one of the age groups. The size of each sector is determined by the size of the population in the age group that sector represents. For example, the sector representing the group between 80 and 99 years of age should be a sector of the circle that covers 3.5 percent of the area of the whole circle. To visualize the corresponding portion of the circle, we write $3.5\% = \frac{3.5}{100} = 0.035 \approx \frac{1}{32}$. We also write the portions corresponding to the remaining groups as fractions with denominator 32. For the group 19 and younger, we write $27.8\% = 0.278 = \frac{(0.278)\cdot(32)}{32} \approx \frac{9}{32}$. In the same way, we find that portions corresponding to the remaining groups can be converted, approximately, to fractions as follows:

Age Range	19 Yrs and Younger	20 to 39 Yrs	40 to 59 Yrs	60 to 79 Yrs	80 to 99 Yrs	100 Yrs and Older
Proportion	$\frac{9}{32}$	$\frac{9}{32}$	$\frac{9}{32}$	$\frac{4}{32}$	$\frac{1}{32}$	0*

(*Note that the number of people age 100 and older is not really 0, but $\frac{(0.000207)\cdot(32)}{32} \approx \frac{0.006}{32}$ is practically 0. In fact, sometimes when we approximate, the sum of the proportions will not equal 1 because of roundoff errors.)

To draw the pie chart, we divide a full circle into two halves, then cut each half into two equal parts, so each is a quarter of the circle. We then cut each of those in eight equal parts to obtain sectors of size $\frac{1}{32}$ of the whole circle, as shown in the following figure:

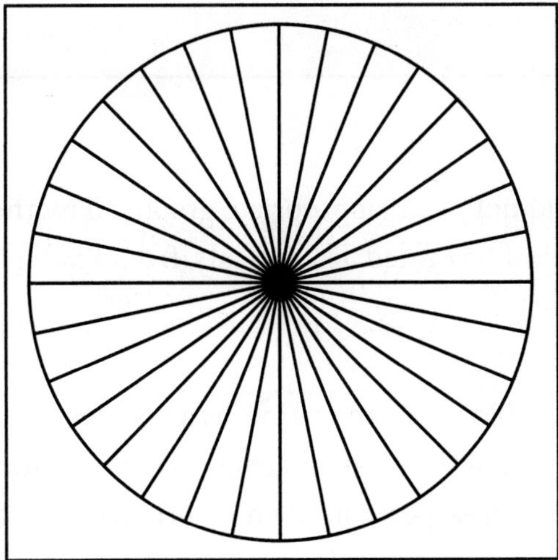

Now we mark how many $\frac{1}{32}$ portions correspond to each population group and label each sector. Here is the pie chart we obtained:

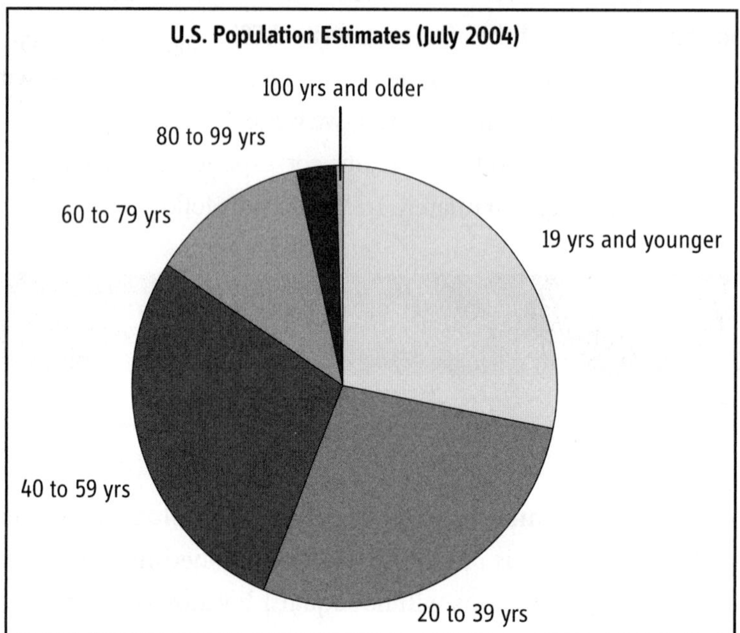

A pie chart is not appropriate to represent all data. A pie chart provides a good representation only when the data represent parts of the same "whole" and consist of a small number of categories or data that can be grouped into a small number of categories.

Example 1.6

Explain why a pie chart would not be an appropriate graphical method to show the variable "audience rating" of top-rated TV specials, from Example 1.1.

Solution

Although the quantitative variable "audience rating" is a percentage, we wouldn't be interested in what proportion of the total percentage is represented by the audience rating of each single show. These percents are not portions of the total audience at the same time.

Histograms

Another type of graph that is useful for visualizing the distribution of quantitative variables is the **histogram**; that is, a histogram shows how the data are distributed.

Quantitative variables such as winter travel spending may take on many different values in a range of, for example, $0 to $10,000 or more. To make sense of the data, we often group the data into **classes**. If the data are given as raw data, we would first need to determine the groupings or classes and then proceed to count how many data values fall into each class. Here is a step-by-step procedure for constructing a histogram for a quantitative variable:

1. Divide the range of data into classes of equal width. We usually choose between 5 and 20 classes, depending on how many cases we are working with. We also want to specify the classes so each data value falls into exactly one class.
2. Count the number of data values that fall into each class.
3. To draw the histogram, first construct a horizontal axis and mark the scale for the variable being graphed. On this scale, mark the boundaries for each class, using consistent measurements.
4. On the vertical axis, mark a scale for the counts for each class.
5. Draw a bar for each class, with the base of the bar covering the class on the horizontal axis and the height of the bar determined by the number of data values in the class. Bars for adjacent classes will touch one another (unlike the bar graph, where bars are generally separated by a space).

In a histogram, we can see the overall pattern and spread of the data. Because bars are of equal width, the area of each bar is determined by its height, and all the data are fairly represented.

Example 1.7

The following table, taken from *Time Almanac 2004,* gives a list of 12 well-known U.S. universities, along with the percent acceptance rate for applicants wishing to be admitted to the university.

College or University	Percent Accepted
Harvard University	11
Yale University	16
Princeton University	12
Johns Hopkins University	32
New York University	29
MIT	16
Duke University	26
Carnegie Mellon University	36
George Washington University	49
Northwestern University	33
American University	72
Cornell University	31

Create a histogram for these data and describe what the graph shows.

Solution

The data values range from 11 to 72, so if we choose classes 10 units wide, starting at 10, we will have seven different classes: 10 to 19, 20 to 29, 30 to 39, 40 to 49, 50 to 59, 60 to 69, and 70 to 79. Now we count the number of data values that fall in each class. The result is shown in the following table:

Class	Number of Universities
10 to 19	4
20 to 29	2
30 to 39	4
40 to 49	1
50 to 59	0

Class	Number of Universities
60 to 69	0
70 to 79	1

Now we create the graph with seven adjacent bars, one for each class, 10 units wide and with a height equal to the number of data values in the class.

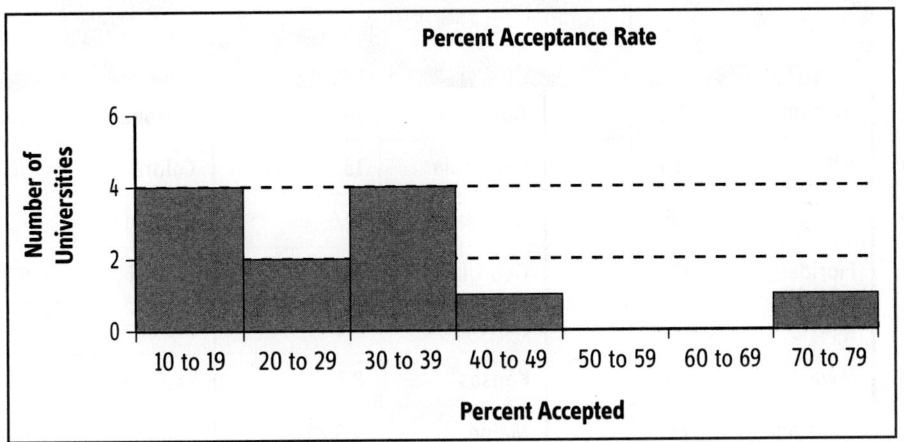

This histogram shows that most of the universities in the given list accept between 10 and 39 percent of the applicants. Only two universities in the list accept more than 40 percent, one of which accepts more than 70 percent of its applicants.

Note that in Example 1.7, we labeled the first class "10 to 19" and the second class "20 to 29." In this case, because all the data values were integers, we knew that no data value would fall between 19 and 20. In general, to allow for data values that are not necessarily all integers, we label the classes in such a way that every real number is in one of the classes. Instead of "10 to 19" and "20 to 29," these classes would be labeled "10 to 20" and "20 to 30." Because these names do not indicate whether the value 20 is considered in the first or the second class, we need to clarify this either on the histogram itself or in a separate sentence. Example 1.8 shows how this can be done.

Example 1.8

The following table, from *The New York Times Almanac 2004,* page 261, contains a list of the states and District of Columbia, along with the percent change in population from the census of 1990 to that of 2000. Construct a histogram for these data.

State	Percent Change	State	Percent Change	State	Percent Change
Alabama	10.1	Alaska	14.0	Arizona	40.0
Arkansas	13.7	California	13.8	Colorado	30.6
Connecticut	3.6	Delaware	17.6	Dist. Of Col.	-5.7
Florida	23.5	Georgia	26.4	Hawaii	9.3
Idaho	28.5	Illinois	8.6	Indiana	9.7
Iowa	5.4	Kansas	8.5	Kentucky	9.7
Louisiana	5.9	Maine	3.8	Maryland	10.8
Mass.	5.5	Michigan	6.9	Minnesota	12.4
Mississippi	10.5	Missouri	9.3	Montana	12.9
Nebraska	8.4	Nevada	66.3	New Hamp.	11.4
New Jersey	8.9	New Mexico	20.1	New York	5.5
North Caro.	21.4	North Dak.	0.5	Ohio	4.7
Oklahoma	9.7	Oregon	20.4	Penns.	3.4
Rhode Is.	4.5	South Caro.	15.1	South Dak.	8.5
Tennessee	16.7	Texas	22.8	Utah	29.6
Vermont	8.2	Virginia	14.4	Washington	21.1
West Va.	0.8	Wisconsin	9.6	Wyoming	8.9

Solution

The first thing we need to do is identify the range of the data. The smallest percent change is –5.7 while the largest is 66.3. We divide that range into suitable classes or intervals of equal length. We could use convenient intervals of 10 units. If we start with –10 as the left boundary of the first interval, the first interval would be from –10 to 0; the second interval would be from 0 to 10, and so on. To ensure that each point is in exactly one interval, we include the right endpoint of each interval in the interval and do not include the left endpoint. Thus, the first interval is: –10.0 < percent change ≤ 0.0; the second interval is: 0.0 < percent change ≤ 10.0; the third interval is: 10 < percent change ≤ 20.0; the next interval is: 20.0 < percent change ≤ 30.0; the next interval is: 30.0 < percent change ≤ 40.0; the next interval is: 40.0 < percent change ≤ 50.0; the next interval is: 50.0 < percent change ≤ 60.0; and the final interval is: 60.0 < percent change ≤ 70.0. Note that we would not need to start our first interval at –10. We could have used –6, or even –5.8, but –10.0 was a convenient choice. All the intervals need to be of equal width and each data value must be in one and only one interval. (Choosing our intervals the way we did ensures this, but it is not the only possibility.) Now we tally the number of data values that fall into each interval:

Interval (Percent Change)	Number of Data Values (States) in Interval
-10.0 < percent change ≤ 0.0	1
0.0 < percent change ≤ 10.0	25
10.0 < percent change ≤ 20.0	13
20.0 < percent change ≤ 30.0	9
30.0 < percent change ≤ 40.0	2
40.0 < percent change ≤ 50.0	0
50.0 < percent change ≤ 60.0	0
60.0 < percent change ≤ 70.0	1

We sketch the histogram as shown next. (Note that if we had chosen intervals of width 8 or 5 the histogram would have looked a bit different.) We want to look at the general shape of the histogram, its center, and see how spread out the histogram is to get an idea of the general data pattern.

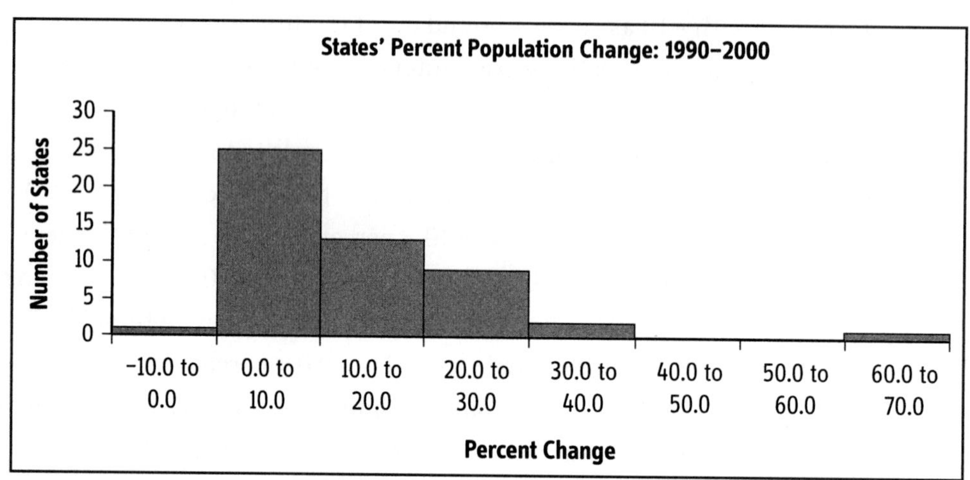

In Example 1.8, we analyzed data on the percent change in population. What does this mean? For example, the first table in Example 1.8 gives the percent change in population for Alabama from 1990 to 2000 as 10.1 percent. This means that Alabama's population in the year 2000 is 10.1 percent greater than its population in 1990. In the following example we show how the percent change is calculated.

Example 1.9

Data from the 1990 census show that the populations of Alabama and California were 4,040,583 and 29,760,021, respectively.

a. According to the 2000 census the population of Alabama in 2000 was 4,447,100. Verify that the percent change in Alabama's population from 1990 to 2000 is 10.1 percent.

b. Use the information given in the first table of Example 1.8 to find the population of California in 2000.

Solution

a. To find the percent change in Alabama's population from 1990 to 2000, we first find the change in population: 2000 population – 1990 population = 4,447,100 – 4,040,583 = 406,517. Next, to find what percentage of the 1990 population this change represents, we divide the change by the population in 1990 and multiply by 100: $\frac{406,517}{4,040,583} \cdot 100 \approx 10.06$. This number, when rounded off to the first decimal, gives a percent change of about 10.1 percent.

b. From the data in the first table of Example 1.8, the percent change in California's population from 1990 to 2000 was 13.8 percent of 29,760,021 (which was the population of California in 1990). So,

$$13.8 = \frac{\textit{change in population}}{29,760,021} \cdot 100, \text{ or } 0.138 = \frac{\textit{change in population}}{29,760,021}$$

Thus the change in population from 1990 to 2000 is: *change in population* = (0.138) · (29,760,021) = 4,106,882.898 ≈ 4,106,883. So the population in 2000 is the sum of the population in 1990 and the change in population or 29,760,021 + 4,106,883 = 33,866,904.

Stemplots

Another type of graph, a **stemplot** or **stem-and-leaf graph**, is often used to display a quantitative variable, particularly if the data set is not too large. This type of graph shows not only the general pattern of the data, as a histogram does, but it also displays all the individual data values. Here are the steps for constructing a stemplot:

1. Divide each data value into two parts: a **stem** consisting of all but the single rightmost digit, and a **leaf**, consisting of the rightmost digit. (For example, if data values represent test scores, the score of 82 would have a stem of 8 and a leaf of 2. If data values represent math SAT scores, the score of 625 would have a stem of 62 and a leaf of 5.)

2. Write the stems in order in a vertical column with the smallest at the top. (We must include all possible consecutive stems, even if there are no values in our data set with that particular stem; otherwise the data are distorted.) We draw a vertical line to the right of the column of stems.
3. Write each leaf in the row to the right of its stem, in increasing order from left to right. Take care to be consistent with the spacing and size of the numbers representing each leaf.

Example 1.10

Create a stemplot for the data on the percent acceptance rate in the following table. (This is the same data used in Example 1.7.) Describe any patterns that emerge.

College or University	Percent Accepted
Harvard University	11
Yale University	16
Princeton University	12
Johns Hopkins University	32
New York University	29
MIT	16
Duke University	26
Carnegie Mellon University	36
George Washington University	49
Northwestern University	33
American University	72
Cornell University	31

Solution

The data values consist of two digits, so we will use the "tens" digit as the stem and the "units" digit as the leaf. We start by listing the stems from 1 to 7, in a vertical column.

1

2

3

4

5

6

7

We add a vertical line after the column of stems and then add the leaves one at a time. The final step is to order the leaves on each stem and produce an ordered stemplot. We also include a title and a key:

Percent Acceptance Rate: 2|5 = 25%

1	1 6 2 6
2	9 6
3	2 6 3 1
4	9
5	
6	
7	2

1	1 2 6 6
2	6 9
3	1 2 3 6
4	9
5	
6	
7	2

The stemplot shows that for this small collection of data values, most of the data fall in the 11 to 49 range. There is one **outlier** (that is, an individual data value that falls outside the overall pattern of the data). In this case, the outlier is much larger than the other values, which in the context of this data, says that one school in the list accepts a much greater percentage of its applicants than the others. The "center" of the data is probably somewhere in the low 30s. From the stemplot, we can see the shape and spread of the data and get a general idea of the center and any outliers; we can also see the actual data values.

Summary

We have seen examples of quantitative and categorical variables and learned how to interpret information given in the four types of graphs.

Bar graphs and pie charts are used to represent categorical data. In a bar graph, each bar represents a category and the height of each bar represents a count or percentage for the category. To make a pie chart sometimes it is necessary to group the data. Each portion of a pie chart shows what percentage each category is of the whole. To find the corresponding portion of a category in a pie chart, we need to use fractions and compute percents.

Histograms and stemplots (stem-and-leaf graphs) are used to display numerical data. A histogram shows the shape and spread of the data and the frequency of data values in determined classes; that is, how many times data values fall in a particular class. A stemplot shows the shape and spread of the data and gives the actual data values.

Explorations

1. For each of the following variables, indicate whether it is quantitative or categorical. Then, identify the individual (or case) and write a sentence explaining why you might be interested in such a variable.

    a. The number of calories in a cup of breakfast cereal

    b. Salaries of last year's college graduates

    c. Preferred brand of cola

    d. Time it takes college freshmen to read a particular editorial

    e. Race of small business owners in Philadelphia

2. Write a paragraph to describe what the following bar graphs show. The graphs accompanied the article "Children of the Gun" that appeared in *Scientific American*, June 2000.

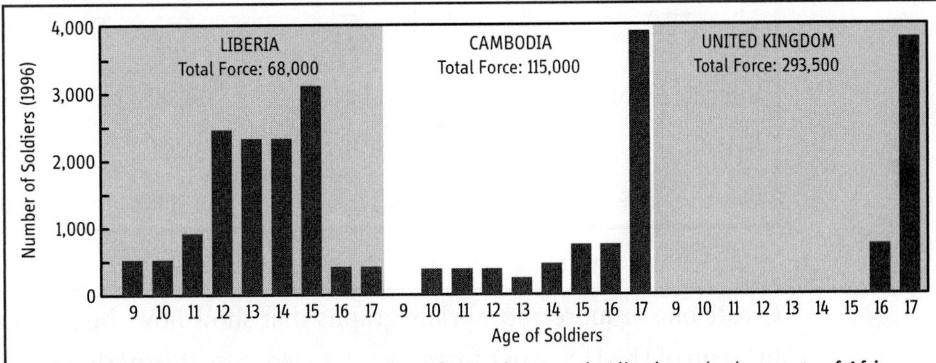

YOUNG CHILDREN make up a large fraction of armed groups in Liberia and other parts of Africa; the percentage is smaller in Cambodia and elsewhere in Asia and in Latin America. Several developed nations recruit 16-year-olds but are now raising the minimum age.

3. Group the data about Liberia's armed forces on the bar graph in Exploration 2 into three age groups. Then use this grouped data to create a pie chart that shows the composition of Liberia's armed forces in the three age groups. Clearly indicate which age groups you are considering and give the percentage of the total armed forces that corresponds to each group.

4. Explain what the pie chart in the solution of Example 1.5 shows about the population of the U.S. in 2004.

5. The following table containing U.S. Bureau of Labor statistics gives the number of workers by race and ethnic origin and gender for the three largest ethnic groups of workers in the U.S. for the years 1996, 2000, and 2005. (Source: *U.S. Department of Labor, Bureau of Labor Statistics,* www.bls.gov.)

Numbers (thousands)	1996	2000	2005
White, non-Hisp.	**102,141**	**108,264**	**111,844**
Men	55,977	59,119	61,255
Women	46,164	49,145	50,589

Numbers (thousands)	1996	2000	2005
Black or African American	12,929	14,444	14,777
Men	6,167	6,741	6,901
Women	6,762	7,703	7,876
Hispanic or Latino	10,996	14,762	17,785
Men	6,655	8,859	10,872
Women	4,341	5,903	6,913

Create one or more appropriate graphs that show how the racial and ethnic makeup of the U.S. work force has changed over this period of time.

6. The following table contains a list of the states and District of Columbia, with the average verbal SAT test scores for high school seniors (for the academic year ending in 2003) and the percentage of high school seniors who take the test. (Source: *New York Times World Almanac 2004,* page 296.)

State	Average Verbal SAT	Percent Taking	State	Average Verbal SAT	Percent Taking
AL	559	10	MO	582	8
AK	518	55	MT	538	26
AZ	524	38	NE	573	8
AR	564	6	NV	510	36
CA	499	54	NH	522	75
CO	551	27	NJ	501	85
CT	512	84	NM	548	14
DE	501	73	NY	496	82
DC	484	77	NC	495	68
FL	498	61	ND	602	4

State	Average Verbal SAT	Percent Taking	State	Average Verbal SAT	Percent Taking
GA	493	66	OH	536	28
HI	486	54	OK	569	8
ID	540	18	OR	526	57
IL	583	11	PA	500	73
IN	500	63	RI	502	74
IA	586	5	SC	493	59
KS	578	9	SD	588	4
KY	554	13	TN	568	14
LA	563	8	TX	500	57
ME	503	70	UT	566	7
MD	509	68	VT	515	70
MA	516	82	VA	514	71
MI	564	11	WA	530	56
MN	582	10	WV	522	20
MS	565	4	WI	585	7
			WY	548	11

a. Make a stemplot of the percentage of high school seniors taking the test.

b. Make another stemplot of the average verbal SAT scores of the states.

c. Describe the shape of each of the distributions of the variables "percent taking" and "average verbal SAT score."

d. Look at the states by the region of the U.S. in which they are located, and make any preliminary observations about the variables "percent taking" and "average verbal SAT score."

7. In the 2005 Electronic Monitoring and Surveillance Survey conducted by the American Management Association, the following information about what employers record and review relative to electronic eavesdropping was obtained. The information was based on responses from 526 companies. Create an appropriate graph to represent these data and explain why you chose the graph you did.

Recorded	Percent of Companies
Telephone conversations	3
Voice-mail messages	7
Computer files	30
E-mail messages	38
Video recording of job performance	6
Telephone use (time spent, numbers called)	31
Computer use (time logged on, keystroke counts)	21
Video surveillance for security	32
Website connections	76

8. The following graph appeared in the *New York Times* (December 19, 1999) and accompanied an article titled "Experts Wonder If Crime Drop Is Near End."

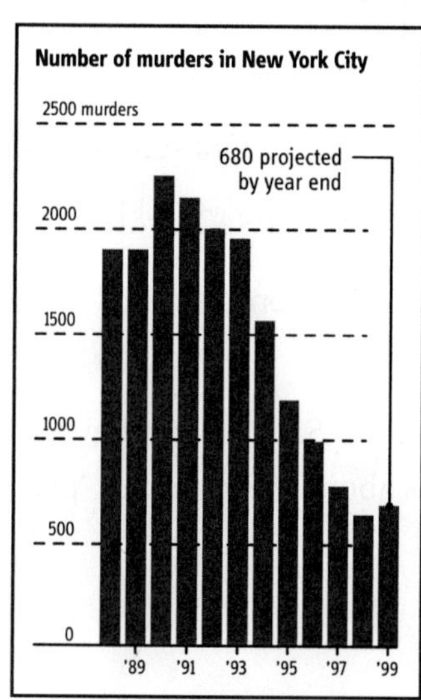

a. Explain in detail what the graph shows about the number of murders in New York City.

b. What was the percent change in number of murders in New York City from 1989 to 1990?

c. What was the percent change in number of murders in New York City from 1993 to 1994?

d. Find data for 2000 to the present and construct a new bar graph with the additional data. Was the crime drop near its end in 1999?

9. Describe the differences and similarities between the histogram created in Example 1.7 using data on acceptance rates in 12 well-known universities and the stemplot in Example 1.10 created using the same data set.

10. The total population of the U.S. was 151,325,798 in 1950, and it increased to 281,421,906 in 2000. What is the percent change in the population from 1950 to 2000?

11. In 1990, the population of Michigan was 9,295,297 and increased by 6.9 percent during the 10-year period 1990–2000. Use this information to find Michigan's population in the year 2000.

12. Use the Internet to find current data on the AIDS epidemic, and create several graphs like those in Example 1.2 to show how the epidemic is continuing to change.

13. From each student in the class, collect the following data, and then create an appropriate graph or chart for each variable and explain what it shows.

a. The yearly salary to the nearest hundred of dollars that they estimate they will earn in their first post-college job

b. Their most preferred leisure-time activity, chosen from among the following: watching a movie; watching television; playing a sport or exercising; reading; going out to eat; playing computer/video games; other

14. Write a summary paragraph explaining when each of the different types of charts or graphs is useful.

15. The number of pieces of "junk mail" in 1978 was 30.5 billion. This represented 31.5 percent of the total number of mail pieces handled by the post office during that year. Find the total number of mail pieces handled by the post office during the year 1978.

16. Use the information given in the graph shown in Example 1.3 to find the total number of mail pieces handled by the post office.

    a. During the year 1983

    b. During the year 1988

    c. During the year 1993

    d. During the year 1998

TOPIC 2

Bivariate Data

OBJECTIVES

After completing this topic, you will be able to

» Construct a scatterplot to describe the relationship between two quantitative variables

» Identify and explain, in context, any trends depicted in a scatterplot

» Describe the mathematical concept of a function and recognize when one variable is a function of another

» Represent functions using words, tables, symbols, and graphs, and move from one mode of representation to another

» Recognize and formulate directly proportional relationships

There are many instances where we want to determine how two variables are related; for instance, we might want to determine how gender and salaries in a particular industry are related. Because gender is a categorical variable and salary is a quantitative variable, we use comparative bar graphs or pie charts to help identify relationships. When we are working with two quantitative variables (that is, bivariate data sets), as in the exploration in Topic 1 involving the average verbal SAT score for each state and the percentage of high school seniors in the state taking the SAT test, we might want to use a **scatterplot** to see how the two variables are related.

In considering the relationship between two quantitative variables, we can sometimes identify one of the variables as the **explanatory variable**, or **independent variable**, and the other as the **response variable**, or **dependent variable**. The response or dependent variable generally *depends on* or is *explained by* the explanatory variable. For example, the time a student spends playing video games (the explanatory variable) might *explain* the student's grade point average (the response variable). A child's height (the response variable) *is explained by* his age (the explanatory variable). Here is a table showing how a particular boy's height changed between two years and five years of age:

Age in Years	2	3	5
Height in Inches	35	38	42.5

(Note that in both of these situations, there are other quantities that might also explain the relationship between age and height. Here we are simplifying the analysis by looking at just one explanatory variable.)

Sometimes we will want to choose values of the explanatory variable and see how the response variable is affected. At other times it might not be obvious which is the explanatory variable and which is the response variable.

When creating a scatterplot, we will use a rectangular coordinate system and plot the explanatory variable on the horizontal axis and the response variable on the vertical axis. We denote points in a rectangular coordinate system as ordered pairs using parentheses, with the explanatory variable as the first coordinate and the response variable as the second coordinate, like this: (explanatory, response). If the choice is not obvious, we might plot either variable on the horizontal axis. The following graph shows a scatterplot of the three points from the boy's age-height table: (2, 35); (3, 38); (5, 42.5).

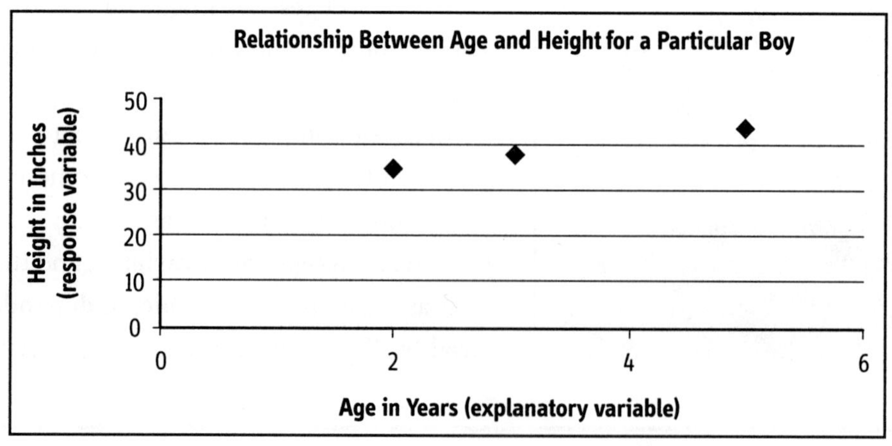

In Example 2.1, we use SAT data to identify the explanatory and response variables, create a scatterplot, and look for trends in the graph.

Example 2.1

Consider the verbal SAT data for states and the percentage of high school seniors in each state who take the test. (A portion of the table that you used in Topic 1, Exploration 6 is shown here.)

State	Average Verbal SAT	Percent Taking Test
AL	559	10
AK	518	55
AZ	524	38
AR	564	6

Create a scatterplot for these bivariate data and identify which variable should be the explanatory variable, which variable should be the response variable, and why you made this choice. Also note any patterns or trends that this plot reveals. Does a high average verbal SAT score for a state mean the state has a sound education system?

Solution

Because the percentage of seniors taking the test might help "explain" the state's verbal SAT score, we chose that as the explanatory variable and plotted it on the horizontal axis. For example, if a lower percentage of seniors took the test in a state, those students might be the higher achieving students, which helps "explain" the state's score. Each point represents one state (that is, one individual in this data set); the percent taking the test is the first coordinate, measured on the horizontal axis, and the average verbal SAT score is the second coordinate, measured on the vertical axis.

On the following graph, the first point, with coordinates (10, 559), represents Alabama with 10 percent of students taking the test and an average verbal SAT of 559. To plot the point, we count 10 units to the right for the first coordinate and 559 units up for the second coordinate. The next point, representing Arkansas with 55 percent of students taking the test and an average verbal SAT of 518, is the second point given on the plot (55, 518).

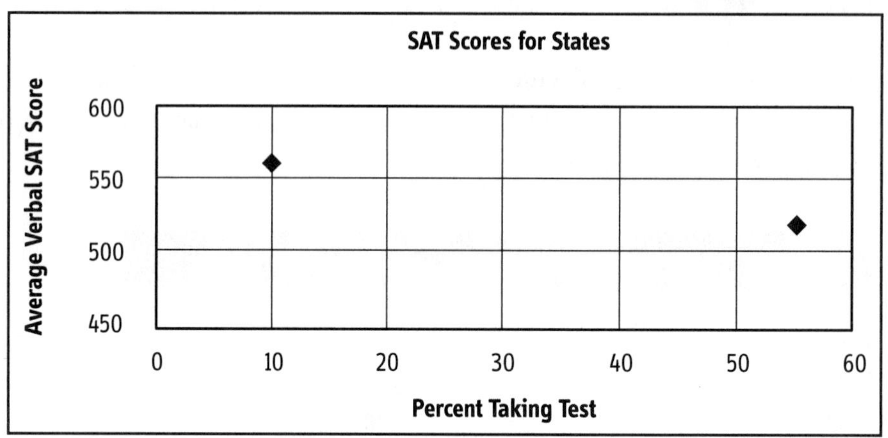

The plot of the data for all states is shown here:

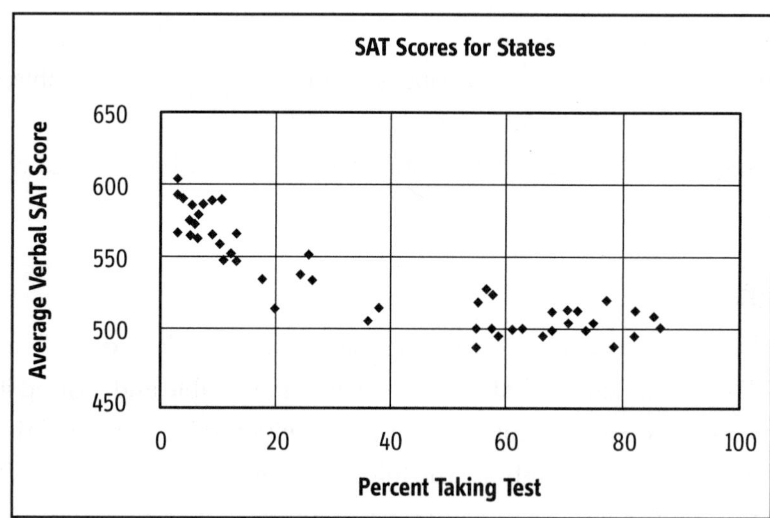

The scatterplot for all the states shows that, in general, higher verbal SAT scores tend to occur in states in which a lower percentage of high school seniors take the test. Lower verbal SAT scores tend to occur in states in which a higher percentage of seniors take the test. So we really cannot say that states with higher verbal SAT scores necessarily do a better job of educating their high school students. A more plausible explanation is that in the states in which a smaller percentage of seniors take the SAT test, the stronger students are the ones who tend to take the test, thus resulting in a higher average verbal SAT score for that state.

Note that the scales used on the two axes in Example 2.1 are not the same. There are some instances in which we will want to use the same scale on both axes and others in which, because of the nature of the variables, we won't. (See Exploration 5 at the end of this topic for an example where we use the same scale on both axes.) In Example 2.1, you should also note that the scale on the vertical axis does not start with 0, but with 450. In this data set, as in many data sets, the values of one or both of the variables are much larger than 0, so the plot shows the relationships much more clearly if the axes intersect at a point different from (0, 0). We need to be sure to mark the axes clearly.

Two variables have a **positive association** if larger values of one variable tend to occur with larger values of the other variable. The variables have a **negative association** if larger values of one variable tend to occur with smaller values of the other variable. From the data in the previous example, we see that states' average verbal SAT score and the percentage of seniors taking the test are negatively associated. This negative association can be seen in the trend of the data in the scatterplot.

Example 2.2

Match the following four descriptions to the following four scatterplots, and explain why you chose the match. Note that in each plot, time is the explanatory variable and the years are given on the horizontal axis. The scale of the response variable is not marked on the vertical axis. Think about the data to match the descriptions to the pictures. Observe if the association between the variable described and year is a positive association, a negative association, or neither.

a. U.S. life expectancy at birth, all races, both sexes (Source: *Time Almanac 2004*, page 192.)

b. U.S. gross federal debt in millions (Source: *Encyclopaedia Britannica Almanac 2004*, page 859.)

c. Active-duty military personnel, excluding reserves on active duty for training (Source: *Time Almanac 2004*, page 396.)

d. U.S. cases of tuberculosis per 100,000 population (Source: *The Centers for Disease Control*, www.cdc.gov/nchstp/tb/surv/Surv.htm.)

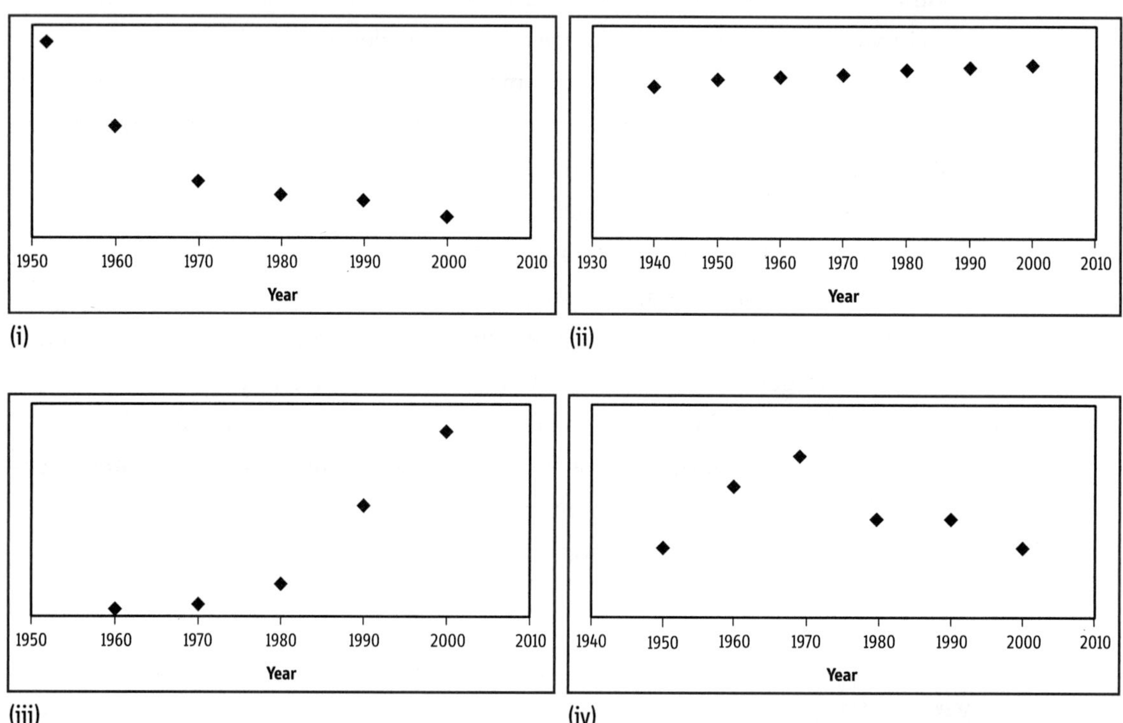

(i) (ii) (iii) (iv)

Solution

Graph (i) shows a decline in the value of the response variable as time increases, which is a negative association. Therefore, this graph represents the cases of tuberculosis per 100,000 population; this decline is a result of better healthcare, improved hygiene, and more advanced treatments for disease. Graph (ii) shows a fairly modest increase (a positive association) and so would be a plot of U.S. life expectancy. Graph (iii) also shows an increase in the response variable as time passes (a positive association), but the increase is more dramatic and thus would be a plot of U.S. gross federal debt over time. Graph (iv) shows an increase, followed by a decrease and would be a plot of U.S. active-duty military personnel. This graph represents an association that is neither positive nor negative. (What accounts for the peak around 1970?)

In mathematics, a general pairing of quantitative variables is called a **relation** or **relationship**. In the next example, we examine a relationship based on the board game Scrabble, to see how the Scrabble point value of a word is related to its length.

Example 2.3

The following table gives the Scrabble point value for each letter in the alphabet.

A = 1	B = 3	C = 3	D = 2	E = 1	F = 4	G = 2	H = 4	I = 1
J = 8	K = 5	L = 1	M = 3	N = 1	O = 1	P = 3	Q = 10	R = 1
S = 1	T = 1	U = 1	V = 4	W = 4	X = 8	Y = 4	Z = 10	

For each of the following ten words, find its Scrabble point value: *case; categorical; chart; data; function; graph; increase; quantitative; scatterplot; variable.* Then make a scatterplot of the variables "number of letters in the word" and "Scrabble point value of the word" and explain what the scatterplot shows.

Solution

Here is a table listing the Scrabble point value for each of the words.

Word	Number of Letters	Scrabble Point Value
Case	4	6
Categorical	11	16
Chart	5	10

Word	Number of Letters	Scrabble Point Value
Data	4	5
Function	8	13
Graph	5	11
Increase	8	10
Quantitative	12	24
Scatterplot	11	15
Variable	8	13

We will plot "number of letters in the word" on the horizontal axis because it *explains* the Scrabble point value. (Another way to think about this is to realize that the "Scrabble point value of a word" *depends upon* the number of letters in it.) The scatterplot shown here reveals that, in general, as the number of letters in a word increases, the Scrabble point value of the word tends to increase as well. However, there are words in our list that are exceptions to this tendency. For example, the word *graph* has 5 letters but its Scrabble point value is 11 points; however, the word *increase* has 8 letters, but it has a smaller Scrabble point value of 10 points.

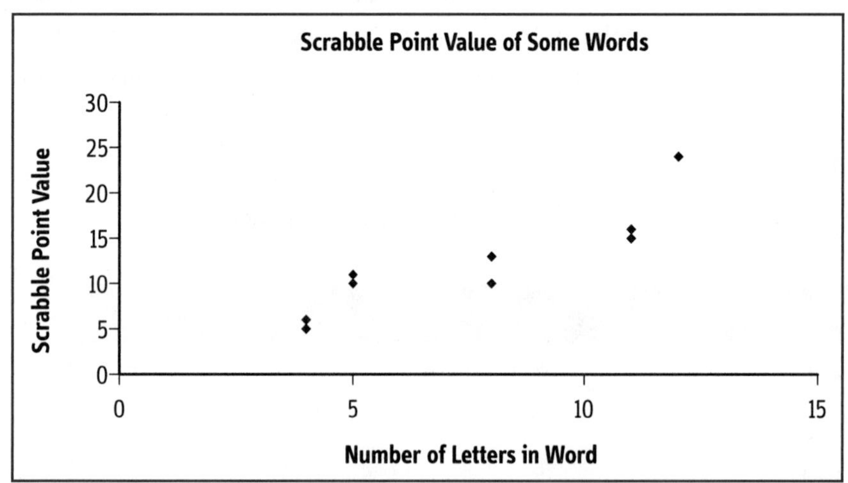

In ordinary language when we say that one entity is a function of another, we mean that the first thing is related to the second and is in some way dependent upon it. As we saw in Example 2.3, the Scrabble point value of a word is related to the length of the word. But if we are using the term *function* mathematically, the Scrabble point value is not a function of the length of the word. In mathematics, we mean something precise when we use the term *function*. A **function** is a relation in which each value of the explanatory variable is paired with exactly one value of the response variable. We will call the explanatory variable x and the response variable y. The variable y = "Scrabble point value of a word" is not a function of the variable x = "number of letters in the word" because there are values of the explanatory variable x that are paired with multiple values of the response variable. For the word *chart*, $x = 5$ and $y = 10$, which results in the point (5, 10), but for the word *graph*, we have $x = 5$ and $y = 11$, resulting in the point (5, 11). This collection of ordered pairs is not a function because two ordered pairs with the same first values (5 in this case) have different second values. Sometimes we will want to take data for which there is not a functional relationship between the explanatory and response variables and *fit* a function to it. We will look at this modeling process in Topic 6.

Functional relationships, or functions, can be represented in various ways. Some functions are represented by **tables** in which we give a list of the allowable values of the explanatory variable x, and for each value of x we give its associated value of y. (Note that because each value of x in the table has exactly one value of y associated with it, there will be one y value for each x value listed, and the x value should not be repeated.) The following table of data supports graph (i) given in Example 2.2. Examining the data, we can see that there is a unique value of y for each x value given in the table.

Year (x)	Cases of Tuberculosis (y)
1953	53.0
1960	30.8
1970	18.3
1980	12.3
1990	10.3
2000	5.8

Functions can also be given using **symbols**. If we let h represent the hours worked in one week at a job that pays $6.15 per hour, then total wages w earned for the week could be represented symbolically as: $w = 6.15h$. We could then show this function relationship in a table for selected values of h as follows (note that w is a function of h because for each h value, there is exactly one w value):

Hours h	5	10	12	15	20	30	40
Wages w	$30.75	$61.50	$73.80	$92.25	$123.00	$184.50	$246.00

If a function is given symbolically as a formula, in the form $y = f(x)$, with y representing the response variable, x the explanatory variable, and $f(x)$ describing how to get y for any particular x, then we can see that for each x put into the formula $f(x)$, we will compute a unique y value. This means that y is a function of x. When representing functions using symbols, we often use letters that suggest the quantities they represent, instead of x and y. For example, we used h to represent hours worked and w to represent total wages. It really doesn't matter what letters we use, but h and w might help us remember the quantity they represent.

Example 2.2 shows four functions represented as **graphs**. Note that on each of these graphs, every value of the year (which is the explanatory variable) that appears on the graph is paired with only one value of the response variable. The graphs in Example 2.2 show functions with a finite number of points. We will look at some additional examples of graphs of functions in which the graph shows an infinite number of points of the function and might appear as a line, a curve, or a series of lines and/or curves.

Words can also be used to describe a function. For example, the energy cost in calories to an individual engaging in an activity such as jogging is affected by a variety of factors that vary from person to person but size is a critical factor. The calories used while doing a particular activity is the response variable, and the individual's weight is the explanatory variable. A 110-pound person, for example, burns approximately 3.4 calories per minute playing table tennis while a 150-pound person burns 4.5 calories per minute and a 190-pound person burns 5.9 calories per minute on the same activity. The next example takes a function given in words and asks us to represent it in table form, as a graph, and in symbols.

Example 2.4

A 150-pound person uses approximately 5.4 calories per minute walking at a normal pace on an asphalt road. Represent the total number of calories used by a 150-pound person as the response variable c, and represent c as a function of the number of minutes spent walking m. Represent this function as a table, in symbols (that is, using the letters c and m and relating them in an equation), and as a graph. (Source: *Setting Your Weight*, George Constable, Editor, Time-Life Books, Alexandria, VA, 1987, page 32.)

Solution

When we describe a function in a table, we need to decide which values of the explanatory variable to include in the table. We choose the explanatory variable to be "minutes spent walking"; we'll use increments of 5 minutes, over an interval from 10 to 50 minutes. We'll then enter the corresponding values for the response variable "total calories used" into the table.

Minutes Spent Walking	10	15	20	25	30	35	40	45	50
Total Calories Used	54	81	108	135	162	189	216	243	270

The table gives total calories expended by a 150-pound person using increments of 5 minutes from 10 minutes to 50 minutes. In symbols, we represent the functional relationship between m and c as: $c = 5.4m$. A graph of this function is shown next:

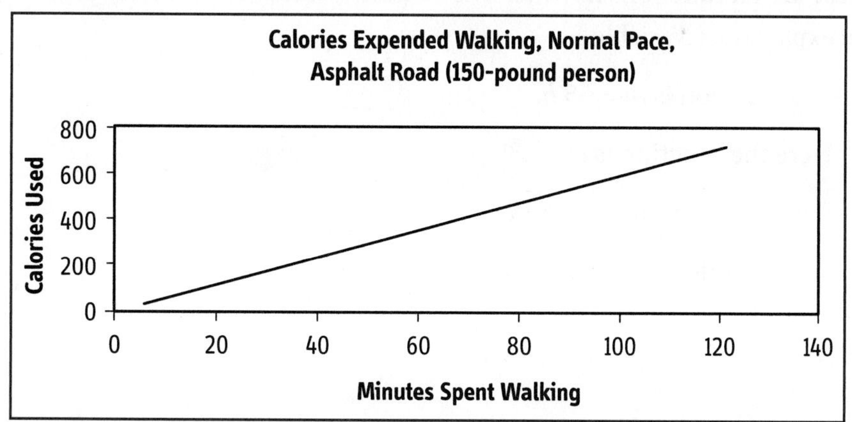

In some functions, such as the one examined in the preceding example, the response variable is equal to a constant multiplied by the explanatory variable. The nature of the relationship determines the constant, called the **constant of proportionality**. In this case, we say that the response variable is **directly proportional** to the explanatory variable. The term *proportional* is used here because the ratios of pairs of response and explanatory variables are constant. In Example 2.4, $\frac{c}{m} = 5.4$ for each pair (c, m) given in the table or represented on the graph. The next example investigates other directly proportional functions.

Example 2.5

For each of the relationships described below, use the indicated letters to represent the variables and write an equation that gives the functional relationship between the variables.

a. Total number of miles m traveled in h hours if traveling at a constant speed of 55 miles per hour

b. Total cost c in dollars of p pounds of bananas if bananas cost \$.39 per pound

c. Length in centimeters c of a ribbon that is i inches long (Recall that an inch is equal to 2.54 centimeters.)

d. Length in inches i of a belt that is c centimeters long

Solution

These are all functions in which the response variable is directly proportional to the explanatory variable.

a. The function is $m = 55\,h$.

b. Here the function is $c = 0.39\,p$.

c. For this function, $c = 2.54\,i$.

d. And for this function, $i = \frac{c}{2.54}$.

Summary

In this topic, we looked at how a scatterplot could be used to show the relationship between two quantitative variables, and we investigated the trends that could be seen using this type of graph. Some pairs of quantitative variables are positively associated, others are negatively associated, and some show no particular relationship. We considered how to express functional relationships between two variables using words, symbols, tables, and graphs. We also explored directly proportional functions; in each directly proportional function involving two variables, the ratio of the variables is a constant.

Explorations

1. Colonial population estimates (in round numbers) are given in the following table for the decades before the establishment of the U.S. Census in 1790. Create a scatterplot for this data table and describe the trends shown by your graph. (Source: *Time Almanac 2004,* page 175.)

Year	Population	Year	Population
1610	350	1700	250,900
1620	2,300	1710	331,700
1630	4,600	1720	466,200
1640	26,600	1730	629,400
1650	50,400	1740	905,600
1660	75,100	1750	1,170,800
1670	111,900	1760	1,593,600
1680	151,500	1770	2,148,100
1690	210,400	1780	2,780,400

2. For each of the following pairs of variables, sketch a rough plot (as a series of points) that could reasonably represent the relationship between the explanatory and response variables. Indicate which is the explanatory variable and which is the response variable and explain why you made the sketch you did. Also determine if your plot shows a positive association between the variables, a negative association, or neither.

   a. Monthly grocery bill of a household; number of people in the household

   b. Hours per day spent watching television; a college student's grade point average

   c. Time it takes to run 100 yards; the runner's age

   d. Number of minutes elapsed since being taken out of the oven; temperature of a pizza

3. For each of the relationships described below, determine if the variables are directly proportional. Write a sentence or two for each to justify your answers.

   a. Total number of words a typist can type and the number of minutes spent typing if he or she types 72 words per minute

   b. The temperature of an asphalt road at several times on a hot summer day and air temperature at those same times

   c. Total number of miles you travel if you average 26 miles per gallon and the gallons of gas used

   d. The total days of vacation earned after y years of work if you earn two vacation days for every three months you work

4. For each student in your class, collect the following data: student's gender; his or her height in inches; his or her hand span measurement, also in inches. (Before collecting these measurements, decide how you will define the hand-span measurement.)

   a. Sketch a scatterplot of the data collected from the class and discuss any trends. Which variable did you choose for the horizontal axis and which variable did you choose for the vertical axis? Does it matter?

   b. Discuss any problems associated with collecting these measurements.

   c. How can you indicate a student's gender on the scatterplot?

5. The following table gives information from a sample of college students: gender; number of children in family of origin; and number of children in their ideal family, in which they may someday be a parent.

Gender	No. of Children in Family of Origin	No. of Children in Ideal Family
F	2	2
M	3	2
M	4	3
F	2	2
F	4	3
M	5	5
F	3	3
F	2	3
F	2	2
F	4	4
M	3	3
F	4	0
F	3	3
F	3	4
F	1	2
M	2	2
M	3	3
M	2	2
M	1	1
M	2	0

a. Sketch a scatterplot of the data collected from the students and discuss any trends. (Use the same scale on both axes.) Which variable did you choose for the horizontal axis and which variable did you choose for the vertical axis? Does it matter?

b. Describe the role of the diagonal line $y = x$ and what it helps you see about the data.

c. Do you think the data for males and the data for females should be considered separately? Why or why not?

6. The following table of data gives hockey star Eric Lindros' career numbers for the eight seasons he played with the Philadelphia Flyers. (Source: *The Internet Hockey Database,* hockeyDB.com.)

Year	Games	Goals	Assists	Points
1993–94	65	44	53	97
1994–95	46	29	41	70
1995–96	73	47	68	115
1996–97	52	32	47	79
1997–98	63	30	41	71
1998–99	71	40	53	93
1999–2000	55	27	32	59
Totals	**486**	**290**	**369**	**659**

a. Sketch a scatterplot of the two variables "number of games played in a season" and "number of points scored by Lindros" and discuss any trends.

b. Which variable did you choose for the horizontal axis and which variable did you choose for the vertical axis? Does it matter?

c. Determine if there appears to be a relationship between goals and assists. What might explain this?

7. Here is a table relating age and height for a particular girl between two years and ten years of age. Create a scatterplot for these data and describe any trends. Is height a function of age?

Age in Years	2	3	4	5	6	8	10
Height in Inches	33	36	37	41	44	49	53

TOPIC 3

Graphs of Functions

OBJECTIVES

After completing this topic, you will be able to

- » Determine when a graph represents a function
- » Use the graph to identify if the values of a function are increasing or decreasing
- » Use the graph to find absolute and relative maximum and minimum values of a function
- » Recognize, from the graph of a function, when the rate of change is increasing or decreasing
- » Apply the information from a graph to help analyze a context

We use functions to model relationships between two variables using available data to better understand the relationship and, in many cases, to predict values. The graph of a function is an important tool for analyzing the function's behavior.

The graph of a function consists of all points with coordinates (a, b) where b is the value of the response or dependent variable that corresponds to the value a of the explanatory or independent variable. In many situations, the values of the independent variable can take on infinitely many values in a given interval (for example when the independent variable is time), so the graph consists of a continuous curve, rather than isolated points. The graph of a function describes the function "at a glance." It provides a quick way of identifying important properties of the function such as whether it is increasing or decreasing and where it peaks. Because each value of the explanatory variable is paired with exactly one value of the response variable, a vertical line drawn anywhere on the graph will intersect the graph in no more than one point. This gives us a test, called the **vertical line test**, to determine whether a graph is that of a function or not.

A function is **increasing** if the values of the response (dependent) variable increase when the corresponding values of the explanatory (independent) variable increase. A function is **decreasing** if the values of the response variable decrease when the corresponding values of the explanatory variable increase.

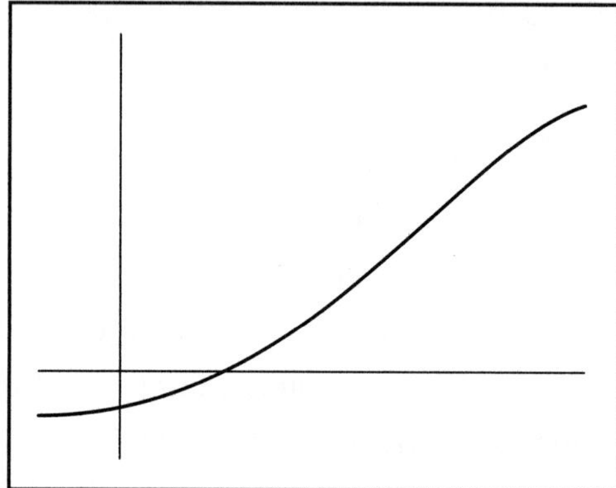

Increasing Function

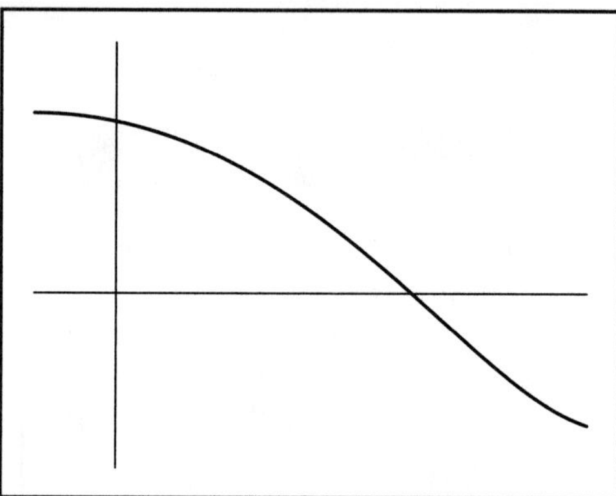

Decreasing Function

Because the independent variable is represented on the horizontal axis with values increasing to the right, and the dependent variable is represented on the vertical axis with values increasing upward, the graph of an increasing function rises when traced from left to right. Similarly, the graph of a decreasing function falls when traced from left to right. Most functions are increasing over some intervals of the independent variable and decreasing over others.

Points of special interest on the graph of a function are the highest and lowest points, which give the (absolute) **maximum** and (absolute) **minimum** values of the function, respectively; that is, they give the largest and smallest values of the response variable.

Example 3.1

The following graph, which appeared in *The Morning Call* on October 6, 1999, represents the national student loan default rate from 1987 to 1997. Use the graph to answer the following questions.

Defaulting on student loans

The national student loan default rate fell to 8.8 percent in 1997, the lowest point since the government started calculating the rate. Here is a look at the rate since 1987.

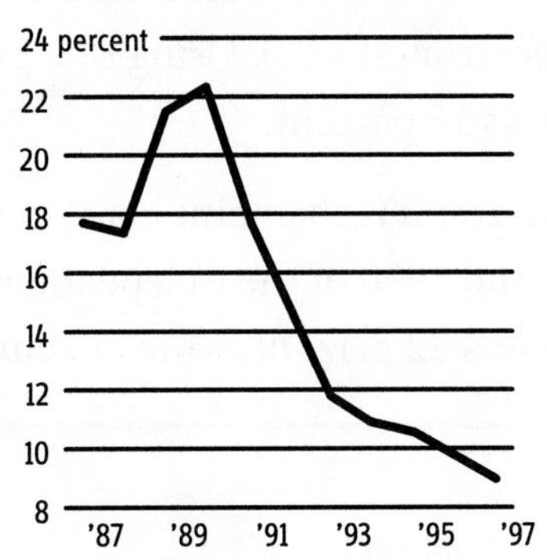

Note: The 1997 rate represents the borrowers whose first loan repayments came due in 1997 and who defaulted sometime before the end of the 1998 fiscal year.

Source: U.S. Department of Education **AP**

a. Is this the graph of a function? Why?

b. During which years was the default rate increasing?

c. During which years was the default rate decreasing?

d. What was the default rate in 1991?

e. What was the maximum default rate between 1987 and 1997? When did it occur?

Solution

a. Because any vertical line intersects this graph at no more than one point, the graph is that of a function.

b. We estimate the interval of time when the default rate is increasing by reading the time interval on the horizontal axis over which the graph of the function rises from left to right: We estimate that the default rate was increasing from 1988 to 1990.

c. The default rate was decreasing from 1987 to 1988 and from approximately 1990 to 1997.

d. The default rate in 1991 is given by the value of the dependent variable (represented in the vertical axis) that corresponds to the value 1991 of the independent variable (represented on the horizontal axis). Using this graph we estimate that the default rate in 1991 was 16 percent.

e. Observing that the highest point of the graph is the point corresponding to the value 22 of the dependent variable and 1990 of the independent variable, we see that the maximum default rate was 22 percent, which occurred in (approximately) the year 1990.

In addition to the maximum and minimum, some points of special interest on the graph of a function are those where the function changes from increasing to decreasing, and those where the function changes from decreasing to increasing. These are "turning points." The value of the response variable at a point where the function changes from increasing to decreasing is a **relative maximum** (or **local maximum**) value of the function. The value of the response variable at a point where the function changes from decreasing to increasing is a **relative minimum** (or **local minimum**) value of the function. The graph here shows a function with two relative maximum values (M_1 and M_2) and one relative minimum (m).

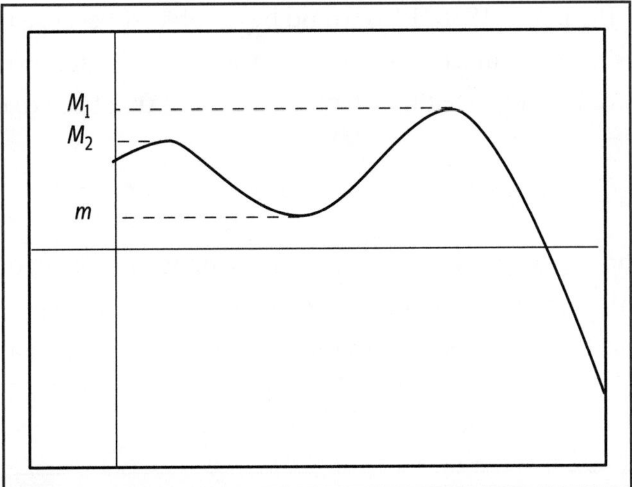

Relative Minimum Value: m. Relative Maximum Values: M_1 and M_2

Example 3.2

Using the function in Example 3.1, give the maximum value, the minimum value, and any relative maximum or minimum values that exist. Also indicate the year in which they occur. Explain why these values are of interest.

Solution

The maximum value of the function is approximately 22 percent, as given in the solution of Example 3.1d. It occurred in 1990, which means the highest default rate occurred in 1990 when approximately 22 percent of borrowers defaulted. This value of 22 percent is also a relative maximum, and a point where the function changes from increasing to decreasing. For the government analyst, this point is of interest because it indicates a trend change and might indicate broader economic improvements. There is no other relative maximum.

The minimum value is approximately 8.8 percent (in 1997), and the only relative minimum is approximately 17.5 percent (in 1988). Note that 8.8 is the minimum value for the time interval considered on the graph, but it is not a relative minimum because the function does not change from decreasing to increasing at the point (1997, 8.8). The point (1988, 17.5) would be of interest because the default rates stopped decreasing and started to increase. Because this is an unwanted change, the analyst would study what may have occurred at that time to cause such a change.

The intervals of the horizontal axis over which a function is increasing or decreasing may be quite small. Sometimes we need to overlook small changes to see the general trend. In Topic 6, we will see how to model more formally such a general trend.

Example 3.3

The following graph shows the fluctuations in annual mean temperature in New York City's Central Park for the years 1876 to 2003. Disregarding small oscillations, explain the general behavior of annual mean temperature in Central Park, giving the maximum and minimum values. (Source: *National Weather Service Forecast Office,* www.erh.noaa.gov/okx/climate.html.)

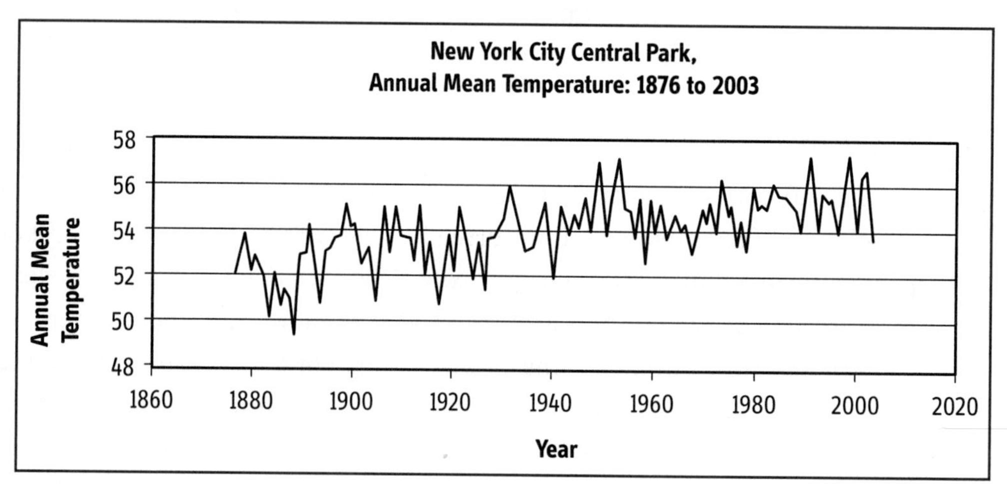

Solution

During the time period from 1876 to 2003, the annual mean temperature in Central Park ranged from a minimum of approximately 49.5 degrees around 1888 to a maximum of 57 degrees in 1998. The relative minimum values are getting larger as time increases, which can be seen if we look at the "dips" in the graph around 1888, 1893, 1904, 1917, 1926, 1940, 1958, 1967, 1978, and several in the 1990s. Similarly, the peaks in the graph also follow a somewhat upward trend.

Another useful piece of information about a function that we can observe from its graph directly is the rate at which the values of the response (dependent) variable are changing per unit change in the explanatory (independent) variable. This tells us whether the function values are increasing or decreasing rapidly or slowly.

If x_1 and x_2 are two values of the explanatory variable and y_1 and y_2 are the corresponding values of the response variable, the **average rate of change** of y per unit change in x over the interval from x_1 to x_2 is the ratio $\frac{y_2 - y_1}{x_2 - x_1}$. We will refer to this quantity as the rate of change of y from x_1 to x_2.

Example 3.4

Estimates of cigarette consumption in the U.S. and of the numbers of cigarettes exported each year are given in the following table. (These data appeared in *The Wall Street Journal Almanac 1999*, page 208.) Compare the rates of change of the two given functions over the intervals of time from

a. 1970 to 1980

b. 1991 to 1993

c. 1994 to 1996

d. 1995 to 1996

Year	U.S. Consumption of Cigarettes	Exports of Cigarettes
1960	484,400,000,000	20,200,000,000
1970	536,400,000,000	29,200,000,000
1980	631,500,000,000	82,000,000,000
1990	525,000,000,000	164,300,000,000
1991	510,000,000,000	179,200,000,000
1992	500,000,000,000	205,600,000,000
1993	485,000,000,000	195,500,000,000
1994	486,000,000,000	220,200,000,000
1995	487,000,000,000	231,100,000,000
1996	487,000,000,000	243,900,000,000
1997	480,000,000,000	217,000,000,000

Solution

a. Because 536,400 million cigarettes were consumed in the U.S. in 1970 and 631,500 million were consumed in 1980, the rate of change in cigarette consumption from 1970 to 1980 was $\frac{631{,}500 - 536{,}400}{1980 - 1970} = \frac{95{,}100}{10} = 9{,}510$ million cigarettes per year. The rate of change in number of cigarettes exported during the same interval of time was $\frac{82{,}000 - 29{,}200}{10} = 5{,}280$ million cigarettes per year. Both the consumption and the export figures increased during this period, with the rate of increase in national consumption being almost twice the rate of increase in exports.

b. From 1991 to 1993, the rate of change in number of cigarettes consumed in the U.S. was $\frac{485{,}000 - 510{,}000}{1993 - 1991} = \frac{-25{,}000}{2} = -12{,}500$ million cigarettes per year. The negative sign here reflects the fact that the number of cigarettes consumed decreased. The rate of change in the number of cigarettes exported during the same period of time was $\frac{195{,}500 - 179{,}200}{2} = 8{,}150$ million cigarettes per year. During this two-year period, consumption decreased at a rate of

12,500 million cigarettes per year while exports increased at a rate of 8,150 million per year.

c. During the two-year period from 1994 to 1996, national cigarette consumption increased at a rate of 500 million cigarettes per year while exports increased at a rate of 11,850 million cigarettes per year.

d. From 1995 to 1996, national consumption remained the same (the rate of change was $\frac{0}{1} = 0$), while exports increased at a rate of 12,800 million per year.

Note that when the function is decreasing, the rate of change is negative because if $x_1 < x_2$ then $x_2 - x_1$ is positive and $y_2 - y_1$ is negative; that is, the values of the response variable decrease when the values of the explanatory variable increase. If the function is increasing, then $y_2 - y_1$ and $x_2 - x_1$ are both positive when $x_1 < x_2$ and so the rate of change is positive.

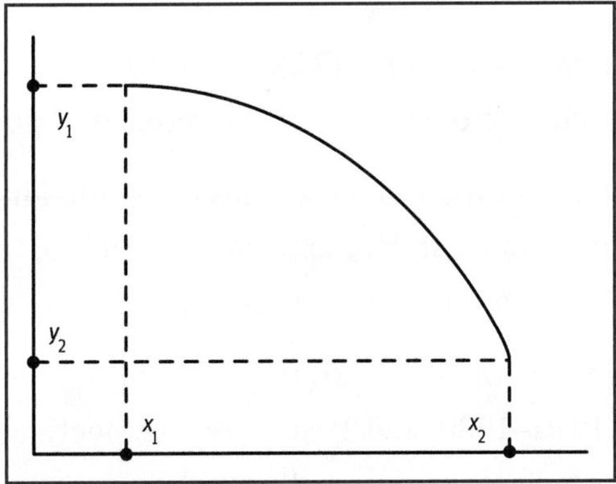

$y_2 - y_1$ is negative.

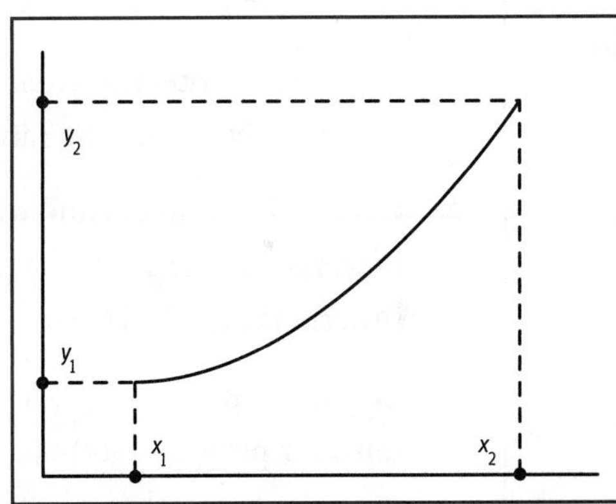

$y_2 - y_1$ is positive.

We can compare rates of change over different intervals by observing how steep the graph of the function is over each interval. If the rate of change is positive, a steeper graph means the rate of change is greater. We know this because for the same interval length $x_2 - x_1$, greater values of $y_2 - y_1$ give greater values of the quotient $\frac{y_2 - y_1}{x_2 - x_1}$.

Example 3.5

The following graphs are of the two functions given in the table in Example 3.4. Use the graph to answer the following questions.

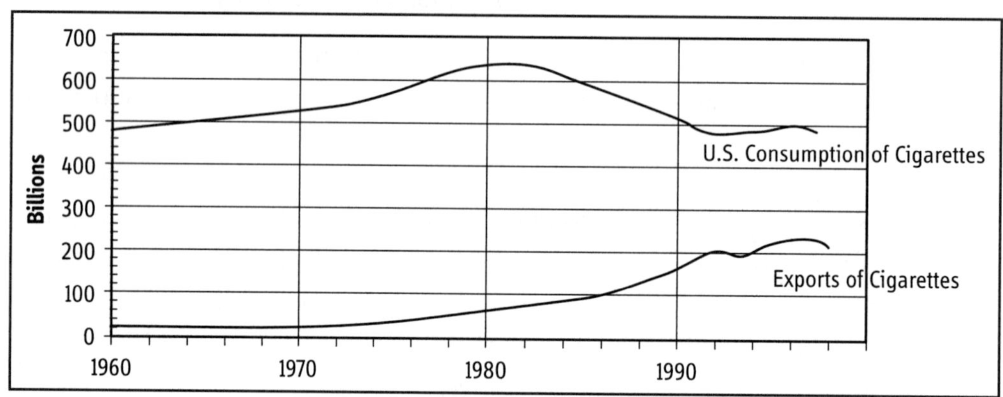

a. Give a time interval when the rate of change of U.S. cigarette consumption was negative while the rate of change of exports of cigarettes was positive.

b. Give an interval of time when the number of cigarettes consumed in the U.S. increased faster than the number of cigarettes exported. Which of the two functions has the larger rate of change over this interval?

c. Let R_1, R_2, R_3 be the rates of change of U.S. cigarette consumption during the ten-year periods 1960–1970, 1970–1980, and 1980–1990, respectively. Without calculating these rates, write them in descending order.

Solution

a. The graph shows cigarette consumption decreased while exports increased from 1980 to approximately 1992. An interval when the rate of change of U.S. cigarette consumption was negative and rate of change of exports was positive is, for example, the interval 1980–1992 (or any smaller interval contained within that interval).

b. The number of cigarettes consumed increased faster than the number of cigarettes exported during the interval 1970–1980. During that interval, the graph of the cigarette consumption function is steeper than the graph of the exports function; therefore, the rate of change of cigarette consumption is greater than the rate of change of exports.

c. R_3 is the smallest since it is negative because the graph is decreasing over the interval 1980–1990. The other two rates are positive, and R_2 is greater than R_1 because the graph is steeper over the interval 1970–1980 than it is over the interval 1960–1970. Placing the rates in descending order we have: R_2, R_1, R_3.

We conclude this topic with another observation we can make from the graph of a function. The way the graph is curved, upward or downward, indicates whether the *rate of change* of the function is increasing or decreasing.

We say that a function is **concave upward** when its graph is a curve bent upward. These are two such graphs:

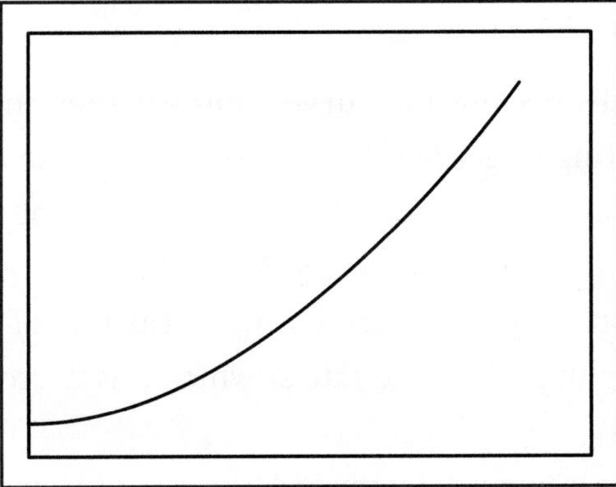

Increasing and Concave Upward

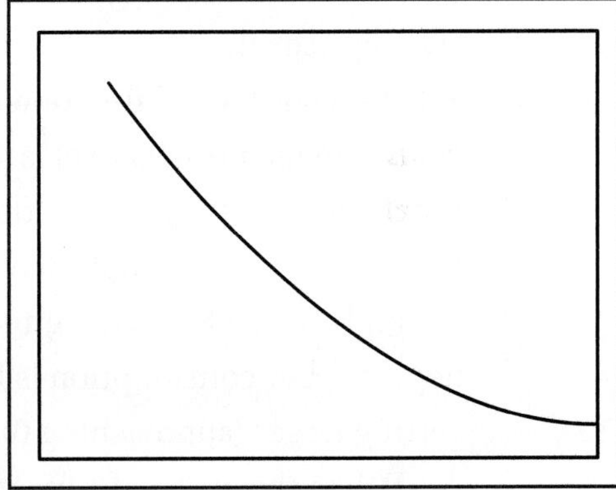

Decreasing and Concave Upward

The following two graphs are curved downward. The functions they represent are **concave downward**.

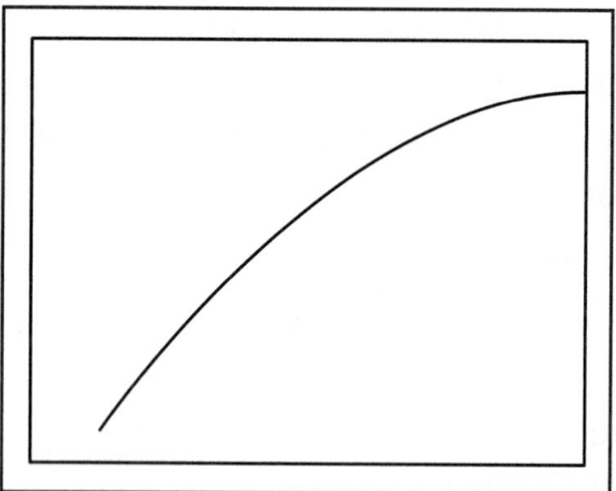

Increasing and Concave Downward

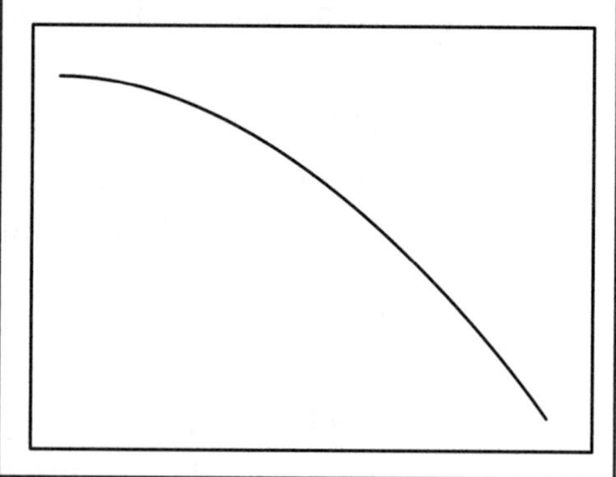

Decreasing and Concave Downward

If the graph of a function is curved upward, then the rate of change of the function is increasing. In the graph in Example 3.5, the graph of the function showing the number of cigarettes exported is increasing and curved upward over the interval 1980–1990. If we look at the rates of change over intervals of equal length, say two-year intervals, within the 1980–1990 period, we can see that these rates are positive and getting larger as we move to the right because the graph is getting steeper.

If the graph of a function is decreasing and curved upward, then the rate of change (which is negative) is getting larger (or "less negative") because the graph is getting less steep. This is the case, for example, for the function that shows U.S. cigarette consumption over the time interval 1986–1993 (see the graph in Example 3.5), where cigarette consumption is decreasing so the rate of change is negative, but consumption is leveling off, so the rate at which it is decreasing is getting larger (approaching 0).

When the graph of a function is curved downward, the rate of change is decreasing. If the function is increasing and the graph curves downward, then the rate of change is positive and getting smaller; that is, the graph is rising but less steeply as we move from left to right. When the function is decreasing and curved downward, then the rate of change is negative and getting smaller (more negative); that is, the curve is falling more steeply as we move from left to right.

In many situations, it is important to look at whether the rate of change is increasing or decreasing; for example, if during an epidemic the number of new cases of sick people is increasing at an increasing rate, health officials would see it as a good sign when the rate of increase of number of new cases starts to decrease. This would mean that their control methods are working and that the end of the epidemic is nearer.

Example 3.6

The following graph, which appeared in *The New York Times* on September 28, 1999, gives the wholesale price of a 14.1-inch liquid crystal display screen (flat-screen used in computer notebooks, videogames, and increasingly in desktop monitors) from the first quarter of 1997 to the second quarter of 1999 and the projected price until the second quarter of 2000. Describe the changes in price and explain why the projected portion of the graph seems to agree with the previous portion.

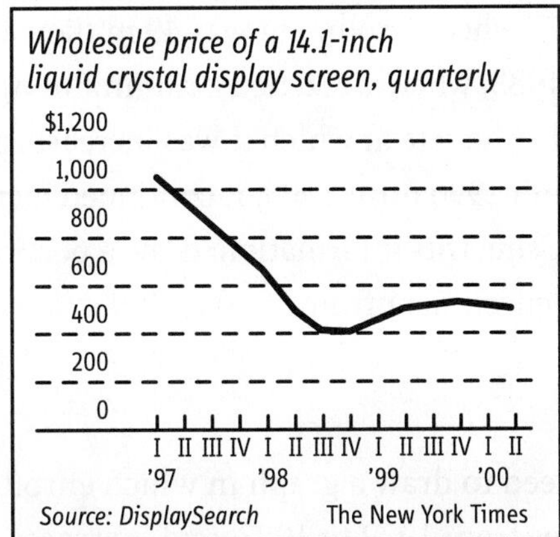

Solution

The price of the 14.1-inch liquid crystal screen was approximately $1,000 at the beginning of 1997 and decreased at a constant rate of approximately $100 per quarter during the year. During the first quarter of 1998, the price had a slightly sharper decline, but during the second and third quarters, the rate of change in price increased causing the decline in price to slow down. The price was lowest at the end of 1998 and then began to increase at a rate of about $100 per quarter. The rate of increase decreased during the first part of 1999, making it reasonable to expect that the price would level off and then start to decline again, continuing the previous trend.

In the following example, we construct a graph from a verbal description of trends.

Example 3.7

The number of students enrolled each year in elementary and secondary schools in the U.S. is given by a function that satisfies the following: the number of students increased from 40 million in 1960 to 51.2 million in 1970. From 1970 to 1983, enrollment decreased, slowly until 1977 when enrollment was 49 million and then decreased more rapidly from 1977 to 1983, when it reached a minimum of 45 million. Enrollment then increased rapidly (the rate at which it was increasing was also increasing) to 51.5 million in 1996. From 1996 until 2003, it continued increasing, but with a decreasing rate of change. Using this information, draw a possible graph of the function that describes student enrollment from 1960 to 2003.

Solution

To satisfy the given description, we need to draw a graph in which enrollment increases from 1960 to 1970, decreases from 1979 to 1983, and increases again from 1983 to 2003. There is a local maximum of 51.2 million in 1970 and a local minimum of 45 million in 1983. Based on the data given, other points that we need to include are (1960, 40), (1977, 49), and (1996, 51.5). To satisfy the other conditions, we draw a function that is concave downward from 1970 to 1983 with a decrease that is not so rapid at first, but gets more rapid. Indicating that the rate of increase from 1983 to 1996 is increasing, we draw the curve concave upward; that is, curved upward. Because the rate of increase decreases from 1996 to 2003, the graph is concave downward but still increasing from 1996 to 2003.

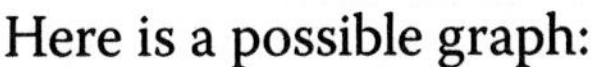

Here is a possible graph:

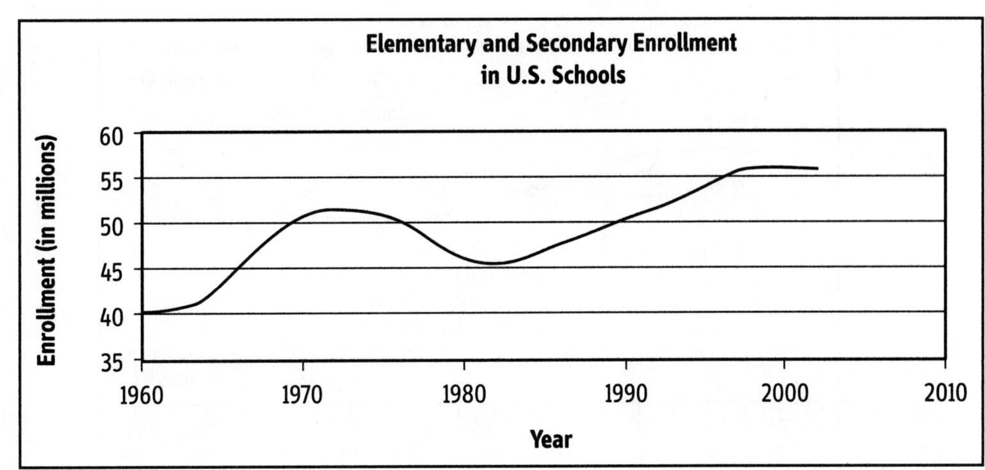

Summary

In this topic, we analyzed graphs of functions to identify relative (local) maximum values, relative (local) minimum values, intervals where the function is increasing and where it is decreasing, where the graph is concave upward and where it is concave downward. We interpreted these characteristics, easily seen on the graph of a function, in terms of the values of the function and the function's rates of change. We also looked at practical implications.

Explorations

1. The following graph (from the article "CBS News May Face More Cuts" that appeared in *The New York Times* on September 9, 1999) represents the television audience of nightly news programs from three major broadcast networks from 1980 to 1999. Use the graph to answer the following questions.

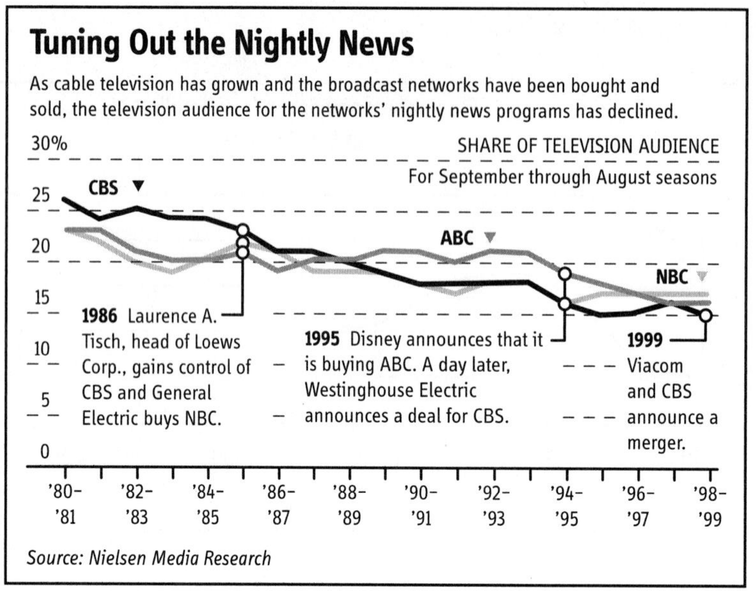
Tuning Out the Nightly News
As cable television has grown and the broadcast networks have been bought and sold, the television audience for the networks' nightly news programs has declined.
30%
SHARE OF TELEVISION AUDIENCE
For September through August seasons
25
CBS
20
ABC
15
NBC
10
5
0
1986 Laurence A. Tisch, head of Loews Corp., gains control of CBS and General Electric buys NBC.
1995 Disney announces that it is buying ABC. A day later, Westinghouse Electric announces a deal for CBS.
1999 Viacom and CBS announce a merger.
'80-'81
'82-'83
'84-'85
'86-'87
'88-'89
'90-'91
'92-'93
'94-'95
'96-'97
'98-'99
Source: Nielsen Media Research

a. Give an interval of time when all three networks were losing viewers.

b. Give an interval over which the audiences of two of the three networks were increasing while the audiences were decreasing for the third network.

c. Give an interval of time when the audiences of two of the three networks were decreasing while the audiences of the third network were increasing.

d. For the function that gives the percentage of viewers of "NBC Nightly News," give each relative minimum and the year it occurred.

e. For the same function as in part d, give each relative maximum and the year it occurred.

2. The following table gives the U.S. budget surpluses or deficits for the years 1955 through 2002 and estimated future values of surpluses or deficits for 2003. Surpluses and deficits are given in millions of dollars. (Source: *The World Almanac and Book of Facts 2004*, page 117.)

Year	Surpluses or Deficits (−)	Year	Surpluses or Deficits (−)	Year	Surpluses or Deficits (−)
1955	-2,993	1971	-23,033	1987	-149,661
1956	3,947	1972	-23,373	1988	-155,151
1957	3,412	1973	-14,908	1989	-153,319
1958	-2,769	1974	-6,135	1990	-220,469
1959	-12,849	1975	-53,242	1991	-269,492
1960	301	1976	-73,719	1992	-290,340
1961	-3,335	1977	-53,644	1993	-255,306
1962	-7,146	1978	-59,168	1994	-203,102
1963	-4,756	1979	-40,162	1995	-163,917
1964	-5,915	1980	-73,808	1996	-107,331
1965	-1,411	1981	-78,936	1997	-21,957
1966	-3,698	1982	-127,940	1998	70,039
1967	-8,643	1983	-207,764	1999	124,360
1968	-25,161	1984	-185,324	2000	236,917
1969	3,242	1985	-212,260	2001	127,021
1970	-2,842	1986	-221,140	2002	-157,666
				2003	-374,000

a. Give the rate of change of the surplus/deficit function from 1955 to 1960, from 1965 to 1970, from 1975 to 1980, and from 1985 to 1990.

b. Use the information obtained in part a of this exploration to decide whether the graph of the function will generally rise or fall over the interval of times mentioned. Does the information tell you that the graph will always rise or fall over each of those intervals? Explain.

c. What is the rate of change from 1998 to 2002? How does it compare with the rate of change over the period 1994 to 1998?

3. The following graph gives the surplus/deficit function from 1975 to 2003 (with an estimate for 2003).

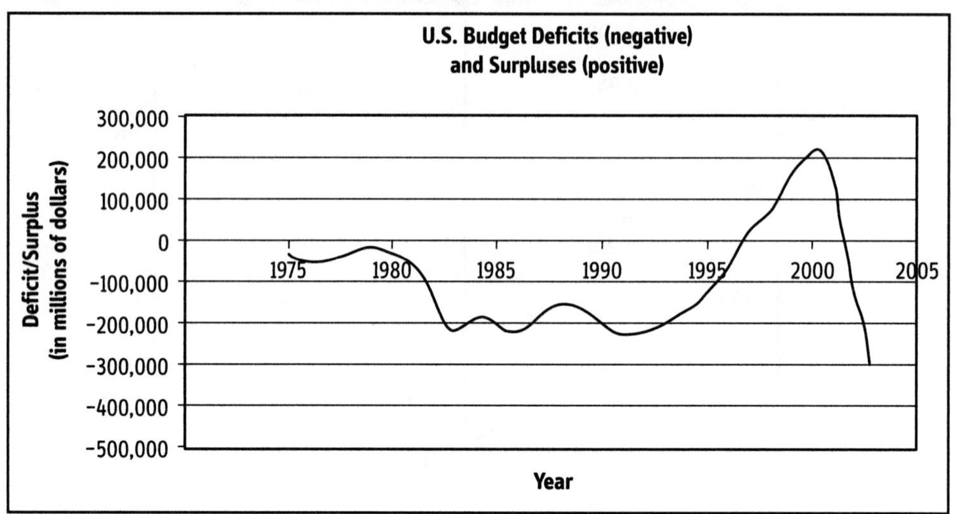

a. Give the intervals where the surplus/deficit is increasing and the intervals where it is decreasing.

b. Give the minimum and the maximum value of the function and the time when it occurred (or when it was expected to occur).

c. Give a relative maximum and a relative minimum (other than the values you gave in part b of this exploration) and the time each occurred. Explain what these values show about the surplus/deficit.

d. Give an interval where the function is increasing at an increasing rate and explain how you can tell this from the graph.

e. Give an interval where the function is decreasing at an increasing rate (so the rate of change of the function is becoming less negative). Explain what this shows about the surplus/deficit.

f. Assuming the prediction for 2003 is correct, what do you predict will happen with the values of the surplus/deficit function? Give a reason for your answer.

4. Explain what this graph from Example 3.5 shows overall about U.S. consumption and exports of cigarettes.

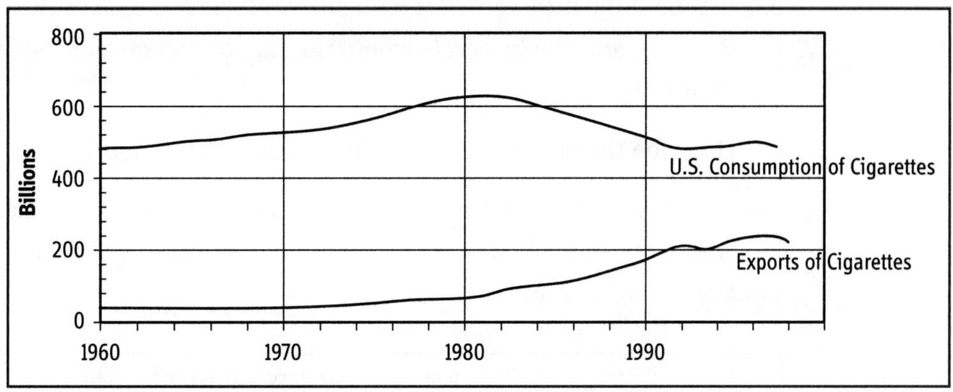

5. The following graph represents the federal debt from the years 1960 to 2002, with an estimated value for 2003. (Source: *The World Almanac and Book of Facts 2004,* page 117.)

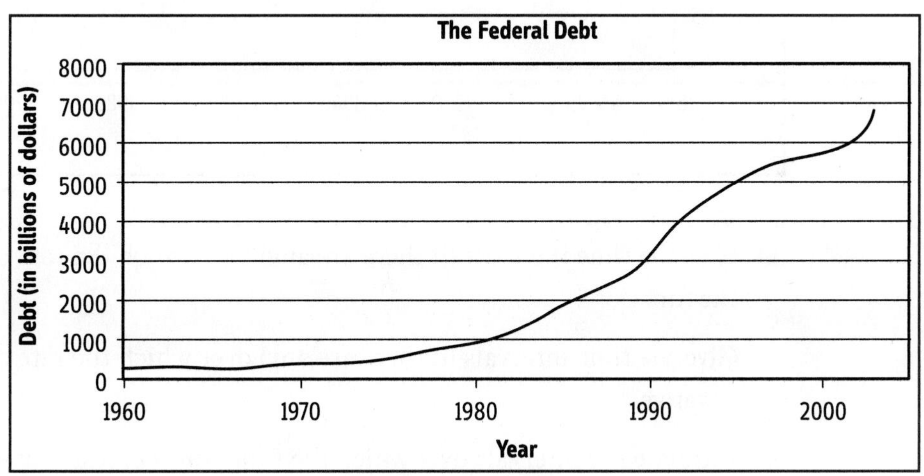

a. Give the time intervals (if there are any) over which the rate of change of the debt function is positive.

b. Give the time intervals (if there are any) over which the rate of change is negative.

c. Estimate the intervals over which the function is concave upward and those intervals where it is concave downward.

d. Estimate the intervals over which the rate of increase of the federal debt is growing, and the intervals where the rate of increase of the federal debt is shrinking.

e. Describe the changes in the values of the federal debt.

6. This graph shows the percentage of U.S. elementary and secondary school students who were enrolled in private schools. (Source: *The World Almanac and Book of Facts 2004*, page 292.)

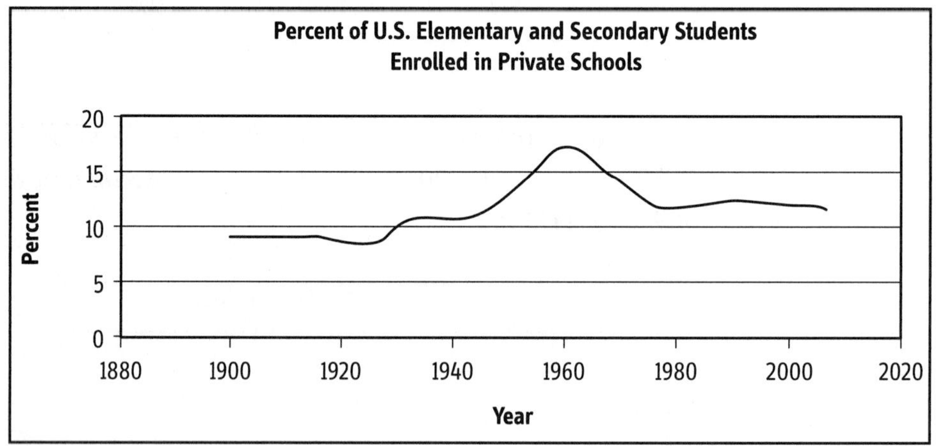

a. Give the time intervals (if there are any) over which the rate of change of the function is positive.

b. Give the time intervals (if there are any) over which the rate of change is negative.

c. Estimate the intervals over which the function is concave upward and those intervals over which it is concave downward.

d. Estimate the intervals over which the rate of increase of the percentage of private school students is growing, and the intervals over which the rate of increase of the percentage of private school students is shrinking.

7. The National Football League average salary was approximately $20,000 in 1960 and rose slowly but at an increasing rate. In 1975, the average salary was approximately $50,000 and in 1980, approximately $100,000. The rate at which salaries continued to rise increased quite sharply until 1991 when the average salary was approximately $780,000. Then, the average salary increased at a small constant rate until 1997, when it started to increase at a rate of approximately $100,000 per year for several years. Draw a possible graph of the function that represents the National Football League average salary.

8. In the 1983–1984 academic year, the number of students per computer in U.S. public schools was 125. The number of students per computer decreased to 75 in the 1984–1985 academic year and continued to decrease to 4.9 students per computer in the 2001–2002 academic year. Suppose we also know that the rate at which the number of students per computer decreased each year was increasing during this time period. Draw a possible graph of the function that represents the number of students per computer in U.S. public schools during this time period.

9. The following table gives the yearly Major League Baseball (MLB) television revenue from 1976 to 1996. (Source: Haupert, Michael. "The Economic History of Major League Baseball." *EH.Net Encyclopedia*, edited by Robert Whaples. August, 2003. http://eh.net/encyclopedia/article.haupert.mlb.)

Year	MLB Television Revenue (millions of $)	Year	MLB Television Revenue (millions of $)
1976	50.01	1986	321.60
1977	52.21	1987	349.80
1978	52.31	1988	364.10
1979	54.50	1989	246.50
1980	80.00	1990	659.30
1981	89.10	1991	664.30
1982	117.60	1992	363.00

Year	MLB Television Revenue (millions of $)	Year	MLB Television Revenue (millions of $)
1983	153.70	1993	616.25
1984	268.40	1994	716.05
1985	280.50	1995	516.40
		1996	706.30

a. Find the rate of change of TV revenue for each of the following four-year periods: 1976–1980, 1980–1984, 1984–1988, 1988–1992, and 1992–1996.

b. Give the change in TV revenue for each one-year period between 1984 and 1988. Are any of these numbers equal to the rate of change in TV revenue over the four-year period 1984–1988? What is the relationship between the change per year and the rate of change over the four-year interval?

c. Suppose that the rate of change of TV revenue over the four-year period 1996–2000 equals the annual change for each year during that period. What would the graph of the function over that period of time look like? Explain.

10. The next table gives the average ticket price of major-league baseball games from 1995 to 2002. (Source: Haupert, Michael. "The Economic History of Major League Baseball." *EH.Net Encyclopedia,* edited by Robert Whaples. August, 2003. http://eh.net/encyclopedia/article.haupert.mlb.) Use this information to answer the following questions.

Year	Average Ticket Price ($)
1995	10.76
1996	11.32
1997	12.06
1998	13.58

Year	Average Ticket Price ($)
1999	14.45
2000	16.22
2001	17.20
2002	17.85

a. Give the rate of change of ticket price over each one-year period between 1995 and 2002.

b. When was the average ticket price decreasing? When was it increasing?

c. Use the data obtained in part a of this exploration to decide where the graph of the average ticket price function will be concave upward and where it will be concave downward.

d. Graph the average ticket price function. Does your graph agree with your answers for parts b and c? Explain.

11. The following graph shows the top major-league baseball salary for each of the years 1988 to 2004. Write a paragraph to explain what this graph shows. In particular, identify when the graph is concave upward and when it is concave downward and explain what that tells you about the top major-league baseball salaries during that time period. Add any other information you can get from the graph.

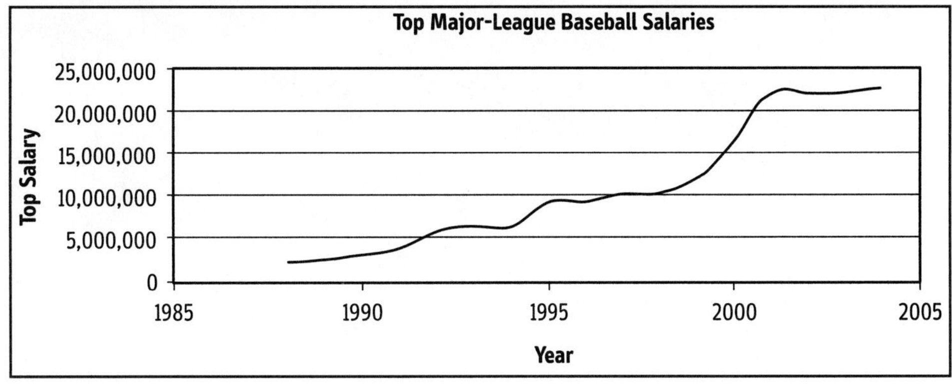

TOPIC 4

Multiple Variable Functions

OBJECTIVES

After completing this topic, you will be able to

- Identify multiple explanatory variables that influence a response variable
- Work with multiple variable functions using words, symbols, numbers in tables, and graphs
- Recognize how the response variable changes when all but one explanatory variable is held constant
- Analyze various quantities relating to real-world applications and understand how they are modeled by multiple variable functions

In previous topics, we looked at situations where we assumed, for simplicity, that one variable (the response or dependent variable) was determined completely by just one other variable. What we often do in examples like these is assume that other variables do not affect the value of the response variable (at least not enough to matter). We will now look at situations in which one response variable depends on several explanatory variables, and we will examine the nature of this dependence in a variety of situations.

We first explore some examples in which a response variable depends on several explanatory variables.

Example 4.1

For each of the following response variables, identify at least three explanatory variables that influence the given variable, and for each explanatory variable identified, determine whether it is a categorical variable or a quantitative variable.

a. The cost of a 1000-mile car trip

b. The length of time a traffic light is set to remain yellow

c. The time it takes to travel from Boston to Miami

d. A college student's grade point average

Solution

There are many variables that may influence each response variable; we give some possibilities but encourage other answers as well.

a. Three variables that influence the cost of a 1000-mile car trip are the price of gasoline, the kind of car driven, and how much the driver pays in road tolls. The kind of car is a categorical variable while the price of gasoline and amount paid in tolls are quantitative variables.

b. The length of time a traffic light is set to remain yellow depends on the speed limit of the road on which it is located, how many other traffic lights there are nearby on the same road, and how heavy the traffic load is on the road. The first two variables are quantitative. How heavy the traffic load is on the road would be a quantitative variable if we measured "how heavy the traffic load is" in terms of how many cars typically travel on the road. If we measured "how heavy the traffic load is" by using categories such as light, medium, and heavy, then it would be a categorical variable.

c. The time it takes to travel from Boston to Miami is a function of the mode of transportation, when you travel (morning, afternoon, evening, or night), and how many stops you make. "How many stops" is a quantitative variable; the other two are categorical.

d. Explanatory variables that influence a college student's grade point average are how much time the student spends studying, what types of courses he or she is enrolled in, and how much time he or she spends watching television or playing video games. "Types of courses" is a categorical variable; the other two are quantitative variables.

Functional relationships in which there are multiple explanatory variables can be given in the same four ways that functions with one explanatory variable are given: using words, tables, symbols, and charts or graphs. The following examples illustrate these modes.

Example 4.2

An adult's basal metabolic rate (BMR) is roughly equivalent to the number of calories burned in a day. The formula (which can be found on www.health.gov) to help people between the ages of 20 and 90 calculate their BMR can be divided into the following six steps:

1. Multiply your weight in pounds by 4.4 (women) or 6.2 (men).
2. Multiply your height in inches by 4.7 (women) or 12.7 (men).
3. Add the answers from 1 and 2.
4. Multiply your age in years by 4.7 (women) or 6.8 (men).
5. Subtract the answer for step 4 from step 3.
6. Add 655 (women) or 666 (men).

The final answer is your BMR.

a. Compare the BMR for a man and a woman, each of whom is 5 feet 7 inches tall, weighs 145 pounds, and is 34 years old.

b. Write two symbolic formulas to calculate BMR, one for men and one for women.

Solution

a. We first need to convert 5 feet 7 inches to inches; 5 feet is 60 inches, so the man and woman are 67 inches tall. For the woman, we perform the following computations:

1. 4.4 * 145 = 638
2. 4.7 * 67 = 314.9
3. 638 + 314.9 = 952.9
4. 4.7 * 34 = 159.8
5. 952.9 – 159.8 = 793.1
6. 793.1 + 655 = 1448.1

For the man, the steps are similar:

1. 6.2 * 145 = 899
2. 12.7 * 67 = 850.9
3. 899 + 850.9 = 1749.9
4. 6.8 * 34 = 231.2
5. 1749.9 – 231.2 = 1518.7
6. 1518.7 + 666 = 2184.7

The man's BMR is higher because most men have larger muscles. These numbers are averages, of course, and are based on studies using large numbers of men and women. In addition to the amount of lean tissue in one's body affecting BMR, an individual's activity level will also affect his or her BMR.

b. We'll use the six steps to put together a general symbolic formula for women and another one for men.

$$\text{BMR for women} = 4.4 * \textit{weight} + 4.7 * \textit{height} - 4.7 * \textit{age} + 655$$

$$\text{BMR for men} = 6.2 * \textit{weight} + 12.7 * \textit{height} - 6.8 * \textit{age} + 666$$

In writing formulas, it is common to use just one letter to represent variables and to omit the sign for multiplication. Letting w represent weight in pounds, h represent height in inches, a represent age in years, and B represent BMR, we can write the formulas as

$$B_{\text{women}} = 4.4w + 4.7h - 4.7a + 655$$

$$B_{\text{men}} = 6.2w + 12.7h - 6.8a + 666$$

(Note that we use the subscript on B for gender.)

For quantities that are functions of more than one variable, we sometimes want to investigate the behavior of one variable at a time as we hold the other variables constant.

Example 4.3

a. Graph the BMR for a woman who is 5 feet 6 inches tall and weighs 130 pounds, as a function of age. Describe what happens to her BMR for every ten years she ages. Does this description change for a woman of different height and weight?

b. Graph the BMR for a man who is 5 feet 10 inches tall and weighs 150 pounds, as a function of age. Describe what happens to his BMR for every ten years he ages. Does this change for a man of different height and weight?

Solution

a. If we let $w = 130$ and $h = 66$, the formula for B_{women} becomes

$$B_{\text{women}} = 4.4 * 130 + 4.7 * 66 - 4.7 * a + 655, \text{ or } B_{\text{women}} = 1537.2 - 4.7a$$

Here is the graph of this function:

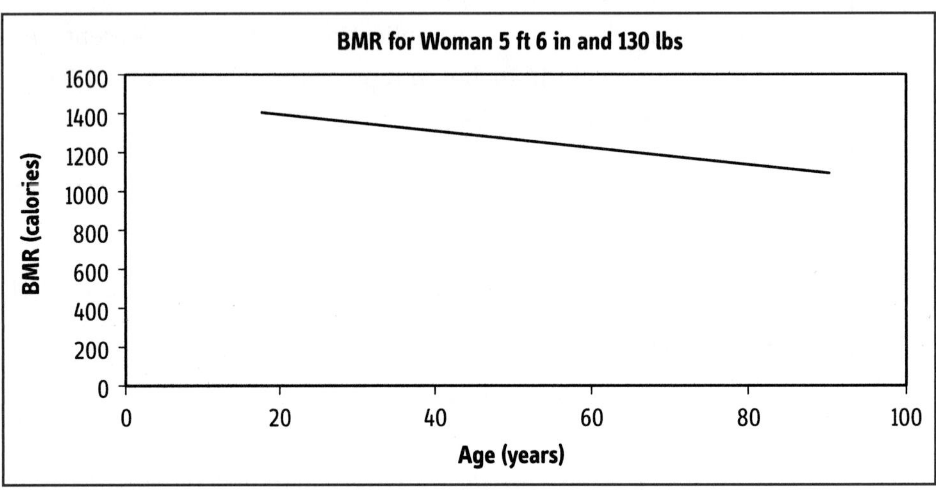

As the woman ages from 30 to 40, her BMR changes from 1396.2 to 1349.2, a decrease of 47. Additional computations will show the same decrease as she ages from 43 to 53, and for any ten-year change in age. If the woman were a different height and weight, the constant term in the formula (1537.2) would be different, but the other term ($-4.7 * a$) would remain the same. Therefore, the change in BMR for any ten-year period would still be a decrease of 47. This is a linear function (the graph is a straight line) and -4.7 is the slope of the line. It tells us that for every one-year increase in age, the BMR decreases by 4.7. And, as we saw, for a ten-year increase in age, BMR decreases by $4.7 * 10 = 47$. (Linear functions are discussed in more detail in Topic 5.)

b. For a man who is 5 feet 10 inches tall and who weighs 150 pounds, the formula for BMR is

$$B_{\text{men}} = 6.2 * w + 12.7 * h - 6.8 * a + 666$$
$$= 6.2 * 150 + 12.7 * 70 - 6.8 * a + 666, \text{ so } B_{\text{men}} = 2485 - 6.8a$$

Here's the graph:

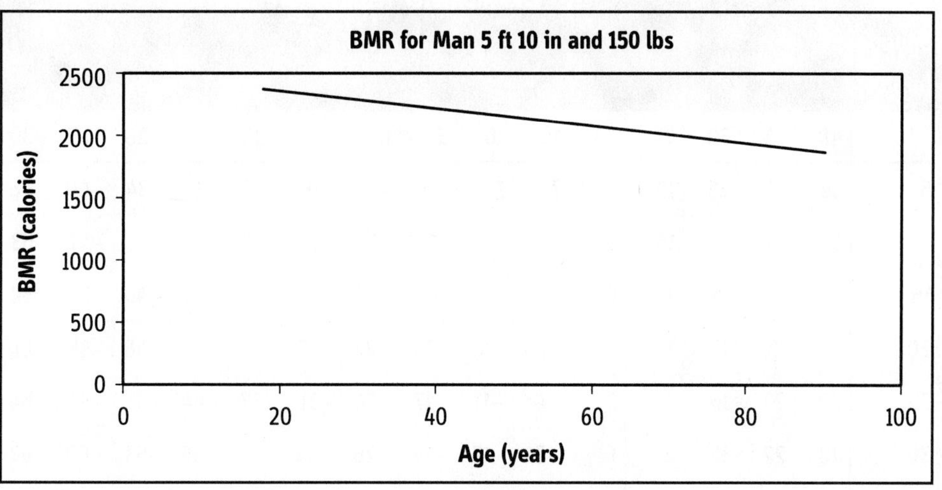

For each ten year increase in age, a man's BMR decreases by 68. This will be true for a man who is 5 feet 10 inches tall and weighs 150 pounds, as well as for any other man, using this model.

Often in winter months, weather reporters will give temperature and wind chill equivalent temperature because how cold we feel when we go out into winter weather depends not only on the outside temperature, but also on wind speed. The term *wind chill* is used to describe the equivalent temperature felt by exposed skin for a particular temperature of the surrounding air and wind speed. The following table gives wind chill equivalent temperatures for various combinations of air temperature, in degrees Fahrenheit, and wind speeds, in miles per hour. Note that the effect of the wind is negligible if its speed is less than 4 miles per hour, and wind speeds greater than 45 miles per hour do not significantly affect body heat further. (Source: *Mount Washington Observatory*, www.mountwashington.org/weather/wind-chill.html.)

Wind Chill Equivalent Temperature

Wind Speed (mph)	Thermometer Temperature (degrees F)																	
	40	35	30	25	20	15	10	5	0	−5	−10	−15	−20	−25	−30	−35	−40	−45
5	36	31	25	19	13	7	1	−5	−11	−16	−22	−28	−34	−40	−46	−52	−57	−63
10	34	27	21	15	9	3	−4	−10	−16	−22	−28	−35	−41	−47	−53	−59	−66	−72
15	32	25	19	13	6	0	−7	−13	−19	−26	−32	−39	−45	−51	−58	−64	−71	−77
20	30	24	17	11	4	−2	−9	−15	−22	−29	−35	−42	−48	−55	−61	−68	−74	−81
25	29	23	16	9	3	−4	−11	−17	−24	−31	−37	−44	−51	−58	−64	−71	−78	−84
30	28	22	15	8	1	−5	−12	−19	−26	−33	−39	−46	−53	−60	−67	−73	−80	−87
35	28	21	14	7	0	−7	−14	−21	−27	−34	−41	−48	−55	−62	−69	−76	−82	−89
40	27	20	13	6	−1	−8	−15	−22	−29	−36	−43	−50	−57	−64	−71	−78	−84	−91
45	26	19	12	5	−2	−9	−16	−23	−30	−37	−44	−51	−58	−65	−72	−79	−86	−93
50	26	19	12	4	−3	−10	−17	−24	−31	−38	−45	−52	−60	−67	−75	−81	−88	−95
55	25	18	11	4	−3	−11	−18	−25	−32	−39	−46	−54	−61	−68	−75	−82	−89	−97
60	25	17	10	3	−4	−11	−19	−26	−33	−40	−48	−55	−62	−69	−76	−84	−91	−98

Example 4.4

a. Suppose wind speed is 25 miles per hour (mph). Graph wind chill equivalent temperature as a function of thermometer temperature and describe this graph.

b. Suppose the temperature is fixed at 30 degrees Fahrenheit and the wind speed is increasing. If wind speed increases by 10 mph from 15 to 25 mph, how much does the wind chill equivalent temperature change? If wind speed increases from 35 to 45 mph, how much does the wind chill equivalent temperature change?

c. Suppose the temperature is 20 degrees Fahrenheit. Graph the wind chill equivalent temperature as a function of wind velocity and describe this graph.

d. Suppose the weather forecaster says that the wind chill equivalent temperature is –43 degrees Fahrenheit and the temperature is –10 degrees Fahrenheit. How fast is the wind blowing?

Solution

a.

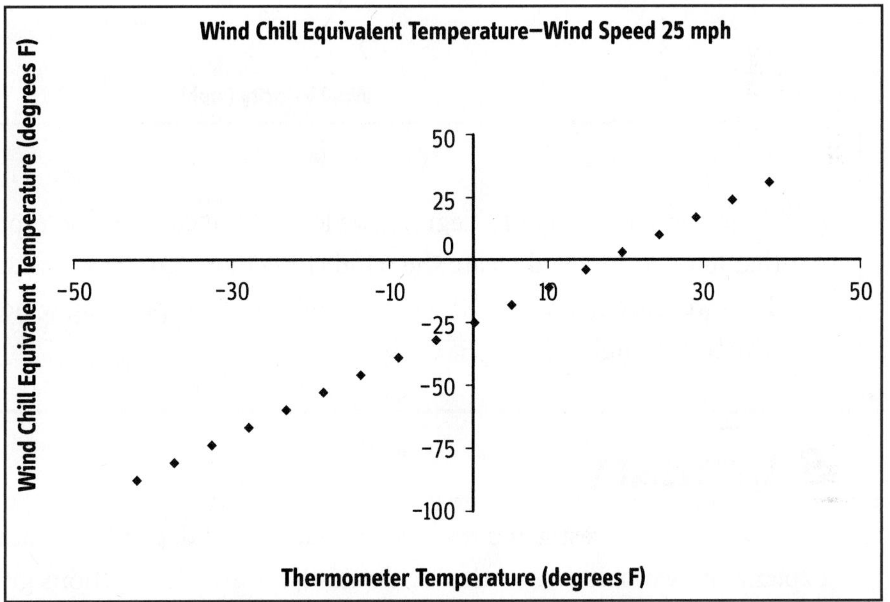

The graph shows a fairly steady and constant increase in wind chill equivalent temperature as the thermometer temperature increases over the interval from –45 to 40 degrees Fahrenheit.

b. If wind speed increases from 15 to 25 mph, the wind chill equivalent temperature drops from 19 to 16 degrees so it decreases by 3 degrees. If wind speed increases from 35 to 45 mph, the wind chill equivalent temperature decreases by 2 degrees.

c. The graph shows that as wind velocity increases, the wind chill equivalent temperature decreases if the temperature is fixed at 25 degrees Fahrenheit.

The decreases are greater for the lower wind velocity values and level off as wind velocity increases.

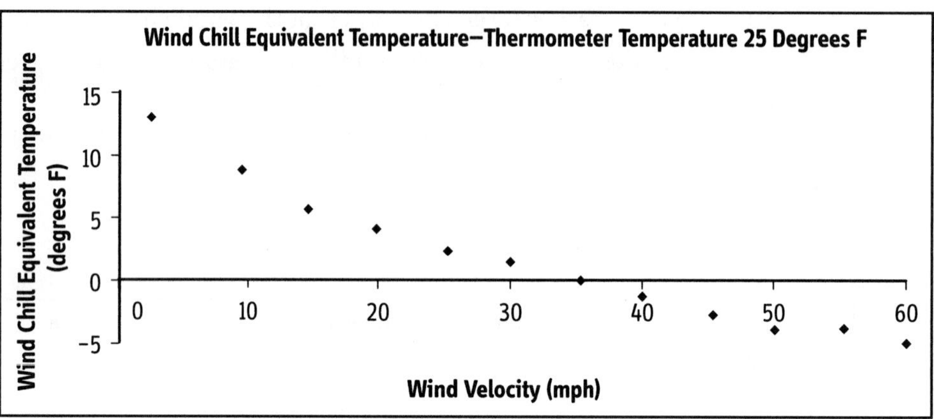

d. If the temperature is –10 degrees, we look at the column corresponding to that temperature and locate the wind chill equivalent temperature of –43. Then we trace that row to the left to find the corresponding wind velocity, which is 40 mph.

Summary

In this topic, we investigated response variables that depend on more than one explanatory variable. We worked with multiple variable functions given in words, in symbols, as graphs, and in table format. We explored basal metabolic rate (BMR) and wind chill equivalent temperature to see how the response variable is affected when all but one explanatory variable is held constant. We identified a linear relationship between BMR and age, which sets the stage for Topic 5.

Explorations

1. In each of the following situations, identify at least three explanatory variables that influence the given response variable. For each explanatory variable, identify if it is quantitative or categorical.

a. The amount of time a student spends writing a research paper for a particular course in which he or she is enrolled

b. A one-month electric bill for a family's residence

c. The amount of profit made by a business during a one-year period

2. The estimated number of calories burned in one day also depends on a person's activity level. If a person is sedentary, his or her activity factor is approximately 1.3; if moderately active, the activity factor is 1.4; if very active, the activity factor is 1.7. Then,

   Total calories needed in one day = BMR * activity factor

   Find the daily calorie needs for the woman and man described in Example 4.2 if they are sedentary. By how much will their calorie needs change if they are moderately active? By how much will their calorie needs change if they are very active?

3. One way to find out whether your weight is "reasonable" for your height is to calculate your body mass index (BMI). A BMI of 20 to 25 is normal. BMI is a function of height and weight and is determined as follows: To calculate a person's BMI, divide his or her weight in kilograms by the square of his or her height in meters. (To convert into kilograms, multiply pounds by 0.4536; to convert into meters, multiply inches by 0.0254.) Thus, you can write an equation to calculate BMI:

$$\text{BMI} = \frac{0.4536 * \textit{wt. in pounds}}{(0.0254 * \textit{ht. in inches})^2}$$

   a. Describe how this equation can be used to calculate BMI for a friend who does not understand the description of BMI.

   b. Abraham Lincoln was 6 feet 4 inches in height and weighed about 180 pounds, according to one historical account. From pictures you have seen of Abraham Lincoln, do you think his BMI would fall in the normal range? Calculate Lincoln's BMI.

c. Find the BMI for a man who weighs the same amount as Lincoln did but who is 5 feet 8 inches tall. Comment on how this compares to Lincoln's BMI.

d. A person who is 5 feet 8 inches tall reduces his or her weight from 200 pounds to 140 pounds. Create a table that contains weights at ten-pound intervals and find this person's BMI at each of those weights. By how much is BMI reduced when weight is reduced from 200 to 170 pounds? By how much is BMI reduced when weight is reduced from 170 to 140 pounds? What can you say about the formula to calculate BMI for fixed height?

4. For improving cardiovascular fitness, exercise physiologists recommend exercising so your heart rate is in the "target zone." The upper and lower limits of this target zone can be represented as a function of two variables: a person's age and resting heart rate (in beats per minute). These limits are found as follows:

Lower limit = (220 – *age in years* – *resting heart rate*) * 0.6 + *resting heart rate*

Upper limit = (220 – *age in years* – *resting heart rate*) * 0.8 + *resting heart rate*

a. Find the upper and lower limits of the target zone for a 20-year-old individual whose resting heart rate is 68 beats per minute.

b. How do these limits change for a 40-year-old individual with the same resting heart rate of 68 beats per minute?

c. Compare the upper and lower limits of the target zone for two 20-year-old individuals with resting heart rates of 65 and 75 beats per minute, respectively.

5. A person's blood alcohol level is a function of his or her gender, body weight, number of drinks consumed, and amount of time spent drinking. The following two tables, obtained from the Pennsylvania State Police, PA Liquor Control Board, give the approximate blood alcohol level for various values of these variables. Note that one drink is 1¼ ounces of 80 proof liquor, 12 ounces of beer, or 5 ounces of table wine.

Male	Body Weight in Pounds							
Drinks	100	120	140	160	180	200	220	240
0	.00	.00	.00	.00	.00	.00	.00	.00
1	.04	.03	.03	.02	.02	.02	.02	.02
2	.08	.06	.05	.05	.04	.04	.03	.03
3	.11	.09	.08	.07	.06	.06	.05	.05
4	.15	.12	.11	.09	.08	.08	.07	.06
5	.19	.16	.13	.12	.11	.09	.09	.08
6	.23	.19	.16	.14	.13	.11	.10	.09
7	.26	.22	.19	.16	.15	.13	.12	.11
8	.30	.25	.21	.19	.17	.15	.14	.13
9	.34	.28	.24	.21	.19	.17	.15	.14
10	.38	.31	.27	.23	.21	.29	.17	.16

Female	Body Weight in Pounds								
Drinks	90	100	120	140	160	180	200	220	240
0	.00	.00	.00	.00	.00	.00	.00	.00	.00
1	.05	.05	.04	.03	.03	.03	.02	.02	.02
2	.10	.09	.08	.07	.06	.05	.05	.04	.04
3	.15	.14	.11	.10	.09	.08	.07	.06	.06
4	.20	.18	.15	.13	.11	.10	.09	.08	.08
5	.25	.23	.19	.16	.14	.13	.11	.10	.09
6	.30	.27	.23	.19	.17	.15	.14	.12	.11
7	.36	.32	.27	.23	.20	.18	.16	.14	.13
8	.40	.36	.30	.26	.23	.20	.18	.17	.15
9	.45	.41	.34	.29	.26	.23	.20	.19	.17
10	.51	.45	.38	.32	.28	.25	.23	.21	.19

a. Describe how blood alcohol levels compare for a male and a female, each weighing 160 pounds, for increasing numbers of drinks; that is, describe the effect of the variable "gender" on blood alcohol levels for a weight level of 160 pounds. Use appropriate graphs.

b. On the same set of axes, sketch a graph of blood alcohol percentage as a function of number of drinks consumed for an "average" woman weighing 120 pounds and an "average" man weighing 160 pounds. Describe the graphs.

c. For a male consuming four drinks, sketch a graph of blood alcohol level as a function of weight. Repeat for a female who consumes four drinks, and explain what the graphs show.

6. A student's final grade in a course is often a function of various student performance measures such as homework, class participation, hour exams, and final exam, as well as the weights the professor assigns to each of these measures. Suppose one professor uses the system:

   Final grade = 0.4 * *hour exam average* + 0.25 * *final exam score* + 0.25 * *homework average* + 0.1 * *class participation grade*

   and a second professor uses the system:

   Final grade = 0.6 * *hour exam average* + 0.15 * *final exam score* + 0.15 * *homework average* + 0.1 * *class participation grade*

   a. Which professor's grading system would you prefer if your grades were as follows: *hour exam average* = 84; *final exam grade* = 68; *homework average* = 90; *class participation grade* = 95?

   b. If you scored 10 points higher on the final exam, what effect would that have on your final grade under each professor's grading system?

   c. If you scored 6 points lower on your hour exam average, what effect would that have on your final grade under each professor's grading system?

   d. Suppose you earned the following grades in a class: *hour exam average* = 78; *final exam grade* = 88; *homework average* = 70; *class participation grade* = 80. What weights would you assign to each of these components to set up a grading system? Assume each component must count for at least 5 percent of the grade (that is, have a weight of at least .05), and explain why you chose the weights you did.

7. Just as low temperatures and high winds together can reduce effective temperature, the combination of high temperatures and high humidity can produce adverse conditions. The following equation gives one model for the apparent temperature (AT), sometimes called *heat index*, as a function of air temperature in degrees Fahrenheit and relative humidity as a decimal between 0 and 1.

   $$\text{AT} = 2.70 + 0.885 * \textit{air temperature} - 78.7 * \textit{relative humidity} + 1.20 * \textit{air temperature} * \textit{relative humidity}$$

   a. Suppose the air temperature is 90 degrees Fahrenheit. Sketch a graph of the apparent temperature as a function of relative humidity. Describe your graph and what it tells you about humidity levels when the temperature is 90 degrees.

   b. Suppose the air temperature is 60 degrees Fahrenheit. Sketch a graph of the apparent temperature as a function of relative humidity. Describe your graph and what it tells you about humidity levels when the temperature is 60 degrees and how this compares to your graph and description in part a of this exploration.

   c. Fill in the following table. Describe how apparent temperature changes when the air temperature increases 15 degrees Fahrenheit from 60 to 75 degrees and from 75 to 90 degrees at 20 percent relative humidity versus 50 percent relative humidity versus 90 percent relative humidity. What is the rate of change of apparent temperature over each of these intervals?

Realtive Humidity

Temperature	.20	.50	.90
60 degrees			
75 degrees			
90 degrees			

TOPIC 5

Proportional, Linear, and Piecewise Linear Functions

OBJECTIVES

After completing this topic, you will be able to

- Identify proportional (directly and inversely), linear, and piecewise linear functions
- Graph directly proportional, linear, and piecewise linear functions
- Write equations to model situations using proportional, linear, or piecewise linear functions
- Find rates of change for functions that are proportional, linear, or piecewise linear

In this topic, we will study several types of functions that establish a fairly straightforward relationship between the response and the explanatory variable—linear and piecewise linear functions. Many situations are modeled by these types of functions.

Directly proportional functions are functions in which the response variable is equal to a constant times the explanatory variable; that is, the response variable is directly proportional to the explanatory variable. In symbols, this is represented as $y = c * x$, where y and x represent the response and explanatory variables, respectively, and c is a constant. You saw examples of such functions in Topic 2 (see Examples 2.4 and 2.5). The graph of a directly proportional function is a (straight) line that passes through the point (0,0). Any function whose graph is a line is a **linear function**. Directly proportional functions are special cases of linear functions. In linear functions, the relationship between the response and explanatory variable is given symbolically by $y = c * x + d$, where c and d are constants.

Example 5.1

A sports nutritionist recommends that exercisers drink two cups of water or other liquid before a race or workout and one cup for every 15 minutes they exercise during a workout to avoid dehydration.

a. Suppose a runner takes 3 hours and 15 minutes to complete a marathon. How many cups of fluid should the runner take in for this event?

b. Let F represent cups of fluid and N represent number of 15-minute intervals spent exercising, and write an equation that can be used to find the recommended number of cups of fluid to consume while exercising.

c. Explain the significance of each of the numbers that appears in the equation and sketch a graph of the equation.

Solution

a. Three hours and 15 minutes corresponds to thirteen 15-minute intervals. So the runner needs thirteen cups (one for each 15-minute interval) plus the two initial cups, which is fifteen cups of fluid.

b. In part a of this example, we computed: *cups of fluid* = 1 × (*number of 15-minute intervals spent exercising*) + 2. We can shorten this equation to: $F = N + 2$.

c. The number 2 in the equation represents the number of cups of fluid recommended at the start of the race; that is, when $N = 0$. The other number that appears in the equation is 1, which is the number of cups of fluid recommended for each 15-minute interval spent exercising. We graph N = "number of 15-minute time intervals spent exercising" on the horizontal axis because it is the variable that "explains" F = "number of cups of fluid recommended."

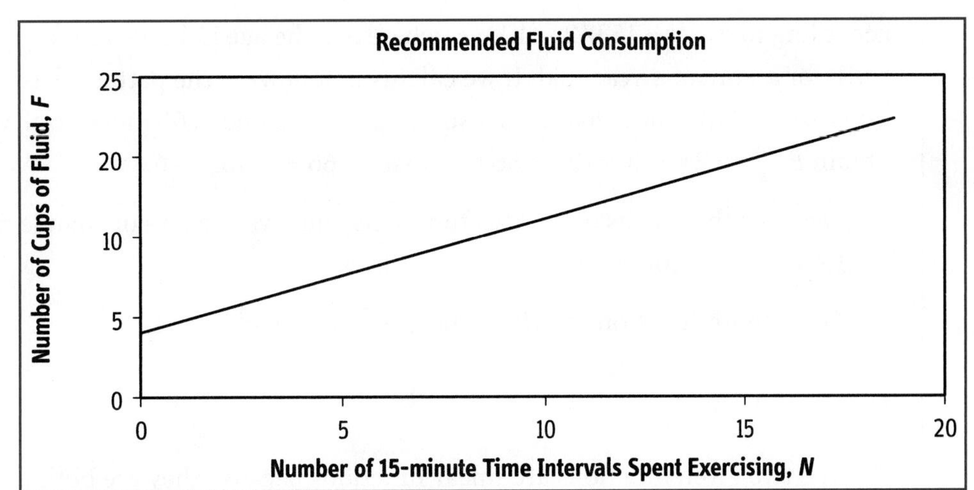

Data that fall in an exact linear pattern are characterized by the fact that no matter which two points you choose, the change in the dependent variable divided by the change in the independent variable will be a constant, called the **slope of the line**. This is the rate of change of values of the function (see Topic 3). This rate of change is given by the number c in the expression $y = c * x + d$. To see that this is true, we compute $\frac{y_2 - y_1}{x_2 - x_1}$, where x_1 and x_2 are two values of the explanatory variable and y_1 and y_2 are the corresponding values of the response variable. Because $y_2 = c * x_2 + d$ and $y_1 = c * x_1 + d$, we have

$$\frac{y_2 - y_1}{x_2 - x_1} = \frac{(c * x_2 + d) - (c * x_1 + d)}{x_2 - x_1} = \frac{c * x_2 + d - c * x_1 - d}{x_2 - x_1} = \frac{c * x_2 - c * x_1}{x_2 - x_1} = \frac{c * (x_2 - x_1)}{x_2 - x_1} = c$$

If the constant c is positive, the function is increasing, so the line rises as we move from left to right. If c is negative, the function is decreasing and therefore the line falls from left to right. The magnitude of c determines how steep the line is; the greater the magnitude, the steeper the line.

In the following example, we compare two linear functions and their rates of change.

Example 5.2

In Example 4.3 in Topic 4, we saw that the basal metabolic rate (BMR) for a woman who weighs 130 pounds and is 66 inches tall is related to the woman's age according to $B_{women} = 1537.2 - 4.7 * a$, where a is the age in years and B_{women} is the BMR for a woman a years old. If we calculate (following the pattern described in Topic 4) the BMR for a man who also weighs 130 lb and is 66 inches tall, we obtain $B_{men} = 6.2 * 130 + 12.7 * 66 - 6.8 * a + 666 = 2310.2 - 6.8 * a$.

a. Describe the graphs of the two functions and give the rate of change per year for B_{women} and for B_{men}.

b. Graph both functions on the same graph, and explain their differences.

Solution

a. The two given functions are linear functions because they are both of the form $B = c * a + d$, where B, the response variable, is B_{women} in one function and B_{men} in the other. The constants are $c = -4.7$ and $d = 1537.2$ for women, and $c = -6.8$ and $d = 2310.2$ for men. The graphs are lines of slope -4.7 and -6.8 for women and men, respectively. The rate of change of B_{women} is -4.7 points per additional year of age and the rate of change of B_{men} is -6.8 points per additional year of age.

b. Here is the resulting graph:

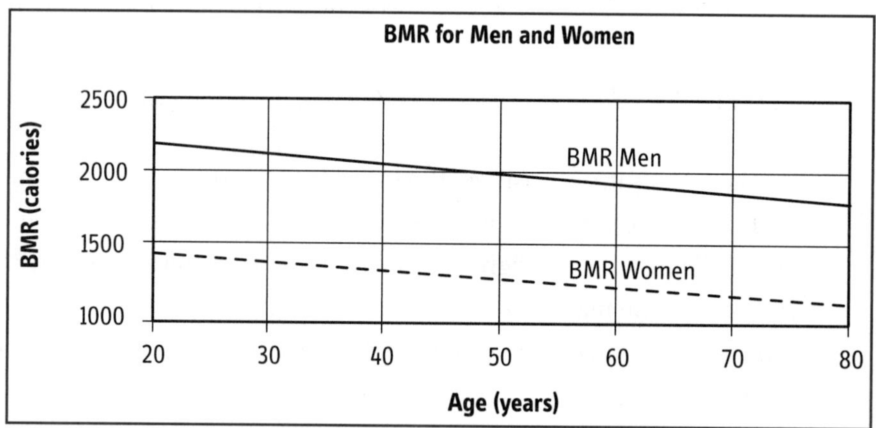

Both lines decline as we move from left to right, which is indicated by the negative slopes. One difference between the two lines is that the graph of the BMR for women is not as steep as that for men. This is because both slopes are negative and the slope of BMR for men is smaller (negative but larger in magnitude) than the slope of the BMR for women.

Linear functions are the only functions with a constant rate of change. To recognize linear functions, we compute the rate of change over several intervals. If we can see that the rate of change is always the same, then we can conclude the function is linear.

Example 5.3

The following is a portion of a table that indicates the postal rate for airmail letters from the U.S. to Canada. (Source: *The New York Times Almanac* 2004, page 167.)

Weight Up to (in ounces)	Rate (in dollars)	Weight Up to (in ounces)	Rate (in dollars)
1.0	0.60	5.0	
2.0	0.85	6.0	
3.0	1.10	7.0	
4.0	1.35	8.0	

a. If we consider only integer numbers of ounces, verify that the table defines the postal rate as a linear function of the letter weight.

b. Assuming the same pattern continues, fill the missing entries in the given table.

Solution

a. We observe from the table that for a 1.0 ounce change in weight, the cost increases by \$0.25. The change in the dependent variable divided by the change in the independent variable is always $\frac{0.25}{1.0} = 0.25$ dollars per ounce, which tells us that the function is linear and its graph is a line with slope 0.25.

b. We now know that the postage and the letter weight are related through a linear equation: postage $= c * (letter\ weight) + d$, where $c = 0.25$. To find the value of the constant d, we note that the postage corresponding to a weight of 1.0 ounce is \$0.60, so $0.60 = 0.25 * 1.0 + d$, hence $d = 0.6 - 1.0 * 0.25 = 0.35$. Thus, the linear relationship between postage and weight, for integer values of letter weight, is given by $postage = 0.25 * letter\ weight + 0.35$.

c. We complete the table using this equation.

Weight Up to (in ounces)	Rate (in dollars)	Weight Up to (in ounces)	Rate (in dollars)
1.0	0.60	5.0	1.60
2.0	0.85	6.0	1.85
3.0	1.10	7.0	2.10
4.0	1.35	8.0	2.35

A function whose graph consists of pieces of different lines is called a **piecewise linear** function. The function that associates income tax owed with salary earned is such a function, as the following example shows.

Example 5.4

The following table gives the year 2003 Federal Tax Rate Schedule for a single individual based on his or her taxable income. (Source: *The New York Times Almanac 2004*, page 164.)

Taxable Income	What You Pay
0 to \$7,000	10% of sum over \$0
\$7,000 to 28,400	\$700 + 15% of sum over \$7,000
\$28,400 to 68,800	\$3,910 + 25% of sum over \$28,400
\$68,800 to 143,500	\$14,010 + 28% of sum over \$68,800

Taxable Income	What You Pay
$143,500 to 311,950	$34,926 + 33% of sum over $143,500
More than $311,950	$90,514.50 + 35% of sum over $311,950

a. Give the tax an individual should pay if his or her taxable income is (i) $5,000 (ii) $21,000, (iii) $54,000, (iv) $250,000, (v) $350,000.

b. Describe symbolically the function that gives the taxes to be paid for a taxable income of i dollars.

c. Graph the function and describe the graph.

d. Give the rate of change in income tax if taxable income changes from $9,000 to $21,000.

e. Give the rate of change in income tax if taxable income changes from $21,000 to $54,000.

f. Give the rate of change in income tax if taxable income changes from $320,000 to $350,000.

Solution

a. (i) Since $5,000 is between 0 and $7,000, the corresponding tax is 10 percent of 5,000, that is (0.10) * $5,000 = $500. (ii) To find the tax that corresponds to a taxable income of $21,000, we need to find 15 percent of the difference 21,000 – 7,000 and add this amount to 700. The tax is then $700 + 0.15 * (21,000 – 7,000) = $2,800. (iii) The tax corresponding to a taxable income of $54,000 is $3,910 + 0.25 * (54,000 – 28,411) = $10,310. (iv) For an income of 250,000, the tax is $34,926 + 0.33 * (250,000 – 143,500) = $70,071. (v) For an income of 350,000, the tax is $90,514.50 + 0.35 * (350,000 – 311,950) = $103,832.

b. To describe this function, we need to use six different expressions, one for each given income range. Note that for an income at the end of a range, we can use either of two rules without ambiguity. For example, if the taxable income is $7,000, we compute the tax using the first line of the tax table to get 0.10 * 7,000 = $700. If we use the second line instead, we get

$\$700 + 0.15 * (7{,}000 - 7{,}000) = \$700 + 0 = \$700$, the same as before. If i represents the taxable income and t the corresponding income tax, the function is as follows:

If i is between 0 and 7,000, then $t = 0.10 * i$

If i is between 7,000 and 28,400, then $t = 700 + 0.15 * (i - 7{,}000)$

If i is between 28,400 and 68,800, then $t = 3{,}910 + 0.25 * (i - 28{,}400)$

If i is between 68,800 and 143,500, then $t = 14{,}010 + 0.28 * (i - 68{,}800)$

If i is between 143,500 and 311,950, then $t = 34{,}926 + 0.33 * (i - 143{,}500)$

If i is over 311,950, then $t = 90{,}514.50 + 0.35 * (i - 311{,}950)$

c. The graph of this function consists of pieces of six different lines. A portion of a line of slope 0.10 lies over the interval on the horizontal axis from 0 to 7,000; a portion of a line of slope 0.15 lies over the interval from 7,000 to 28,400; a portion of a line of slope 0.25 lies over the interval from 28,400 to 68,800; a portion of a line of slope 0.28 lies over the interval from 68,800 to 143,500; a portion of a line of slope 0.33 lies over the interval from 143,500 to 311,950; and a portion of a line of slope 0.35 lies over the values of the horizontal axis beyond 311,950. Here is the graph:

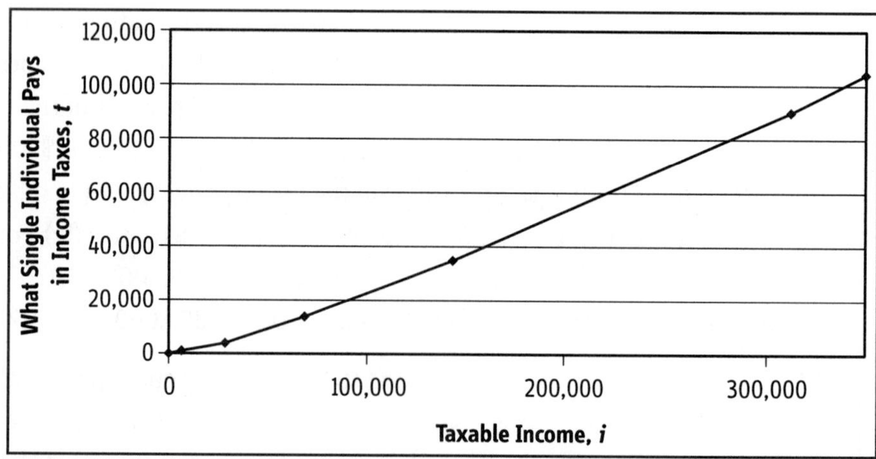

d. The rate of change from \$9,000 to \$21,000 is the rate of change of the function over that interval. This is the slope of the line $t = 700 + 0.15 * (i - 7{,}000)$, or 0.15. (We could also compute this rate by dividing the change in taxes by the change in income, but we have already seen that it equals the slope of the line.)

e. Over the interval from 21,000 to 54,000, the function consists of portions of two different lines, so we need to compute the rate of change by calculating the ratio of the change in the dependent variable divided by the change in the independent variable. The values of t corresponding to the income values of $21,000 and $54,000 are $2,800 and $10,310, respectively; the rate of change is $\frac{(10{,}310 - 2{,}800)}{(54{,}000 - 21{,}000)} \approx 0.228$, or 22.8 cents per dollar change in taxable income.

f. Over the interval from 320,000 to 350,000, the function satisfies the linear equation $t = 90{,}514.5 + 0.35(i - 311{,}950)$. The rate of change is the slope of this line, which is 0.35.

The following is another example of a piecewise linear function. Each piece consists of a horizontal segment with a "jump" between them.

Example 5.5

The instructions on the use of the allergy medicine Triaminic Syrup contain the following dosage table.

Age	Weight	Dosage
Under 6 years	Under 48 lb	Consult a doctor
6 to under 12 yrs	48–95 lb	2 teaspoons
12 yrs to adult	96+ lb	4 teaspoons

a. How much medication should we give a 10-year old boy who weighs $88\frac{1}{2}$ lb? One who weighs 95 lb 7 oz? One who weighs 99 lb 2 oz?

b. Describe in words and graph the function that relates the dosage as the response variable with the weight as the explanatory variable.

Solution

a. The dosage for a 10-year old who weighs $88^1/_2$ lb is 2 teaspoons because his weight falls in the 48–95 lb range. A child who weighs 95 lb 7 oz is between two categories, so the decision is not that clear. We prefer to err on the safe side and will give this child 2 teaspoons. (Some people may decide to round up to the closest pound, and in this case, would administer the higher dose.) The child who weighs 99 lb 2 oz gets 4 teaspoons of medicine because his weight is more than 96 lb.

b. There is no assigned value when the weight is less than 48 lb, so we say that the function is not defined for weights of less than 48 lb. As explained previously, we will administer the higher dose of 4 teaspoons only when the boy's weight reaches 96 lb, so from 48 lb to 96 lb (not included), the assigned value of the response variable is always 2 teaspoons. The corresponding value when the boy's weight is 96 lb or more is 4 teaspoons. We may describe this function in symbols as follows: $d = \begin{cases} 2 \text{ if } 48 \le w < 96 \\ 4 \text{ if } \quad w \ge 96 \end{cases}$ where d represents dose in teaspoons and w represents weight in pounds. The graph of this function consists of all points $(w,2)$ for $w < 96$ and all points $(w,4)$ for $w \ge 96$. Here is the graph:

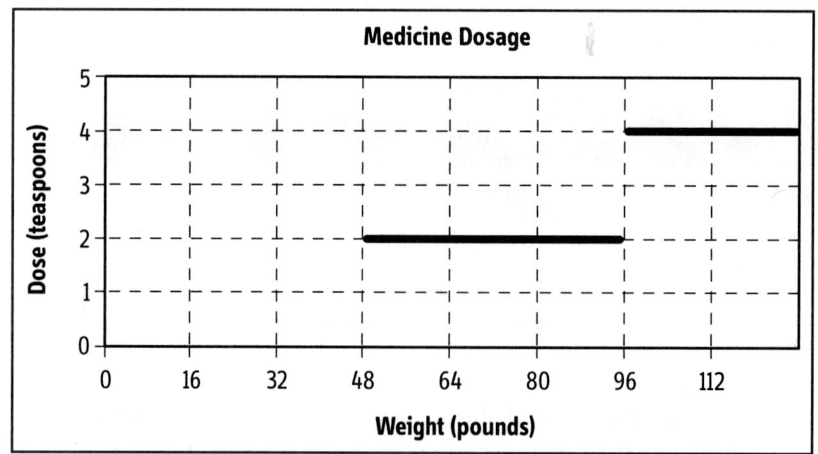

A function that has a symbolic expression $y = \frac{c}{x}$, where c is a constant, y represents the dependent variable, and x represents the independent variable, is an **inversely proportional** function. These are not linear functions as the following example shows.

Example 5.6

The average velocity v of a moving object is given by the distance d traveled divided by the time t elapsed. In symbols, $v = \frac{d}{t}$. The following table gives the average speed of the winners of the Daytona 500 auto race for several years. (Source: *The World Almanac and Book of Facts 2004*, page 955.)

Year	1960	1965	1970	1975	1980	1985	1990	1995	2000
Avg MPH	124.74	141.539	149.601	153.649	177.602	172.265	165.761	141.71	155.669

a. For each of the years given, calculate the time it took the winner to complete the 500-mile race. Create a table that displays the velocity and the corresponding time.

b. Graph the function that gives the time as the response variable and the average velocity as the explanatory variable for the 500-mile race. Include the points obtained in part a of this example.

c. Give the rate of change in time for an increase in average velocity from 130 to 140 mph, from 140 to 150 mph, and from 150 to 160 mph. How does this function differ from a linear function?

Solution

a. Because the distance traveled is 500 miles, the relationship between the distance and the time is given by $v = \frac{500}{t}$ or, solving for t, $t = \frac{500}{v}$, where t represents time in hours and v represents average velocity in miles per hour. So for $v = 124.74$, $t = \frac{500}{124.74} \approx 4.008$. Using this formula, we obtain the following times for the given velocities:

Average Speed	Time
124.74	4.008
141.539	3.533
149.601	3.342
153.649	3.254
177.602	2.815
172.265	2.903
165.761	3.016
141.71	3.528
155.669	3.212

b. This is the graph of the function $t = \frac{500}{v}$. We have included all the points given in the table along with additional ones.

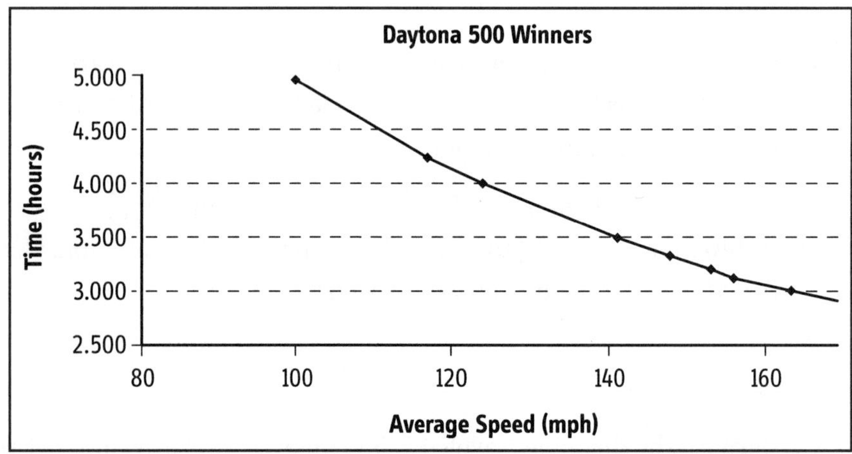

c. The time corresponding to 140 mph is $\frac{500}{140} \approx 3.571$ hrs, and the time corresponding to 130 mph is $\frac{500}{130} \approx 3.846$ hrs. The rate of change in time when the velocity increases from 130 to 140 mph is then $\frac{(3.571 - 3.846)}{(140 - 130)} = \frac{-0.275}{10}$ $= -0.0275$ hrs per each mile per hour change in velocity. This means that the time is reduced on average 0.0275 hrs = (0.0275) × (60) = 1.65 minutes per 1 mph increase over the interval from 130 to 140 mph. Similarly, we calculate the rate of change for an increase in velocity from 140 to 150 mph and from 150 to 160 mph, obtaining approximately −0.0238 and −0.0208 hrs per mph, respectively. We see that the rate of change of this function is not constant, but it decreases as the values of the independent variable increase. This distinguishes it from a linear function, which has a constant rate of change.

Summary

In this topic, we analyzed directly and inversely proportional functions. A directly proportional function has an equation of the form $y = cx$ and is a particular case of a linear function. Inversely proportional functions are given by equations of the form $y = \frac{c}{x}$ and are not linear. We also discussed linear functions, which have equations of the form $y = cx + d$, and piecewise linear functions, which are described by different linear equations over different intervals of x.

Explorations

1. A wireless phone service charges $14.99 as a monthly charge and $0.30 per minute for local calls.
   a. How much would you pay in a month when you made ten local calls of three minutes each?
   b. Let C represent your total monthly cost; assuming you made only local calls, did not receive any calls, and the total number of minutes you were on the phone is m, write the equation to find your monthly bill.
   c. How can you tell that the relationship between C and m follows a linear pattern?
   d. Graph the function that gives the total monthly cost and give the slope of the line.

2. A household electricity bill for one month in which 2,494 kWh (kilowatt per hour) were used includes, among other charges, a distribution charge. The bill reads:

Current Charges
Charges for – UTILITIES

Residential Rate: RS for Jun 9 – Jul 12		
Distribution Charge:		
Customer Charge	6.47	
200 KWH at 1.79600000¢ per KWH	3.59	
600 KWH at 1.59400000¢ per KWH	9.56	
1,694 KWH at 1.47200000¢ per KWH	24.94	
Transmission Charge:		
2,494 KWH at 0.37700000¢ per KWH	9.40	
Transition Charge:		
200 KWH at 1.79800000¢ per KWH	3.60	
600 KWH at 1.59400000¢ per KWH	9.56	
1,694 KWH at 1.47300000¢ per KWH	24.95	
Generation Charge:		
Capacity and Energy		
200 KWH at 4.82600000¢ per KWH	9.65	
600 KWH at 4.23800000¢ per KWH	25.43	
1,694 KWH at 3.88600000¢ per KWH	65.83	
PA Tax Adjustment Surcharge at 0.11818100%	0.23	
Total UTILITIES Charges		$193.21
Pay This Amount No Later Than Aug 2, 2002		**$193.21**
Account Balance		**$193.21**

a. Give the total distribution charge when electricity consumption is (i) 150 kWh, (ii) 755 kWh, (iii) 920 kWh.

b. Describe the function that the utility company uses to find the total distribution charge.

c. Sketch and describe the graph of the function.

d. Give the rate of change of the distribution charge if the electricity consumed changes from 150 to 160 kWh.

e. Give the rate of change of the distribution charge if the electricity consumed changes from 195 to 205 kWh.

f. Give the rate of change of the distribution charge if the electricity consumed changes from 300 to 310 kWh.

3. An aerobics instructor says she drinks one cup (8 ounces) of water before she starts her classes and then she drinks 10 ounces every 10 minutes.

   a. How much water does she drink when she teaches 2 one-hour classes in a row?

   b. Let *W* represent ounces of water and *N* the number of 10-minute intervals the aerobics instructor spends exercising, and write an equation to find the number of fluid ounces the instructor consumes during exercise.

   c. Explain the significance of each of the numbers that appears in the equation and sketch the graph of the function.

   d. Is the aerobics instructor following the advice of the sports nutritionist of Example 5.1?

4. The following is the recommended dose of a junior-strength painkiller, according to the child's weight:

Weight (lb)	Number of Tablets
Less than 48	Consult Physician
48–59	2
60–71	2½
2–96	3
96 and over	4

   a. How much medication should you give a child who weighs (i) 62 lb 3 oz? (ii) 71 lb 9 oz? (iii) 98 lbs? Explain how you obtained your answers.

   b. Describe and sketch the graph of the function that relates the dosage as the response variable with the weight as the explanatory variable.

   c. Find the rate of change of dosage if the child's weight increases from (i) 49 to 62 lb and (ii) from 72 lb 2 oz to 92 lb.

   d. Is the function linear? Is it piecewise linear?

5. The number of violent crimes reported in the U.S. in 1995 was 1,798,790, and decreased to 1,682,280 in 1996. Assume that the function that relates the number of violent crimes as the dependent variable, with the year that the number of crimes occurred as the independent variable, is linear.

    a. Write an equation for the line.

    b. If this equation were valid forever, how many years would it take for violent crime to disappear?

    c. Is the assumption that the function is linear, a reasonable assumption? Explain.

6. The following table contains the winning times in the women's 3,000-meter speed skating races in the Olympic games from 1960 to 1994. The time is given in minutes, seconds, and tenth of seconds. For example, the winner's time in 1960 was 5 minutes and 14.3 seconds or $5 * 60 + 14.3 = 314.3$ seconds.

Year	Winner's Time
1960	05:14.3
1964	05:14.9
1968	04:56.2
1972	04:52.1
1976	04:45.2
1980	04:32.1
1984	04:24.8
1988	04:11.9
1992	04:19.9

    a. For each given year, find the winner's average velocity.

    b. Give the function that relates the winner's time as the explanatory variable with the winner's average velocity as the response variable.

    c. Graph the function from time 0 to 350 seconds.

d. Find the rate of change when the time increases from 314 to 324 seconds and from 334 to 344 seconds.

e. Is the function linear? Explain.

f. Is the function increasing, decreasing, or neither increasing nor decreasing over the interval of time from 0 to 350 seconds.

g. Is the function concave upward or downward over the given interval? What does this say about the velocity function?

7. The table shown in Example 5.3 gives the postage rate for letters whose weights are integers (in ounces). When the weight is not an integer number of ounces, the post office rule is to round up to the nearest ounce to calculate the postage.

   a. What would the postage be for a letter weighing 1.6 ounces?

   b. Sketch a graph of the function for all weights from 0 to 4 ounces. Explain why the horizontal axis is used for the weight of the letter.

TOPIC 6

Modeling with Linear and Exponential Functions

OBJECTIVES

After completing this topic, you will be able to

- Recognize when data have an exact linear relationship
- Visualize a regression line and use an equation of the line to predict values of the response variable for particular values of the explanatory variable
- Determine when some quantity is growing or decreasing at an exponential rate
- Understand the differences between linear and exponential growth

Many times we want to summarize a set of data by constructing a function that in some way "fits" the data. The original data, however, might not represent a function, in the mathematical sense, because at least one value of the explanatory variable appears with two (or even more) different values of the response variable. But the description of the situation or the data leads us to suspect that a straight line or an exponential function fits the situation. By analyzing the data and the circumstances, we evaluate if a linear (straight line) model is appropriate or if some other type of curve would be a better model. In this topic, we look at linear and exponential models as examples. Although there are other ways to model data, we will not address them in this book.

Example 6.1

Each of the following tables represents a specific function. Determine if the relationship between the variables is a straight line.

a. This table represents the total popular votes, in thousands rounded up to the nearest thousand, cast for president of the United States in presidential election years 1968, 1972, and 1976. (Source: *The Federal Register,* www.archives.gov/federal-register.)

Year	1968	1972	1976
Total Votes Cast (in thousands)	74,000	78,000	82,000

b. The following table gives the temperature in degrees Celsius and the corresponding temperature in degrees Fahrenheit.

Temperature (degrees Celsius)	0	25	75	100
Temperature (degrees Fahrenheit)	32	77	167	212

c. The following table represents one model for recommended number of calories consumed per day as a function of a person's weight.

Weight (pounds)	120	140	180
Calories Per Day	1800	2000	2300

Solution

a. We check to see if the slope, determined using successive pairs of points, is constant. We see that $\frac{(78{,}000 - 74{,}000)}{(1972 - 1968)} = 1000$, and $\frac{(82{,}000 - 78{,}000)}{(1976 - 1972)} = 1000$. Therefore, all three points lie on the same line and the relationship represented by these data is linear. Note that this is only a portion of the data representing all presidential elections, and the remaining data probably does not lie on the same line.

b. Again checking the slope between successive pairs of points, we see that $\frac{(77-32)}{(25-0)} = 1.8$, and the quotient of the difference in temperature in degrees Fahrenheit divided by the difference in temperature in degrees Celsius, using any two points in the table, will also be 1.8. For example, $\frac{(77-212)}{(25-100)} = \frac{(-135)}{(-75)} = 1.8$. Recall that it doesn't matter which point is used as the first point in calculating the differences for a rate of change. To calculate the slope of the line joining the first two points in the table, we could also compute $\frac{(32-77)}{(0-25)} = \frac{-45}{-25} = 1.8$.

c. For these data, $\frac{(2000-1800)}{(140-120)} = 10$, but $\frac{(2300-2000)}{(180-140)} = 0.753$, so these data are not linear.

The data in Example 6.1a show that the response variable, "total votes cast," experienced **linear growth** over the time interval from 1968 to 1976; that is, the function values increased at a constant rate over this time interval. In many cases, however, we have a data set such as the one given in Example 6.1c, which is not exactly linear but which can be approximated or "fit" by a linear (straight line) model. The line used most frequently to fit data that have an approximately linear relationship is called the **least-squares regression line**. A computer or calculator can be used to produce the equation of the least-squares regression line, sometimes called the **least-squares line** or the **regression line**. The regression line is the particular line, among all possible lines, that minimizes the sum of the squares of the vertical distances from the data points to the line. We can use the regression line to predict values of the response variable for appropriate values of the explanatory variable.

Example 6.2

The following table gives the median weekly earnings of full-time wage and salary workers 25 years of age and older who have had four years or more of college, for years between 1980 and 2000. (Source: *U.S. Bureau of Labor Statistics*, www.bls.gov.)

Year	Salary ($)	Year	Salary ($)
1980	376	1994	733
1982	438	1995	747
1984	486	1996	758
1986	525	1997	779
1988	585	1998	821
1990	639	1999	860
1992	697	2000	896

a. Create a scatterplot of these data and verify that they do not represent an exact linear relationship.

b. Let s represent weekly salary, in dollars, and let t represent year. Sketch the line: $s = 24.61t - 48{,}352$ on the scatterplot and describe what you see. (This is the regression line for these data. Although there are formulas to find the equation of this line, given a set of data points, we won't discuss those formulas here.)

c. Use the line to predict the median weekly earnings for 2002 for workers with four or more years of college.

Solution

a. The data given in the table do not represent a linear relationship. Checking the rate of change quotients $\frac{\text{change in } s}{\text{change in } t}$ for successive pairs of points, shows that the rate of change is not constant. We get different quotients, for example, $\frac{438-376}{2} = 31$, while $\frac{486-438}{2} = 24$ and $\frac{525-486}{2} = 19.5$. In the scatterplot shown here, the points look fairly close to a line, so it is appropriate to use a line as a model for these data.

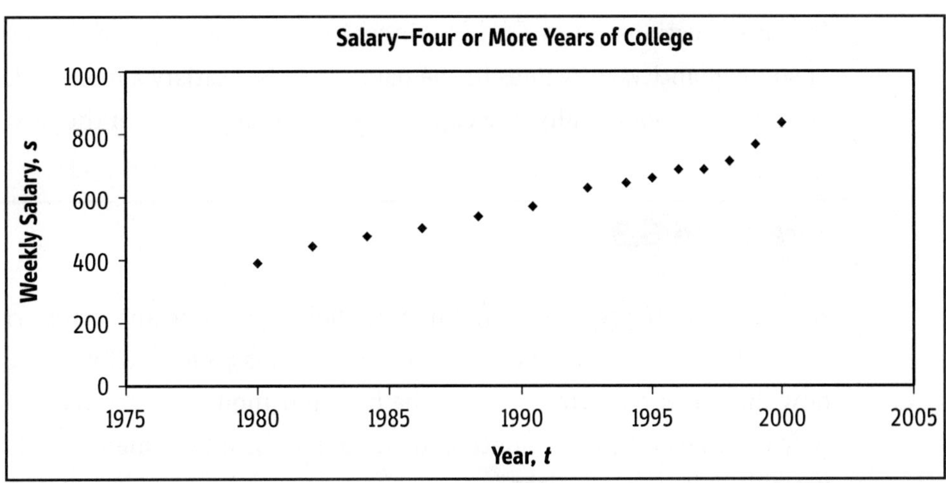

b. We sketched the given line on the following scatterplot. (To do so, we can plot any two points that satisfy the equation of the line and then draw the line through these two points.) We can see that some of the data points lie above the line and some lie below the line. In the equation of the line as given, the slope is 24.61. Note that this number is fairly close to the $\frac{\textit{difference in salary}}{\textit{difference in year}}$ calculated for several pairs of successive points in part a of the solution.

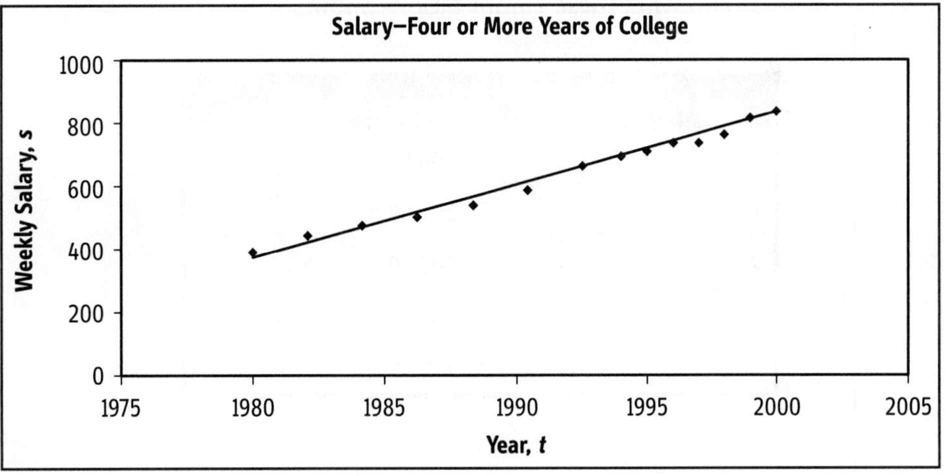

c. To use the line to predict weekly salary s for these workers in 2002, we let the explanatory variable t equal 2002 and compute $s = 24.61t - 48{,}352 = (24.61) \cdot 2002 - 48{,}352 = 917.22$. This is the predicted weekly salary for these workers in 2002. The point (2002, 917.22) would lie on the line if we extended the line on the graph to the year 2002.

In some situations, given a data set in which the values represent growth (or decay) of some quantity, the values in the data set do not satisfy a linear relationship, but are related exponentially. We explore exponential growth in the next example.

Example 6.3

A family is in the process of furnishing their new home and is incurring some debt. At the beginning of the year, their debt was $500; the following table shows how their debt has grown during the first four months of the year. The explanatory variable is "time in months," using one-month increments, and the response variable is "amount of debt."

a. Add two columns to the table. In the third column of the table, compute the rate of change of debt with respect to time over each one month period; that is, compute the difference of the current month's debt minus the previous month's debt. In the fourth column, compute the ratio of the current month's debt to the previous month's debt. (Note that for month 1, the previous month's debt would be the debt of $500 at the beginning of the year.)

b. Explain what these computations show.

End of Month Number	Amount of Debt ($)
0	500.00
1	550.00
2	605.00
3	665.50
4	732.05

Solution

a. The following table shows the two additional columns. To obtain each value in the third column, we compute a difference. (The difference is a rate of change but the denominator of each rate is 1.) To get each value in the fourth column, we compute a quotient.

End of Month Number	Amount of Debt ($)	Current – Previous Debt	Current/ Previous Debt
0	500.00	--	--
1	550.00	50.00	1.1
2	605.00	55.00	1.1
3	665.50	60.50	1.1
4	732.05	66.55	1.1

b. As we can see, the difference of the current month's debt minus the previous month's debt is increasing; the ratio of the current month's debt to the previous month's debt is constant. We could write this relationship as $\frac{\textit{current mo. dept}}{\textit{previous mo. debt}} = 1.1$ or *current mo. debt* = (1.1) × *previous mo. debt*. This means that the family's debt is growing at a rate of 10 percent per month, because *current mo. debt* = (1 + 0.10) × *previous mo. debt*.

The pattern shown in Example 6.3, in which the value of a quantity after the next time interval is a constant greater than 1 times the current value of the quantity, is characteristic of **exponential growth**. A graph can show how the family's debt will grow over a two-year period if it continues to increase by 10 percent per month. Notice that the graph is concave upward, which tells us that the rate of increase is increasing. This is exactly what the third column of the table in Example 6.3 tells us, too.

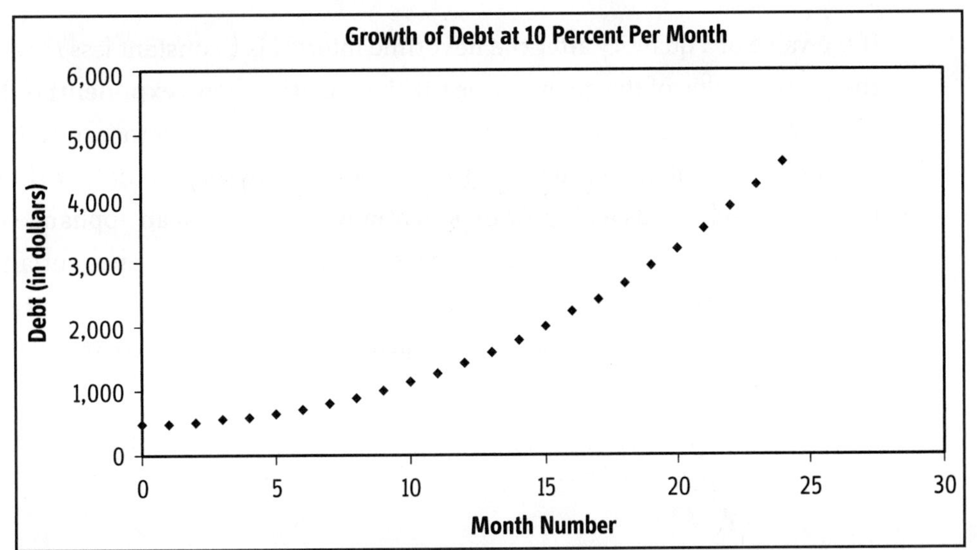

In the next example, we develop a formula that helps explain why this type of growth is called exponential growth.

Example 6.4

Find a pattern and develop a formula that gives the family's debt as described in Example 6.3 for any month *m*.

Solution

The debt in month 0 is \$500 and each succeeding month's debt is 1.1 times the previous month's debt. So let's use d_i to denote the family's debt in month *i*. Then $d_0 = \$500$, and $d_1 = (1.1) \cdot 500$, because the debt in month 1 is 1.1 times the debt in month 0. We know that the debt in month 2 is 1.1 times the debt in month 1, so $d_2 = (1.1) \cdot d_1 = (1.1) \cdot (1.1) \cdot 500 = (1.1)^2 \cdot 500$. We can continue this reasoning, and we see that the debt in month 3 is 1.1 times the debt in month 2: $d_3 = (1.1) \cdot d_2 = (1.1)(1.1)^2 \cdot 500 = (1.1)^3 \cdot 500$. Continuing the pattern, we note that for any value of *m*, the debt in month $m = d_m = (1.1)^m \cdot 500$. We can check to see that the formula is consistent with our graph; for example, in month 20, $d_{20} = (1.1)^{20} \cdot 500 \approx 3363.75$. It looks like the point on the graph is approximately (20, 3400). Notice that in the formula, the explanatory variable *m* appears as an exponent, so the term exponential growth makes sense.

If the value of a quantity after the next time interval is a constant less than 1 times the current value of the quantity, we say the quantity shows **exponential decrease.** Various phenomena grow or decrease exponentially. For instance, interest on investments involves exponential growth. Some analysts predict that the use of the Web to sell goods will grow exponentially. The value of an appliance may decrease exponentially, and radioactive substances decrease exponentially. (This type of decrease is called radioactive decay.) Population growth is another situation in which we often see exponential growth, as the next two examples show.

Example 6.5

The following table gives the estimated population (in millions) of Mexico for the years 1980 to 1986.

Year	Population in Millions
1980	67.38
1981	69.13
1982	70.93
1983	72.77
1984	74.66
1985	76.60
1986	78.59

Verify that the population of Mexico experienced exponential growth during the years 1980 through 1986, and explain why an exponential model for population growth is a reasonable one.

Solution

We check for exponential growth by dividing each year's population by the previous year's population and determining if this value is the same (that is, a constant) for the time period under consideration. For example, $\frac{\textit{population in 1981}}{\textit{population in 1980}} = \frac{69.13}{67.38} \approx 1.026$. Also, $\frac{\textit{population in 1982}}{\textit{population in 1981}} = \frac{70.93}{69.13} \approx 1.026$, and so on. We need to check each pair to make sure. Note that the *difference* in population in successive years is increasing from a growth of 1.75 million between the years 1980 and 1981 to an increase of 1.99 million between 1985 and 1986. Because individuals in the population (whether people or bacteria) reproduce, it is reasonable to predict that a population (unless there are other factors to be considered) will grow exponentially. Therefore, we might expect the population to increase at a rate proportional to the number of individuals now in the population. The population will grow faster and faster as time goes on. This is in contrast to a linear function, which always increases at the same rate.

Example 6.6

A recent newspaper article pointed out that India's current population of approximately 1 billion may double to 2 billion in just another 100 years. The article predicts that India's population by 2050 will be 1.6 billion. Assume that the population growth of India is exponential and that the prediction of India's population in 2050 is correct; use this information to determine what the population of India will be at the end of the next century.

Solution

Using the information given, we assume that the population of India grows from 1.0 billion to 1.6 billion in the 50-year time period from 2000 to 2050. Thus the growth factor is $\frac{\textit{population in 2050}}{\textit{population in 2000}} = \frac{1.6}{1.0} = 1.6$. Therefore, assuming that this growth will continue, (*population in 2100*) = 1.6 × (*population in 2050*) = 1.6 × 1.6 = 2.56 billion. The article goes on to quote one of the authors of the report, whose observations are corroborated by this example: If the growth continues at this rate, India may add even more than 1 billion people in the next 100 years.

Through several examples, we saw that a quantity that grows exponentially grows more quickly than one that grows linearly. The next example shows a situation in which growth is even faster than exponential growth.

Example 6.7

We want to test various brands of chocolate to see which is preferred by more subjects. To be fair, we want to vary the order in which the different brands are presented to the taste testers. If we have one brand to test, clearly there is only one way to arrange the order in which we present it to the testers.

a. How many orders of presentation are there if we have two brands to test? Three brands? Four brands?

b. Fill in the table with the values for number of ways the brands can be presented for $n = 2, 3$, and 4 brands, and describe what you observe in this table.

Number of Brands Tested	Number of Orders of Presentation
1	1
2	
3	
4	
5	120
6	720
7	5,040
8	40,320
9	362,880
10	3,628,800

c. Sketch a graph of the number of orders of presentation as a function of brands tested. Explain why the growth in the number of orders of presentation is not exponential.

Solution

a. This is an example of a counting problem. If we think of lining up in a row the various brands of chocolate to be tested, we are actually counting the number of ways to rearrange objects in a row. If there are two brands, say A and B, either A can be first or B can be first. Thus, there are two orders of presentation. With three brands, say A, B, and C, any one of the three can be presented first. If A is offered first, then either B is next or C is next; if B is offered first, then A or C is next; and if C is offered first, then A or B is next. We can write out all the possibilities (that is, **enumerate** them) and see that there are six orders of presentation with three brands. With four brands, each of the four brands could be presented first. Then for each of those four choices, there are six orders of presentation for the remaining three brands (which we just figured out!). Thus there are 4 × 6 or 24 orders of presentation with four brands.

b. We observe that the number of orders of presentation is increasing rapidly, as the number of brands tested increases.

c.

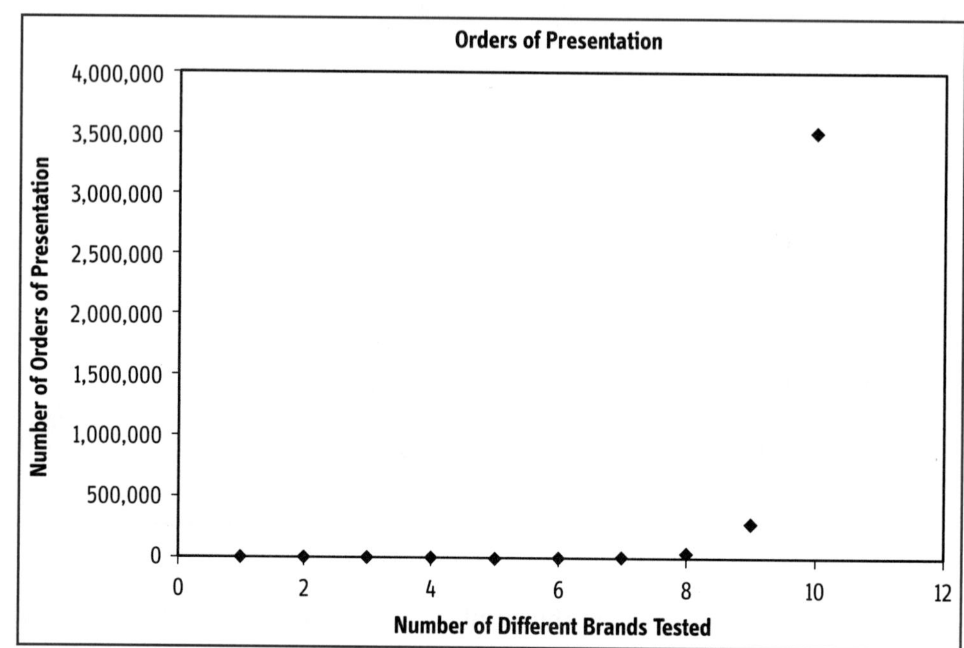

It's difficult to tell from the graph if the growth is exponential or not. We can see that if we connected the points on the graph, the graph would be concave upward. (What does this tell us?) When we look at the ratios of successive values of orders of presentation, we see that those values are increasing and are not constant. (You will be asked to verify this in Exploration 3 later in this topic.) For this to be an exponential function, those ratios must be the same.

Summary

In this topic, we investigated two types of models, linear and exponential models. We investigated how to recognize linear growth and exponential growth and explored some applications of each type of growth. We also saw how linear growth compares to exponential growth, looked at formulas for each type of growth, and examined why the terms *linear* and *exponential* are appropriate for each type of model.

Explorations

1. One linear model for computing an individual's "ideal" weight uses the following equations:

   Ideal weight for man of height h inches: $w_{man} = 160 + 6 \times (h - 65)$

   Ideal weight for woman of height h inches: $w_{woman} = 100 + 5 \times (h - 60)$.

   a. Sketch the graph of each of these equations on the same set of axes with h on the horizontal axis and "ideal" weight on the vertical axis. Describe how the two equations differ.

   b. How do the "ideal" weights compare for a man and a woman who are both 5 feet 7 inches tall?

   c. What is a reasonable interval of values of h for which these equations make sense?

   d. How do you think these models were obtained?

2. The following table gives the median weekly earnings of full-time wage and salary workers 25 years and older who have had one to three years of college. (Source: *U.S. Bureau of Labor Statistics,* www.bls.gov.)

Year	Salary ($)	Year	Salary ($)
1980	304	1994	499
1982	351	1995	508
1984	382	1996	518
1986	409	1997	535
1988	430	1998	558
1990	476	1999	580
1992	485	2000	598

a. Sketch a scatterplot of these data.

b. Let s represent salary and let t represent year. The least-squares regression line that best fits this data set is $s = 13.199t - 25{,}812$. Sketch this line on the graph with the scatterplot and describe what you see.

c. Compare this line with the line given in Example 6.2. What value for salary do you get from each of the lines if you substitute the value 2001 for the variable t, and what do those values tell you?

3. Verify that the ratios of successive values of orders of presentation in Example 6.7 are indeed increasing. Identify the pattern of these increasing values.

4. The following table gives the gross federal debt figures (in millions of dollars) for the U.S. every five years for the years 1945 to 2000. (Source: *U.S. Department of the Treasury*, www.ustreas.gov.)

Year	Gross Federal Debt (million $)
1945	260,123
1950	256,853
1955	274,366
1960	290,525
1965	322,318
1970	380,921
1975	541,925
1980	909,050
1985	1,817,521
1990	3,206,564
1995	4,921,005
2000 (estimate)	5,686,338

a. Sketch a scatterplot of these data. Does the graph appear to be similar to an exponential graph?

b. For each five-year value given in the table, compute the following ratio: gross federal debt in that year divided by the gross federal debt in the previous time period (five years earlier) and record these values. What do you observe about these ratios?

c. Is the growth of the gross federal debt exponential? How would you determine if the growth is exponential?

5. Cars depreciate in value as soon as you take them out of the showroom. A certain car originally cost $25,000. After one year, the car's value is $21,500. Assume that the value of the car is decreasing exponentially; that is, assume that the ratio of the car's value in one year to the car's value for the previous year is constant.

   a. Find the ratio: $\frac{\textit{worth after one year}}{\textit{original worth}}$.

   b. What is the car's value after two years? After ten years?

   c. Approximately when is the car's value half of its original value?

   d. Approximately when is the car's value one-quarter of its original value?

   e. If you continue these assumptions, will the car ever be worth $0? Explain.

6. The U.S. Census Bureau International Database gives the population of the ten most-populated countries in the year 2000 and population predictions for 2050 for these countries: China had 1.2 billion people in 2000 and is predicted to have a population of 1.3 billion in 2050. The U.S. had a population of 284 million in 2000 with a prediction of 420 million by 2050.

   a. Assuming exponential population growth, predict the populations of China and the U.S. in the years 2100 and 2200.

   b. Assuming exponential population growth, approximately when will China's population double?

   c. What factors might make the assumption of exponential growth a faulty assumption?

7. World-wide food supply, or the ability to produce food, generally grows linearly while population tends to grow exponentially, if left unchecked. Explain why this is such a problem and illustrate with graphs.

8. Consider the formula $P = 67.38 \cdot (1.026)^t$. If we let P represent the population of Mexico in year t where t is the number of years from 1980, confirm that this formula gives the same population values as those given in the table in Example 6.5.

   a. Explain where the number 67.38 and the number 1.026 were obtained.

   b. What would the population in 1990 have been if growth had continued in this same pattern?

9. A newspaper article in *The Morning Call* on Friday, July 16, 2004, reported 2003 sales in billions of dollars and number of stores for each of America's ten largest retailers. These data were used to sketch the following scatterplot:

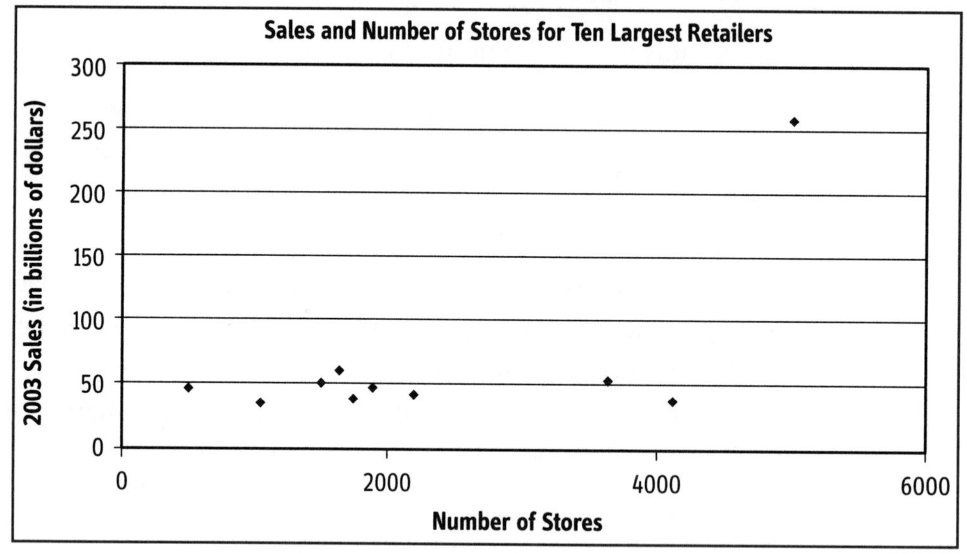

   a. Is a line an appropriate fit for these data? Why or why not?

   b. If you want to fit a line to these data, why should "number of stores" be the explanatory variable and "2003 sales" be the response variable?

   c. Identify any unusual data value(s). What happens if you delete this (or these) value(s)?

TOPIC 7

Logarithms and Scientific Notation

OBJECTIVES

After completing this topic, you will be able to

- » Compare real-world quantities that are measured by logarithmic scales
- » Use properties of logarithms to solve problems
- » Use scientific notation and make estimates involving very large or very small quantities

What do sound levels measured in decibels, intensity of earthquakes measured in points on the Richter scale, and acidity of liquids measured in pH levels, have in common? They are all measurements on a special scale called a logarithmic scale. In this topic, we look at two logarithmic scales and discuss the definition and properties of the common logarithm.

A linear scale, like that on a yardstick, has the property that the distance between the 1- and 2-inch marks is the same as the distance between the 31- and 32-inch marks. With a **logarithmic scale**, the unit steps increase in a multiplicative way. In a similar way, the sequence 1, 4, 7, 10, . . . is linear, or additive, because each term is three more than its predecessor, while the sequence 1, 3, 9, 27, 81 . . . is **multiplicative** because each term is three times its immediate predecessor. Logarithmic scales are used for some quantities that have a large range of variation, such as magnitude of earthquakes and levels of sound.

The Decibel Scale

The human ear can hear sounds that are as much as 100 trillion times louder than the faintest sounds. In the **decibel scale**, the least audible sound, which corresponds to a sound wave of intensity 10^{-12} watts/m^2, is assigned the value of 0; a sound that is 10 times louder than the least audible sound is assigned a value of 10; a sound 100 = 10^2 times louder than the faintest sound is assigned a value of 20; a sound 1,000 = 10^3 times louder than the faintest sound is assigned a value of 30; and so on. Thus, an increase of 10 decibels means a tenfold increase in sound intensity or loudness.

Example 7.1

The sound of normal conversation measures approximately 60 decibels; the noise level inside a subway measures approximately 90 decibels.

a. How many times as loud as the faintest audible sound is normal conversation?

b. How many times as loud as the faintest audible sound is the noise inside a subway?

c. How many times as loud as normal conversation is the noise inside a subway?

Solution

a. Normal conversation at 60 decibels is $10 * 10 * 10 * 10 * 10 * 10 = 10^6$ = 1,000,000; that is, normal conversation is one million times louder than the faintest audible sound.

b. The noise inside a subway is 90 decibels, so this noise is 10^9 = 1,000,000,000 or one billion times as loud as the faintest sound.

c. The difference between the sound of normal conversation and the sound inside a subway is 30 decibels. Because each increase of 10 decibels corresponds to a ten-fold increase in loudness, an increase of 30 decibels corresponds to a sound $10 * 10 * 10 = 10^3 = 1{,}000$ times as loud. So the noise inside a subway is 1,000 times as loud as the sound of normal conversation. (We can obtain the answer in a different way by computing the ratio of the two sounds' loudness relative to the faintest sound, $\frac{10^9}{10^6} = 10^{9-6} = 10^3$.)

The decibel scale is called logarithmic because the decibel measure and the loudness relative to the faintest audible sound are related through logarithms. Logarithms were first introduced by John Napier in 1614 to help simplify complicated calculations. The **common logarithm** (or logarithm with base 10) of a number n is the exponent r where $10^r = n$. The common logarithm, also referred to as the **decimal logarithm** or just **logarithm**, of a number n is denoted by $\log n$.

For example, $\log 1 = 0$ because $10^0 = 1$; $\log 10 = 1$ because $10^1 = 10$; and $\log 0.01 = -2$ because $10^{-2} = \frac{1}{10^2} = \frac{1}{100} = 0.01$. For most numbers n, the computation of $\log n$ must be done using a scientific calculator or computer program. Before this technology was available, scientists and engineers used tables of logarithms or slide rules. For example, if we use the calculator to find the logarithm of $n = 50$, we obtain $\log 50 \approx 1.699$, which means that $10^{1.699} \approx 50$. The function that relates the quantities n as the explanatory variable and r as the response variable through the equation $r = \log n$ is called the **logarithmic function with base 10**.

Example 7.2

Consider the logarithmic function with base 10, $r = \log n$.

a. Use a calculator to complete the following table of values of the function.

n	0.01	0.1	0.5	1	2	5	10	15	20	50	100
$r = \log n$	-2		-0.30								

b. Explain why when trying to evaluate log n for a value of $n \le 0$, the calculator gives an error message.

c. Find the rate of change in r per unit change in n, when n increases from 0.1 to 0.5, from 1 to 5, from 10 to 15, and from 20 to 100.

d. Sketch and describe the graph of the function over the interval 0.01 to 100.

Solution

a. We notice that $0.1 = \frac{1}{10} = 10^{-1}$, so $\log 0.1 = -1$. Also, $100 = 10^2$, so $\log 100 = 2$. We have already seen that $\log 1 = 0$ and $\log 10 = 1$. Using a calculator, we then approximate the remaining entries. Here is the completed table:

n	0.01	0.1	0.5	1	2	5	10	15	20	50	100
r	-2	-1	-0.30	0	0.30	0.7	1	1.18	1.30	1.7	2

b. For any value of r, the value of 10^r is positive. (Note that when r is negative, 10^r means the reciprocal of a positive power of 10, that is $10^r = \frac{1}{10^s}$, where $s = -r$.) Because $r = \log n$ means that $n = 10^r$, n must be positive.

c. The rate of change in r from 0.1 to 0.5 is $\frac{-0.30 - (-1)}{0.5 - 0.1} = \frac{-0.30 + 1}{0.4} = 1.75$ units per unit change in n. In the same manner, we compute the rate of change over the other given intervals. Here are the results:

The rate of change from 1 to 5 is $\frac{0.70 - 0}{5 - 1} = 0.175$.

The rate of change from 10 to 15 is $\frac{1.18 - 1}{15 - 10} = 0.036$.

The rate of change from 20 to 100 is $\frac{2 - 1.30}{100 - 20} = 0.00875$.

From these calculations, we can see that the rate of change decreases as n increases.

d. The following graph was obtained by using the points in the table completed in part a of this example. This function is always increasing (the rate of change is positive) and is concave downward (the rate of change is positive and decreasing, as the calculations in part c of this example show).

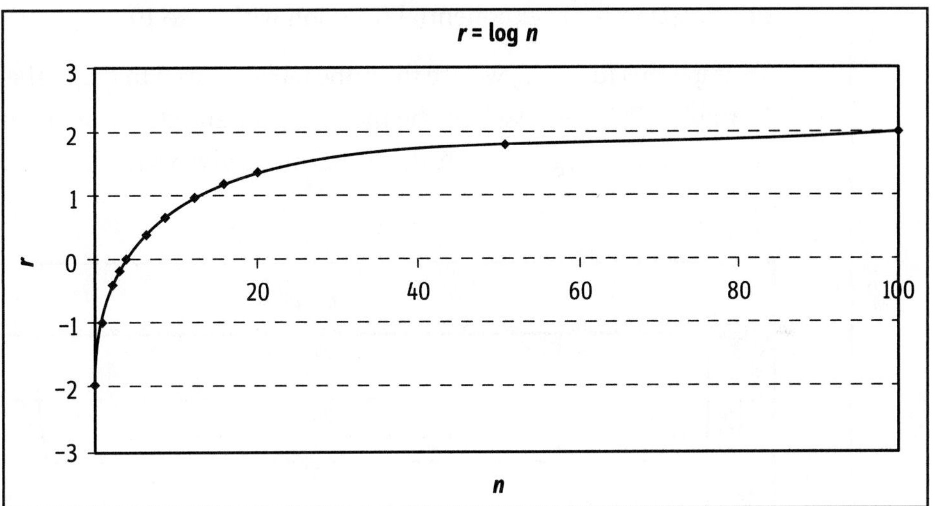

The logarithmic function in Example 7.2 is closely related to the exponential function with base 10. We study this relationship in the following example.

Example 7.3

Consider the function defined by the same equation $r = \log n$ in Example 7.2, but using r as the explanatory variable and n as the response variable.

a. Express n in terms of r and determine what type of function is obtained.

b. Graph the function.

c. Using both horizontal and vertical axes from –2 to 10, sketch both this function and the one in Example 7.2 on the same graph. Explain the relationship between the two graphs.

Solution

a. Because $r = \log n$ means $10^r = n$, n is obtained from r by raising 10 to the power r. The function $n = 10^r$ is an exponential function because the explanatory variable appears as an exponent in the formula. Because the base is 10, this function is the exponential function with base 10.

b. To graph this function, we use the same table we used to graph the function in Example 7.2, but we switch the independent and dependent variables. We can see from the graph that this function is always increasing and is concave upward.

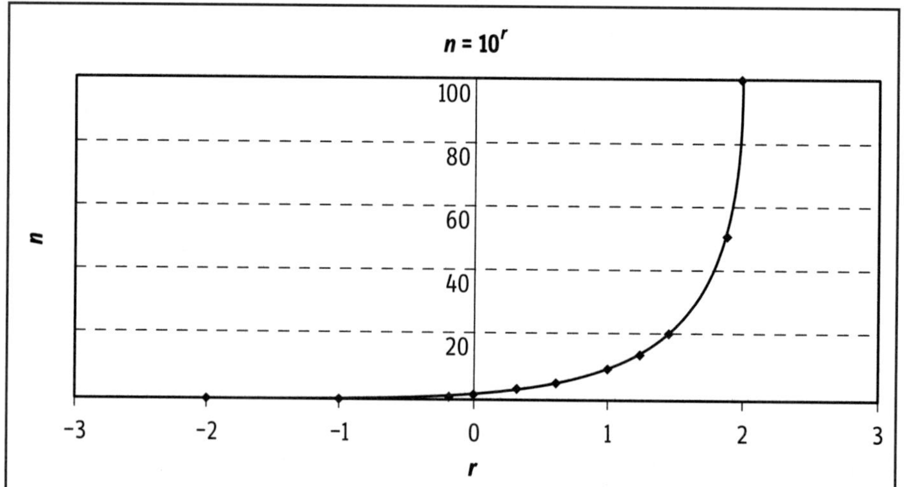

c. For this graph, we will use the names x and y for the independent and dependent variables, respectively. We can see that the graph of each of the functions can be obtained from the other function by reflection across the line $y = x$ (graphed as a dotted line). Thus, for example, the point (0,1) is on the graph of $y = 10^x$ and (1,0) is on the graph of $y = \log x$; the point (0.7, 5) is on the graph of $y = 10^x$ and (5, 0.7) is on the graph of $y = \log x$, and so on.

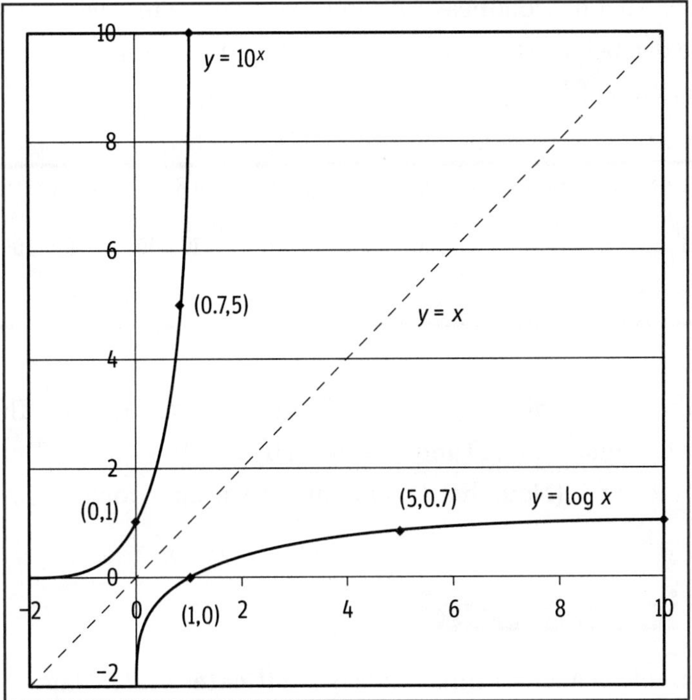

In the next example, we investigate further the connection between logarithms and the decibel scale.

Example 7.4

Use logarithms to find a formula that relates the decibel measure of a sound with the loudness of the sound relative to the loudness of the faintest audible sound.

Solution

We assume that the loudness of the least audible sound, which measures 0 in the decibel scale, is 1. Then a sound that measures 10 decibels has loudness 10, which is 10 times 1; a sound that measures 20 decibels is 100 times louder than the least audible sound, so it has loudness 100, and so on. In the following table, we have included the decibel measure for several sounds, the corresponding loudness L and the logarithm of L.

D	0	10	20	30	40	50	60
L	1	10	100	1,000	10,000	100,000	1,000,000
log L	0	1	2	3	4	5	6

Notice that the quantity log L differs from D by a factor of 10. The relationship between the two quantities, D and L, is given by $D = 10 * (\log L)$. Thus D is directly proportional to log L. (Note that D is not directly proportional to L, but to log L.)

The Richter Scale

Another example of a logarithmic scale is the **Richter scale**. Named after the American geologist Charles F. Richter, it is used to measure the magnitude of earthquakes. Seismologists use records taken by a seismograph to measure the total amount of motion produced by the vibrations during an earthquake. A number derived from this seismograph measurement is assigned to the earthquake and is called the magnitude of the earthquake. A unit change in the Richter scale represents a tenfold increase in energy released by the earthquake. For example, an earthquake that measures 5 on the Richter scale releases 10 times more energy than an earthquake that measures 4. If E represents the relative energy released by an earthquake and r indicates the magnitude of the earthquake on the Richter scale, then $E = 10^r$.

Example 7.5

In 1906, the American continents experienced two major earthquakes. The San Francisco earthquake, which occurred in April of that year, measured 8.3 on the Richter scale. In August of the same year, an earthquake that measured 8.6 on the Richter scale occurred in Valparaiso, Chile. How many times stronger was the Valparaiso earthquake than the San Francisco one?

Solution

The relative energy released by the San Francisco earthquake was $10^{8.3} \approx 199{,}526{,}231.5$, while the relative energy released by the Valparaiso earthquake was $10^{8.6} \approx 398{,}107{,}170.6$. Because $\frac{398{,}107{,}170.6}{199{,}526{,}231.5} \approx 1.995$, the Valparaiso earthquake was almost two times stronger than the San Francisco one.

Logarithms have been used extensively to simplify computations because they have the property that they transform products into sums, quotients into differences, and powers into products. That is, if a and b are two positive numbers, and r is any nonzero number, we have the following properties:

Property (1): $\log ab = \log a + \log b$

Property (2): $\log \left(\frac{a}{b}\right) = \log a - \log b$

Property (3): $\log a^r = r \log a$

Example 7.6

Use values from the table obtained in Example 7.2 to provide examples for each of the three properties just described.

Solution

Because $20 = 2 * 10$, property (1) gives $\log 20 = \log 2 + \log 10 = 0.30 + 1 = 1.30$. Note that $\frac{20}{2} = 10$ and from the table in Example 7.2, we know that $\log 10 = 1$, $\log 20 = 1.30$, and $\log 2 = 0.30$. So by property (2), we have $\log \left(\frac{20}{2}\right) = \log 20 - \log 2 = 1$. We illustrate property (3) with 100. We know that $100 = 10^2$, so $\log 100 = \log 10^2 = 2 \log 10 = 2 * 1 = 2$.

Scientific Notation

Scientific notation is used mostly to express very large or very small numbers. A number is in scientific notation if it is written in the form

$a * 10^r$, where $1 \leq a < 10$ and r is an integer

When a number is written in scientific notation, the logarithm of the number can be related easily to the logarithm of a number between 1 and 10. This concept made it possible, before calculators and computers existed, to use tables of logarithms to calculate products and quotients involving numbers with several digits. Some calculators use scientific notation in their output. For example, 4,260,000,000 would appear as 4.26E9 (or as 4.26e9), meaning $4.26 * 10^9$; and 0.0000258 would appear as $2.58E^{-5}$, meaning $2.58 * 10^{-5}$.

Example 7.7

The distance from a planet to the Sun varies as the planet moves in its orbit.

a. The closest distance (called the perihelion) from Mercury to the Sun is $2.86 * 10^7$ miles and its farthest distance (aphelion) is $4.34 * 10^7$ miles. What is the difference between the two measurements?

b. The farthest distance from Jupiter to the Sun is $5.07 * 10^8$ miles. What is the difference between the farthest distance from Mercury to the Sun and the farthest distance from Jupiter to the Sun?

Solution

a. Because each term has a common factor of 10^7, we can factor it out and then subtract. The difference between the two measurements is $4.34 * 10^7 - 2.86 * 10^7 = (4.34 - 2.86) * 10^7 = 1.48 * 10^7 = 14{,}800{,}000$ miles.

b. For this calculation, we need the same term, 10^7, in each expression so we can factor it out. We first rewrite $5.07 * 10^8$ as $5.07 * 10 * 10^7 = 50.7 * 10^7$. Then, the difference between the farthest distance from Jupiter to the Sun and that of Mercury to the Sun is $5.07 * 10^8 - 4.34 * 10^7 = (5.07 * 10 - 4.34) * 10^7 = (50.7 - 4.34) * 10^7 = 46.36 * 10^7 = 463{,}600{,}000$ miles.

Scientific notation is useful to make estimates when large numbers are involved, because operating with powers of 10 is straightforward in the decimal numerical system that we use. In the next example, we see how to use scientific notation when estimating.

Example 7.8

According to its website (www.mcdonalds.com), McDonald's has 30,000 restaurants in 119 countries around the world. In Australia, for example, there are 725 McDonald's restaurants that serve over 1 million customers daily. Use this information to estimate how many McDonald's hamburgers are consumed each year around the world. State any assumptions you make.

Solution

We assume that, on average, every customer who visits McDonald's eats one hamburger (some might eat chicken or fish or salad, but other customers consume more than one hamburger, so this assumption does not seem unreasonable). To estimate the daily number of customers, we will assume that the average daily number of customers per restaurant all over the world is the same as the average daily number of customers in Australia. This is $\frac{1,000,000}{725} \approx 1,379$ customers per restaurant per day. Because there are 30,000 restaurants around the world, each serving 1,379 hamburgers a day, a rough estimate of daily consumption is

$$(3.0 * 10^4) * (1.4 * 10^3) = 4.2 * 10^7 = 42 \text{ million hamburgers per day}$$

In one year, the number of hamburgers consumed at McDonald's is approximately

$$4.2 * 10^7 * 360 = 4.2 * 10^7 * 3.6 * 10^2 = (4.2 * 3.6) * 10^9 \approx 15 \text{ billion}$$

Summary

In this topic, we defined the common logarithm and reviewed rules of logarithms. We looked at two logarithmic scales: the decibel measure for sounds and the Richter scale for magnitude of earthquakes. We also did some calculations and made estimates using scientific notation.

Explorations

1. Explain why a logarithmic scale is useful when measuring sound.
2. The following graph appeared in *The Morning Call* on October 5, 1999. Use the information on the graph to answer the following questions.

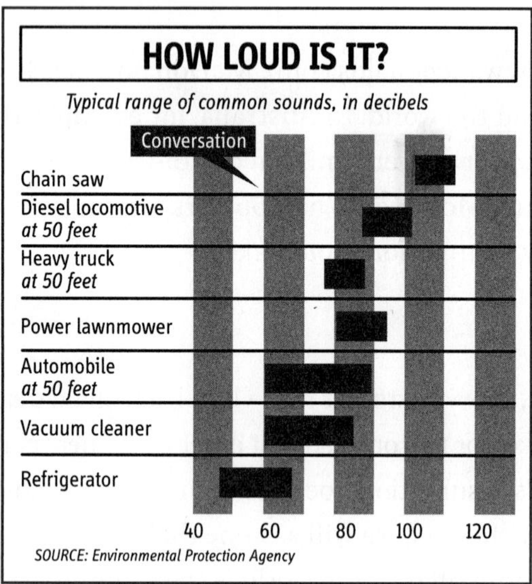

   a. How many times louder is the sound of the most silent chain saw than the sound of the loudest power lawnmower?

   b. How does the sound of an approaching diesel locomotive compare to the sound of normal conversation when the locomotive is 50 feet away? (Consider the full range of possibilities.)

3. Consider the following statement: "An ambulance siren has a noise level of about 100 decibels when heard from 100 feet away. That is about 8 times as loud as normal conversation." Is this statement correct? Explain your answer.

4. Answer the following:

   a. Use the fact that log 2 = 0.30 and the properties of logarithms discussed in this topic to find the following values: log 200, log 0.002, log 16 = log 2^4.

   b. Use a calculator to evaluate the logarithms in part a of this exploration. Did you get the same answer?

5. In February of 1997, northwestern Iran suffered an earthquake of magnitude 6.1 on the Richter scale. In May of the same year, an earthquake of magnitude 7.5 occurred in northern Iran. The first earthquake caused 1,000 deaths and the second earthquake caused 1,560 deaths.

   a. How many times stronger was the second earthquake than the first one?

   b. What was the ratio between the death tolls in the earthquakes?

   c. Name other factors that might increase the death toll of an earthquake other than its magnitude.

6. If an earthquake has magnitude 4.2 on the Richter scale, what is the magnitude on the Richter scale of an earthquake that is 500 times stronger? Explain your answer.

7. Pure water has a pH level of 7.0. The pH scale is a measurement scale that indicates how acidic or basic (that is, alkaline) a liquid is. Each one-unit decrease in pH indicates a ten-fold increase in acidity. Similarly, for pH values greater than 7.0, each one-unit increase indicates an acidity level one-tenth of the next lower integer value.

   a. Black coffee has a pH level of 5. How much more acidic is black coffee than pure water?

   b. Ammonia has a pH of 11. What is its acidity level compared to pure water?

   c. Lemon juice has a pH of 2. How does its pH level compare with that of Milk of Magnesia with a pH of 10?

8. The distance from Pluto to the Sun varies from $2.76 * 10^9$ miles to $4.58 * 10^9$ miles. What is the difference between the closest and farthest distances from Pluto to the Sun?

9. In August 1999, a 36-year-old woman was awarded $23 million dollars by a Texas jury who found that manufacturers of the diet drug Fen-phen were responsible for the woman's serious medical problems. There were 3,100 similar lawsuits against the company at the time. In July 1999, the company had reported second quarter profits of $299 million.

   a. Estimate how much the company would have to pay out if the company loses all those lawsuits.

   b. Estimate how many years the company would take to pay off all that money. State any assumptions you make.

10. An angstrom is a unit of length equivalent to 0.0000001 millimeters.

    a. Write the number 0.0000001 as a power of 10.

    b. An electron is 10^{-12} millimeters in diameter. What is this diameter's length in angstroms?

11. A recent news report stated that, as a result of a recent downturn in the stock market, a man whose worth was $10 billion lost 99 percent of his wealth. How much is he now worth?

TOPIC 8

Indexes and Ratings

OBJECTIVES

After completing this topic, you will be able to

- Analyze trends in several commonly used indexes
- Use and calculate indexes to understand and compare data
- Recognize rating systems as a type of indexing system and investigate what might go into setting up a rating system

Quantitative indexes and rating systems are used to give information about general trends and to allow us to make comparisons and judgments. We'll examine some frequently used indexes and rating systems and look at how to use them and what goes into setting them up.

The **Dow Jones Industrial Average (DJIA)** is a well-publicized index that reflects the value of stock prices. The DJIA includes 30 stocks that represent a variety of industries—financial, food, technology, retail, heavy equipment, oil, chemical, pharmaceutical, consumer goods, and entertainment. The DJIA is not a simple average but is adjusted to take into account the changes in price associated with stock splits in each of the included companies. The average is calculated by summing the prices of the 30 stocks and then dividing by a constant called the divisor that is adjusted periodically. (Originally, the divisor was 30, making the DJIA a simple average, but it has been reduced over the years and is currently less than 1.)

Example 8.1

The following graph of the Dow Jones Industrial Average (DJIA) shows the daily closing values from mid-July 2003 through mid-July 2004. (The graph appeared in *The New York Times* on July 21, 2004.) Explain what trends this index shows for the time period covered in the graph. Why do you think the DJIA index uses 30 stocks rather than all stocks?

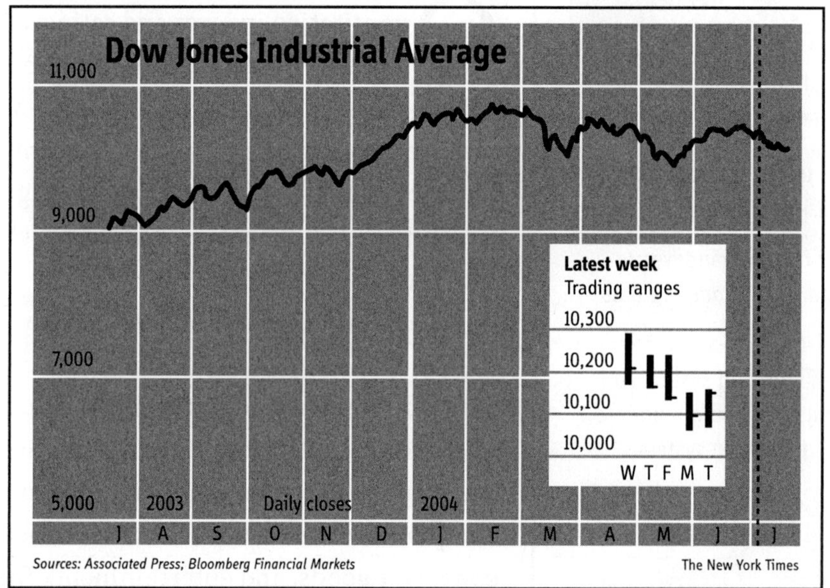

Solution

The index shows that the stock market experienced a gradually increasing trend through early March 2004 (with slight dips at the end of September, end of October, and in late November.) Prices then declined a bit and subsequently rose slightly during the last days of March. The DJIA remained fairly steady through April, decreased somewhat from the end of April to mid-May but returned to the previous level during the second half of May. It remained fairly steady through the month of June and started to decrease again during the first part of July 2004. Using 30 well-chosen, representative stocks (that is, a representative **sample** of the stock market) gives an accurate reflection of what is happening in the stock market as a whole, without having to include all stocks.

The **Consumer Price Index (CPI)** is an index number determined and published each year by the Bureau of Labor Statistics. It is used as one measure of inflation and measures the price of a market basket of approximately 300 goods and services purchased by consumers. (The market basket is weighted using the following percentages: 39.6 percent housing, 16.3 percent food and drinks, 17.6 percent transportation, 5.6 percent medical care, 4.9 percent apparel and upkeep, 6.1 percent entertainment, 5.5 percent education and communication, 4.3 percent other.) The CPI is not the only index used as an indicator of inflation, but it is a commonly used one. In the next example, we look at the CPI over time and then examine how we can use it to understand changes in the value of a dollar over time.

Example 8.2

The following table gives the Consumer Price Index for the years 1970 through 2003.

Year	1970	1971	1972	1973	1974	1975	1976	1977	1978	1979	1980	1981
CPI	38.8	40.5	41.8	44.4	49.3	53.8	56.9	62.6	65.2	72.6	82.4	90.9
Year	1982	1983	1984	1985	1986	1987	1988	1989	1990	1991	1992	
CPI	96.5	99.6	103.9	107.6	109.6	113.6	118.3	124	130.7	136.2	140.3	
Year	1993	1994	1995	1996	1997	1998	1999	2000	2001	2002	2003	
CPI	144.5	148.2	152.4	157.6	160.3	163	166.6	172.2	177.1	179.9	184	

Sketch a graph of the CPI over the years given in the table and explain what the graph shows.

Solution

A scatterplot of the CPI over the years 1970 to 2003 appears next. This graph shows that the increase in CPI from 1970 through 2003 was fairly steady with slightly larger increases around 1980 and slightly smaller increases since 1990.

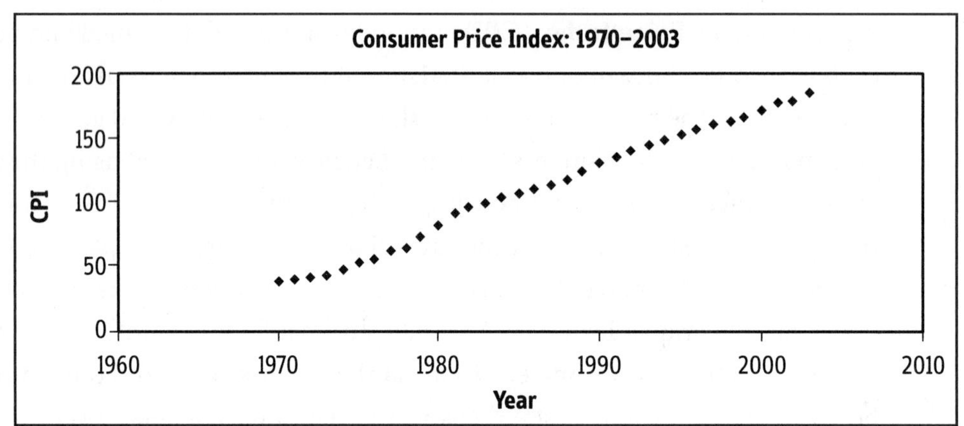

Index numbers such as the CPI allow us to make comparisons. When we look at the cost of something over time, in order to understand the nature of the changes in cost, we might want to look at the increases and prices in dollar units that reflect actual buying power at one particular time. We can re-express the costs in dollar units, called **constant dollars**, which adjust the monetary units for inflation.

To convert a monetary amount from **current dollars** at *time A* to an amount with the same buying power at another time, say *time O*, the ratio of the monetary amount to the CPI must be the same for both times; that is,

$$\frac{\text{Dollars at time A}}{\text{CPI at time A}} = \frac{\text{Dollars at time O}}{\text{CPI at time O}}$$

We can write this ratio as:

$$\text{Dollars at time O} = \text{Dollars at time A} * \frac{\text{CPI at time O}}{\text{CPI at time A}}$$

Example 8.3

The following table shows the federal minimum hourly wage rate from 1974 to 2004 and the years in which it increased during that time period. (Note that the minimum wage from 1998 to mid-2004 remained at the 1997 level.)

Year	1974	1975	1976	1978	1979	1980	1981	1990	1991	1996	1997
Wage ($)	2.00	2.10	2.30	2.65	2.90	3.10	3.35	3.80	4.25	4.75	5.15

a. By what percentage did the CPI increase from 1974 to 1997?

b. By what percentage did the minimum wage increase from 1974 to 1997?

c. What do the values calculated in parts a and b of this example tell you?

d. Convert the 1974 (*time A*) minimum wage into constant 2000 (*time O*) dollars. Repeat for the minimum wage for each of the other years given in the table and record in a new table.

e. Create a graph of the minimum wage in constant 2000 dollars for the years 1974 to 1997 and describe what happened to minimum wage rates over the years 1974 to 1997.

Solution

a. We recall how to calculate percent increase. We take the amount of increase in CPI and divide it by the CPI at the time from which we measured the increase, 1974; then we multiply by 100 to convert to percentage. The CPI increased from 49.3 to 160.3, which is an increase of 111 units. We then divide this by 49.3 and convert it to a percentage: $\left(\frac{111}{49.3}\right) \times 100\% \approx 225.2\%$. So the CPI increased by about 225 percent from 1974 to 1997.

b. The minimum wage rate increased from \$2.00 to \$5.15, which is an increase of \$3.15. We divide by 2.00 to get $\left(\frac{3.15}{2.00}\right) \times 100\% = 157.5\%$.

c. These calculations show that the CPI increased by a much higher percentage than the minimum wage rate did, which means that the minimum wage rate did not increase as fast as the cost of market basket goods.

d. With *time A* as 1974 and *time O* as 2000, we convert the minimum wage *dollars at time A* of \$2.00 to its dollar value in 2000 (i.e. at *time O*) using the following formula:

$$\textit{Dollars at time O} = \textit{Dollars at time A} * \frac{\textit{CPI at time O}}{\textit{CPI at time A}}$$

$$\text{1974 minimum wage in 2000 dollars} = 2.00 \times \left(\frac{172.2}{49.3}\right) \approx 6.99$$

For the 1975 minimum wage in 2000 dollars, we consider *time A* to be 1975 and use 2.10 as *dollars at time A* and also use the CPI in 1975 as *CPI at time A*. We then get 1975 minimum wage in 2000 dollars = $2.10 \times \left(\frac{172.2}{53.8}\right) \approx 6.72$. Using a similar calculation for each of the years in the table, we get these values:

Year	1974	1975	1976	1978	1979	1980	1981	1990	1991	1996	1997
Wage	2.00	2.10	2.30	2.65	2.90	3.10	3.35	3.80	4.25	4.75	5.15
Wage 2000$	6.99	6.72	6.96	7.00	6.88	6.48	6.35	5.01	5.37	5.19	5.53

e. The graph shows that the minimum wage, measured in 2000 dollars, was higher than its 2000 value during the years 1974 through 1981. Minimum wage measured in 2000 dollars declined from 1978 through 1990. In 1991 it rose again, but by 1996, even the increase was not enough to "keep up with inflation."

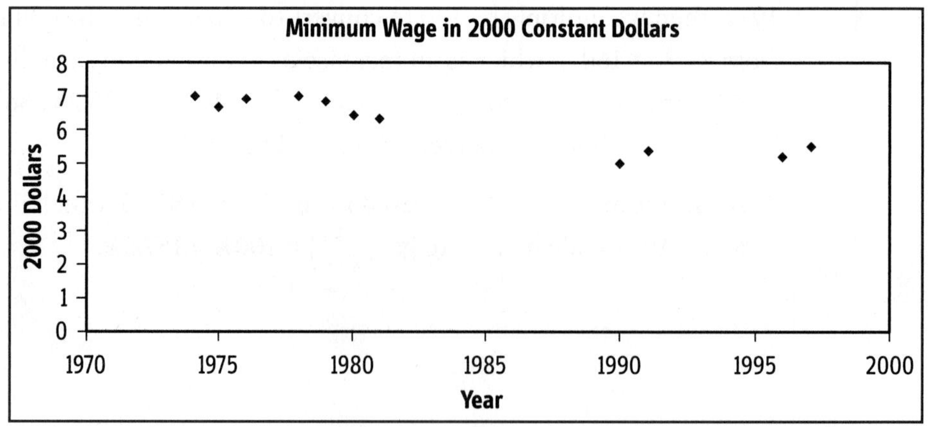

The **Consumer Confidence Index (CCI)** is a measure of consumers' optimism about the state of the economy; it is based on a sample of 50,000 U.S. households and is designed to reflect people's feelings about general business conditions, employment opportunities, and their own income prospects. The next example looks at how this index has changed over time.

Example 8.4

The Consumer Confidence Index is set up so that the index number for 1985 is 100, and the other years' CCI values are given relative to the 1985 index of 100. The following table gives CCI values for the years 1970 through 1997. (Source: *The Wall Street Journal Almanac 1999*, page 135.)

Year	1970	1971	1972	1973	1974	1975	1976	1977	1978	
CCI	89.6	80.4	103.3	98.3	70.9	74.5	94.3	97.9	106.0	
Year	1979	1980	1981	1982	1983	1984	1985	1986	1987	
CCI	91.9	73.8	77.4	59.0	85.7	102.3	100.0	94.7	102.6	
Year	1988	1989	1990	1991	1992	1993	1994	1995	1996	1997
CCI	115.2	116.8	91.5	68.5	61.6	65.9	90.6	100.0	104.6	125.4

a. Sketch a graph of these data and describe what the graph shows.

b. From approximately 1989 to 1992, the graph of CCI is decreasing and concave upward. What does the shape and direction of the graph tell us?

c. Does there appear to be any relationship between the CCI and the CPI?

Solution

a. A graph of the CCI appears next. The CCI goes up and down throughout the time period under consideration, reaching its maximum value in 1997 and its minimum value in 1982. For most years, the CCI is below the 1985 level of 100.

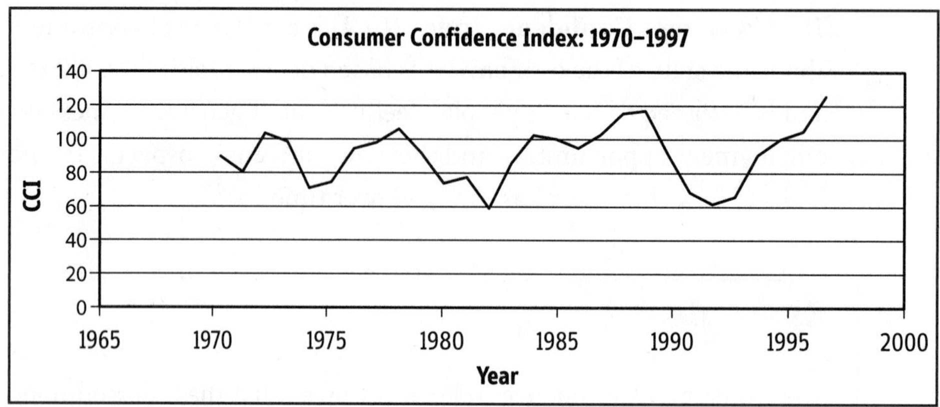

b. Because the graph is decreasing, the CCI is going down; that is, consumers' optimism is decreasing. However, because the graph is concave upward, even though consumer confidence is decreasing, the rate at which it is changing is increasing, so it appears as though consumers' outlook is improving. A look at the graph for the years after 1992 confirms this.

c. There doesn't appear to be a link between CCI and CPI. The following graph shows both CCI and CPI.

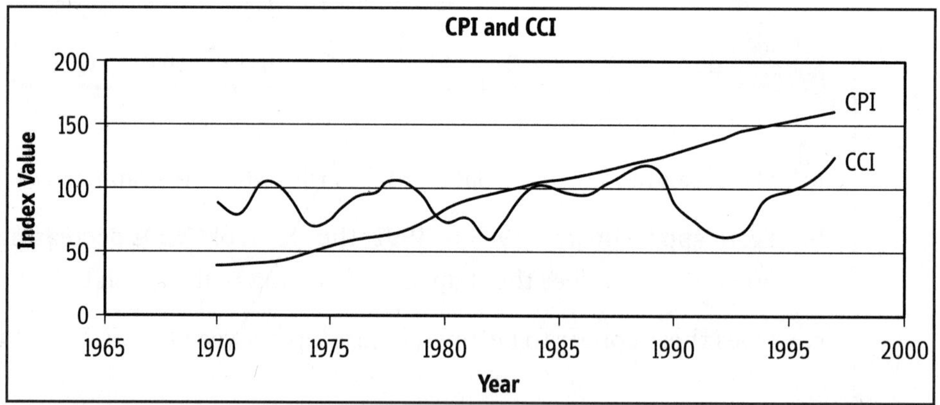

The **Fog Index** is one measure of the reading difficulty of a passage of text and is sometimes used by newspaper and magazine editors. (There are other measures of reading difficulty.) The Fog Index takes into account word count, sentence length, and word size. For our purposes, we will define "big" words as words with three or more syllables. We'll also use the following conventions: proper names do not count as "big" words; compound words that are combinations of easy words, such as "everything," and words in which one of the three syllables is formed from a suffix such as "ed" or "ing" do not count as "big" words; however, "big" words that are repeated count each time they are used. The Fog Index (FI) is defined and calculated using the following formula:

$$FI = \left(\frac{\textit{number of words}}{\textit{number of sentences}} + 100 * \frac{\textit{number of "big" words}}{\textit{number of words}} \right) * 0.4$$

A FI value of 9 purportedly indicates ninth-grade reading level; a value of 12 indicates twelfth-grade level; 14 is college-sophomore level.

Example 8.5

Find the Fog Index for the following passage from the article, "Who Will Name the Next Supreme Court Justice?" that appeared in *The Wall Street Journal* on May 21, 2000. Explain how the FI is calculated.

> When David M. O'Brien, a government professor at the University of Virginia, took his students on a tour of the Supreme Court recently, they ended up in a private question-and-answer session with a justice who abruptly turned to presidential politics. He blurted out several hot-button issues involving federalism, and said the future composition of the court would dramatically affect the outcome of those cases.
>
> "Vote carefully," the justice, a Republican appointee, implored the students assembled in a stately white oak conference room.
>
> To this justice, and to interest groups on the left and the right, the 2000 presidential campaign is not so much about whether Al Gore or George W. Bush makes it to the White House. It's about whom Mr. Gore or Mr. Bush would put on the Supreme Court, where vacancies are likely, if not in the next four years, then certainly in the next president's potential second term. Three of the nine justices are age 70 or older.

Solution

Before we begin, we need to agree on some criteria for "counting words." We will count a hyphenated word (for example, question-and-answer) as one word and an initial in someone's name or a number will count as a word. Words with an apostrophe will also count as one word. (This follows the conventions used in the word-processing package Microsoft Word.) Thus, the word count for the three paragraph passage is 161. If we disregard words such as "recently" and "appointee" that reach three syllables because of a suffix and disregard proper names, we count 16 "big" words and only 6 sentences. (The "big" words are government, professor, presidential, politics, several, federalism, composition, dramatically, Republican, assembled, conference, interest, presidential, vacancies, president's, potential.) Therefore, we calculate the FI as follows:

$$FI = \left(\frac{161}{6} + 100 * \frac{16}{161}\right) * 0.4 \approx (26.83 + 9.94) * 0.4 \approx 14.7$$

To really understand this index, however, we need to look at the FI for a variety of passages. The first term of the FI involves $\frac{\textit{number of words}}{\textit{number of sentences}}$, which is the average number of words per sentence. The second term is the percentage of "big" words in the passage. These two figures are added together and then we multiply that sum by a factor of 0.4.

A rating system can be considered a type of indexing scheme; rating systems are often set up to compare movies, cities, colleges, and other institutions. Sometimes these rating systems organize places or things into categories and other times rating systems rank order things. Movies are rated by movie industry insiders according to the following categories: G, PG, PG-13, R, or NC-17. Each year, *U.S. News and World Report* releases its ratings of colleges and universities based on diverse criteria. The magazine does not publish an actual index number for each college or university, but they rank order the colleges based on their system. We encounter rating systems when various organizations rate, for example, the best cities for walking or the best "family-friendly" companies.

To use these rating systems' indexes appropriately, we need to consider the information included in the rating as well as the reliability of that information.

In the next example, we consider what quantities should be included in particular rating systems.

Example 8.6

a. A 268-page report "The State of the Child in Pennsylvania" rates counties on child well-being, using factors such as infant deaths and juvenile delinquency figures. What other quantities would be appropriate to include in a rating system for well-being of children?

b. *U.S. News and World Report* publishes an issue each year that rates the nation's colleges. What variables do you think should be included in such a rating scheme?

c. *Money* magazine rates cities on "livability." What variables do you think should be included in such a rating scheme?

Solution

a. In addition to the variables mentioned, the number of children on welfare, the number of children living in families with incomes below the poverty level, and the proportion of children who finish high school might be included in the rating. (You might think of other variables to include, as well.)

b. In rating colleges, the proportion of students who graduate in four years, "student satisfaction" with the college's social atmosphere, the proportion of students receiving financial aid, and the proportion of students who pursue graduate or professional school, might be taken into consideration. (Note that we could create either a simple or a fairly sophisticated rating system. Also, different individuals might think certain variables are more or less important.)

c. A partial list might include: unemployment rate; tax rate; accessibility to professional sports teams; and distance to and size of shopping malls.

In addition to choosing what to rate, sometimes there is a problem with how to define and measure certain variables. For example, a person's "unemployment status" is one such variable. Is a person who is contented in the role of homemaker employed, unemployed, or neither? What about part-time workers looking for full-time employment? What about someone who has no job, is discouraged about looking for a job, and who has given up looking for one? We will consider some of these questions further in the Explorations.

Summary

In this topic, we investigated several indexes that are used to show trends over time, like the Dow Jones Industrial Average, the Consumer Confidence Index, and the Consumer Price Index. We also used the Consumer Price Index to look at changes in minimum wage over time. The Fog Index was studied as one way to measure the reading difficulty of a passage of text. Finally, we discussed how rating systems could be viewed as a type of index and looked at what factors should be included in particular rating systems.

Explorations

1. To track increases in cost and to allow for comparisons, the Fan Cost Index (FCI) for various professional sports leagues has been computed for the past several seasons. This index includes the cost of four average-price tickets, four small soft drinks, two small beers, four hot dogs, parking for one car, two game programs, and two souvenir caps. The following tables show the FCI for four professional sports leagues. (Source: *Team Marketing Report,* www.teammarketing.com.)

MLB Season	FCI ($)	NFL Season	FCI ($)
1994	96.41	1994	180.33
1995	97.55	1995	199.22
1996	103.07	1996	208.48
1997	107.26	1997	227.54
1998	115.06	1998	241.89
1999	121.76	1999	256.09

MLB Season	FCI ($)	NFL Season	FCI ($)
2000	132.44	2000	279.90
2001	140.63	2001	278.73
2002	145.21	2002	290.41
2003	151.19	2003	301.75

NHL Season	FCI ($)	NBA Season	FCI ($)
1994–95	195.79	1994–95	182.72
1995–96	203.31	1995–96	192.18
1996–97	219.26	1996–97	204.38
1997–98	234.45	1997–98	216.55
1998–99	238.91	1998–99	241.84
1999–00	258.42	1999–00	268.45
2000–01	264.81	2000–01	281.72
2001–02	239.24	2001–02	244.44
2002–03	240.43	2002–03	254.88
2003–04	253.65	2003–04	261.26

a. Discuss the FCI, including if it is a reasonable measure of costs.

b. Look at the pattern of FCI values for the NHL and NBA from 2000–01 to 2003–04 and compare with the values for the previous four-year period. What do you observe?

c. Determine whether the cost of attending sporting events has outpaced inflation.

d. Explain the increases or decreases in the FCI for the different sports during the 2000–03 time period.

2. The following table lists first-class postage rates and the years in which they have increased:

Year	1919	1932	1958	1963	1968	1971	1974	1978	
Postage in Cents	2	3	4	5	6	8	10	15	
Year	1981	1983	1988	1991	1995	1999	2001	2002	2006
Postage in Cents	18	22	25	29	32	33	34	37	39

a. Use the CPI values in Example 8.2 to convert the 1974 postage rate into constant 1971 dollars.

b. Repeat for the other dates in the table from 1971 to 2002. Describe what these values show about postage rates over the years 1971 to 2002.

3. The following graph (which appeared in *The New York Times* on July 28, 2004) gives the monthly values of the Consumer Confidence Index (CCI) from July 2003 to July 2004. Describe what the graph shows.

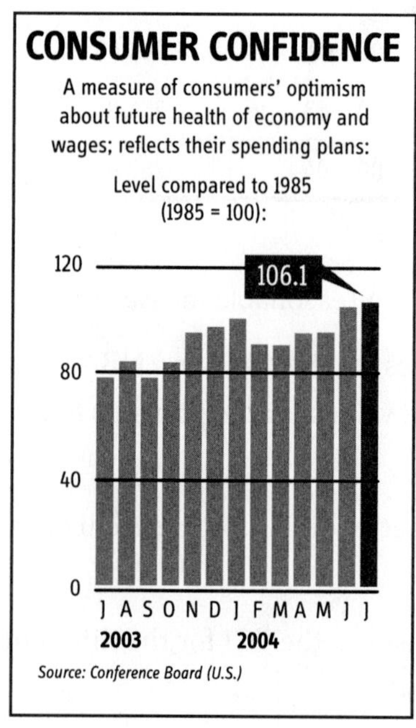

4. The monthly values of the Consumer Price Index (CPI) from January 2003 to June 2004 are given in the following table. Sketch a graph of these data and describe what the graph shows and how the CPI compares with the CCI over the same time period. (See Exploration 3.)

CPI	Jan.	Feb.	Mar	April	May	June	July	Aug.	Sept.	Oct.	Nov.	Dec.
2003	181.7	183.1	184.2	183.8	183.5	183.7	183.9	184.6	185.2	185.0	184.5	184.3
2004	185.2	186.2	187.4	188.0	189.1	189.7						

5. Describe as precisely as possible how you would measure the following items and indicate any problems that might arise.
   a. A person's employment status
   b. The length of a person's foot
   c. A person's sensitivity toward others
   d. How many calories a person consumes in one day
6. Think about the quantities used for computing the Fog Index of a passage of text. What other quantities might be used to evaluate the reading difficulty of a passage of text?
7. Find an editorial from a recent newspaper.
   a. Compute the Fog Index for the editorial. Indicate any problems or questions that arose in computing the index.
   b. How would the Fog Index for this section change if there were twice as many sentences?
   c. How would it change if there were five more big words? Twice as many big words?
8. Find a short passage from a children's book and compute the Fog Index for the passage. Find a passage from a paper you have written for a class and compute the Fog Index for the passage. Compare these values with the Fog Index computed in Example 8.5.

9. Use the rating systems we looked at in Example 8.6 to answer the following questions.

   a. What additional variables, other than the ones mentioned in the solution to Example 8.6a, could you include in a system that rates counties on child well-being?

   b. What five variables do you think would be most important to include in a system that rates colleges and universities? Give reasons to support your answers.

   c. What additional variables, other than those listed in the solution to Example 8.6c, should be included in a system that rates cities on "livability"?

10. The following data, gathered by the National Center for Education Statistics, U.S. Department of Education, shows expenditures in unadjusted and constant dollars (in billions) and expenditures per full-time-equivalent student in constant dollars in U.S. public four-year degree-granting institutions of higher education from 1987–88 to 2001–02. Explain in detail what the table shows, including any trends.

Year	Expenditures in Billions: Unadjusted Dollars	Expenditures in Billions: Constant 2001-02 Dollars	Dollars per Student, in Constant 2001–02 Dollars
1987–88	$60.1	$92.5	$21,039
1988–89	65.3	96.1	21,330
1989–90	70.9	99.5	21,529
1990–91	76.7	102.1	21,537
1991–92	81.3	104.9	21,868
1992–93	86.1	107.6	22,429
1993–94	89.7	109.3	22,930
1994–95	94.9	112.4	23,668
1995–96	97.9	112.9	23,733
1996–97	103.1	115.6	24,244

Year	Expenditures in Billions: Unadjusted Dollars	Expenditures in Billions: Constant 2001-02 Dollars	Dollars per Student, in Constant 2001-02 Dollars
1997-98	109.2	120.3	24,986
1998-99	115.2	124.7	25,608
1999-2000	124.9	131.4	26,581
2000-01	134.6	137.0	27,256
2001-02	139.9	139.9	27,118

11. A recently released poll rated the top 20 "party" colleges and universities in the country. What variables do you think should be included in such a rating and what importance would you place on each of these variables in such a rating?

12. The following graph, created using data from the Missouri Economic and Information Center (http://www.misssourieconomy.org) shows the value of the Purchasing Managers' Index from August 2003 to July 2005.

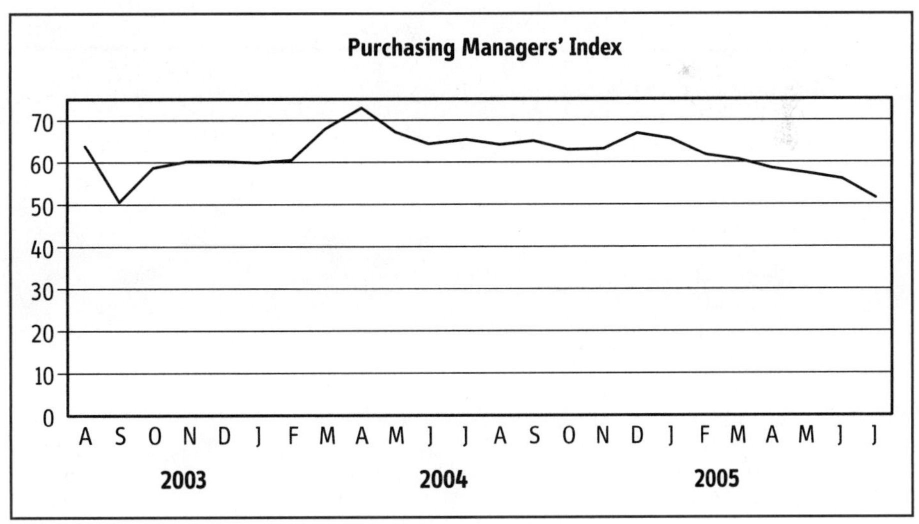

a. Use the Web to find how this index is calculated.

b. Describe what the graph shows in this context.

TOPIC 9

Personal Finances

OBJECTIVES

After completing this topic, you will be able to

- Use the terminology associated with personal financial management
- Analyze relevant formulas to compute simple and compound interest
- Understand ordinary annuities and use the accumulated savings formula
- Apply the loan payment formula to understand and analyze installment loans and credit card loans

In previous topics and activities, we investigated several problems dealing with personal financial management. For example, we looked at the growth of debt in conjunction with exponential functions, and we investigated credit card balances in an activity about multiple variable functions. In this topic, we will consider the interest received on savings and the costs and payments associated with various types of loans in more detail.

In order to understand loans in general, we first consider the concept of **interest**. Interest is money that the borrower pays for using the lender's money. If money is lent under a **simple interest** agreement, the loan is for a fixed time period and the borrower repays the money borrowed plus interest at the end of that time period. Simple interest, denoted I, is computed using the formula:

$$I = p \times r \times t$$

where p is the amount of money borrowed, called the **principal**, r is the interest rate (expressed as a decimal), and t is the duration of the loan. The interest rate and the duration of the loan must be expressed in the same time increment. This means that if the interest rate is given in rate per year, t must be expressed in years; if the interest rate is expressed in rate per month, t must be given in months.

Example 9.1

An electronics store is offering a laptop computer for \$2000, with two payment options. Under the first option, there is no payment for the first nine months, but the \$2000 must be repaid at the end of that time, at an annual simple interest rate of 19.8 percent. Under the second option, the borrower must pay \$400 initially and then pay \$1875 at the end of one year. Compare the two options by finding the total cost of the first option and the annual interest rate for the second option.

Solution

To compute the total cost under the first option, we use time in years because we are given an annual interest rate. Then the amount of time t is $\frac{9}{12}$ of a year and an annual interest rate of 19.8 percent means $r = 0.198$. So, the interest paid is $I = p \times r \times t = \$2000 \times 0.198 \times \frac{9}{12} = \$297$; the total cost of the computer is \$2297. To find the interest rate of the second offer, we use the amount of the one-year loan, \$1600, to first find the amount of interest paid: $I = \$1875 - \$1600 = \$275$. Then we use the simple interest equation $I = p \times r \times t$, with $I = 275$, $p = 1600$, and $t = 1$ to solve for r: $275 = 1600 \times r \times 1$. Thus, $r = \frac{275}{1600} = 0.171875$ and the annual interest rate is approximately 17.2 percent. Therefore, the interest rate under the second option is less than with the first option. The total cost of the computer using the second option is \$400 + \$1875 = \$2275, which is \$22 less than the cost under the first option.

In finding simple interest, we saw that the interest was computed only on the principal. Interest that is computed on the principal and any accumulated interest is called **compound interest**. Interest paid on savings accounts is normally compound interest. In the next examples, we develop a method for finding compound interest and then compare how interest accrues using different compounding periods.

Example 9.2

A student saves \$2000 from his summer job and wants to invest it. If he invests it in an account at an annual interest rate of 3 percent compounded quarterly, and doesn't add or withdraw any money during the year, how much will he have in the account at the end of one year?

Solution

The annual interest rate is $r = 0.03$. Because interest is paid quarterly, the interest rate paid each quarter is one fourth of the annual interest rate: $\frac{r}{4} = \frac{0.03}{4} = 0.0075$. At the end of the first quarter, the amount in the account (account balance) is $2000 + 0.0075 \times 2000 = (1 + 0.0075) \times 2000 = 2015$. At the end of the second quarter, the account balance is the amount at the end of the first quarter, plus the interest on that amount: $2015 + 0.0075 \times 2015 = (1 + 0.0075) \times 2015$. Recall from the previous calculations that $2015 = (1 + 0.0075) \times 2000$. So, $(1 + 0.0075) \times 2015 = (1 + 0.0075) \times (1 + 0.0075) \times 2000 = (1 + 0.0075)^2 \times 2000 \approx 2030.11$. We can use a table to record our findings:

Quarter	Interest Paid This Quarter	Balance End of This Quarter	Formula for Balance End of This Quarter
1	$0.0075 \times 2000 = 15$	\$2015	$\left(1 + \frac{r}{4}\right)p = \left(1 + \frac{0.03}{4}\right)(2000)$
2	$0.0075 \times 2015 = 15.11$	\$2030.11	$\left(1 + \frac{r}{4}\right)^2 p = \left(1 + \frac{0.03}{4}\right)^2(2000)$
3	$0.0075 \times 2030.11 = 15.23$	\$2045.34	$\left(1 + \frac{r}{4}\right)^3 p = \left(1 + \frac{0.03}{4}\right)^3(2000)$
4	$0.0075 \times 2045.34 = 15.34$	\$2060.68	$\left(1 + \frac{r}{4}\right)^4 p = \left(1 + \frac{0.03}{4}\right)^4(2000)$

The balance at the end of one year is \$2060.68.

The previous example illustrates the following general **compound interest formula** for finding the amount A of money in an account after t years, if the annual interest rate, also called the **annual percentage rate**, is denoted APR and n is the number of compounding periods per year:

$$A = \left(1 + \frac{APR}{n}\right)^{nt} \times p$$

In the next example, we use this formula to compare the amount of money in an account if the interest is compounded quarterly, monthly, or daily. Quarterly means interest is compounded four times per year, so $n = 4$; monthly means interest is compounded $n = 12$ times per year. There are three methods for calculating daily compounding. Historically, accountants used $n = 360$, assuming there are 12 months each year with 30 days. In the modern technological world, many banks now use $n = 365$. And some banks use what's called "365/360," which uses the APR divided by 360 but then compounds it 365 times a year. Here we will use $n = 365$ if interest is compounded daily.

Example 9.3

A student has a choice of three banks at which to invest her \$2000 of summer earnings. All three banks offer an annual interest rate of 3 percent; however, the first bank's interest is compounded quarterly, the second bank's interest is compounded monthly, and the third bank's interest is compounded daily. If she doesn't deposit or withdraw any money for two years, how much money would she have in the account at the end of two years if she had invested her money at each of the banks?

Solution

All three banks have an APR of 3 percent and in all three cases $t = 2$ years. If interest is compounded quarterly, $n = 4$; if interest is compounded monthly, $n = 12$; and if interest is compounded daily, $n = 365$. So, if interest is compounded quarterly, $A_{\text{quarterly}} = \left(1 + \frac{APR}{n}\right)^{nt} \times p = \left(1 + \frac{0.03}{4}\right)^{(4)(2)} 2000 = \left(1 + \frac{0.03}{4}\right)^{8} 2000 \approx 2123.20$. If interest is compounded quarterly, the amount after two years, rounded to the closest cent, is \$2123.20. If interest is compounded monthly, $A_{\text{monthly}} = \left(1 + \frac{APR}{n}\right)^{nt} \times p = \left(1 + \frac{0.03}{12}\right)^{(12)(2)} 2000 = \left(1 + \frac{0.03}{12}\right)^{24} 2000 \approx 2123.51$. If interest is compounded monthly, the amount after two years, rounded to the closest cent, is \$2123.51. If interest is compounded daily, $A_{daily} = \left(1 + \frac{APR}{n}\right)^{nt} \times p = \left(1 + \frac{0.03}{365}\right)^{(365)(2)} 2000 = \left(1 + \frac{0.03}{365}\right)^{730} 2000$ ≈ 2123.67. So if interest is compounded daily, the amount after two years, rounded to the closest cent, is \$2123.67. Using these formulas, we see that there are only small differences in the total amount in each account after two years.

The power of compounding is more evident if we consider a time period longer than two years, as the next example shows.

Example 9.4

Since the 18th century, the U.S. government has had ordinances that established a system for the sale of publicly owned land. A 1785 ordinance allowed purchase of a "section" of land, minimum size 640 acres, at no less than \$1.00 per acre. Suppose, instead of purchasing a 640-acre section of land for \$640 in 1785, a family had invested that money at 5 percent *APR* and didn't add to or withdraw from the account. How much would the account be worth in 2005, if interest is compounded

a. Annually?

b. Daily?

Solution

In both scenarios, the money is invested from 1785 to 2005, so it would be invested for $t = 220$ years.

a. If interest is compounded annually, $n = 1$, and we use the formula,

$$\left(1 + \frac{APR}{n}\right)^{nt} \times p = \left(1 + \frac{0.05}{1}\right)^{(1)(220)} 640 = (1.05)^{220} 640 \approx 29{,}364{,}713.60$$

So, if interest is compounded annually, the account would be worth \$29,364,713.60.

b. If interest is compounded daily, $n = 365$, so

$$\left(1 + \frac{APR}{n}\right)^{nt} \times p = \left(1 + \frac{0.05}{365}\right)^{(365)(220)} 640 = (1.000136986)^{80300} 640 \approx 38{,}290{,}593.48$$

So, if interest is compounded daily, the account would have \$38,290,593.48, almost 9 million dollars more than if interest is compounded annually!

When interest is compounded at periods of time less than one year, the actual gain in the account at the end of one year is not the same as the annual percentage rate times the initial principal, but it is slightly higher. In Example 9.2, we found that the actual gain in the \$2000 account after one year when the interested is compounded quarterly was \$60.68, but $0.03 \times 2000 = 60$, so after one year, the interest actually collected is higher than 3 percent of the initial principal. To find what percentage of the initial principal the actual interest collected represents, we divide the interest gained by the original principal: $\frac{60.68}{2000} = 0.03034$, or 3.034 percent. This percentage is called the **annual percentage yield** and is denoted by *APY.* Banks usually list both the *APR* and the *APY.* For this account, the bank would state $APR = 3$ percent and $APY = 3.034$ percent.

Note that in general, we can calculate the annual percentage yield as the ratio $\frac{A-p}{p}$, where A is the amount in the account after one year, p is the original principal, and $A - p$ is the interest earned on the principal p invested at an annual interest rate of APR in an account in which interest is compounded n times per year. Given APR and n, we can use any value of p to find A and then use that A and p to find APY. In particular, using $p = 1$ may make the calculations simpler, as we'll see in the next example.

Example 9.5

A grandfather would like to be able to give his grandson a car when the grandson graduates from college in four years. He decides he will need \$25,000 at that time.

a. How much will the grandfather need to invest now in a 4-year account that has an annual interest rate of 5.2 percent compounded monthly in order to have \$25,000 in four years?

b. What is the annual percentage yield on this account?

Solution

a. We use the general formula for finding the amount A of money in an account after t years if the annual interest rate is APR and n is the number of compounding periods per year: $A = \left(1 + \frac{APR}{n}\right)^{nt} \times p$, with the principal p as the quantity that we need to determine. In this example, $A = 25{,}000$, $r = .052$, $n = 12$, and $t = 4$.

Thus our equation is $25000 = \left(1 + \frac{0.052}{12}\right)^{(12)(4)} \times p$ or $25000 \approx (1.0043333)^{48}p$. So $p = \frac{25000}{1.0043333^{48}} \approx 20{,}314.33$; the grandfather needs to invest \$20,314.33 in the account in order to have \$25,000 in four years.

b. With an *APR* of 5.2 percent compounded monthly, if \$1 is invested for one year, the amount in the account after one year is $A = \left(1 + \frac{0.052}{12}\right)^{12} \times 1 \approx 1.053257$. Thus the increase in the \$1 account is $1.053257 - 1 = 0.053257$, and the *APY* is $\frac{0.053257}{1}$, or approximately 5.326 percent.

Suppose we want to save for something in the future (as the grandfather in Example 9.5 did) but we don't have a lump sum of money to invest initially. Instead, we want to add to our account in smaller amounts on a regular basis. There are a variety of savings plans that promote this kind of savings, such as Individual Retirement Accounts (IRAs), Keogh plans, and employee pension plans. A series of fixed, regular payments is called an **annuity**. Leases and rental payments, as well as savings plans in which you deposit a specified amount at fixed intervals of time, are examples of annuities. There are two types of annuities: an ordinary annuity and an annuity due. An **ordinary annuity** is one in which the payments are required at the end of each time period. With an **annuity due**, payments are required at the beginning of each time period. We'll concentrate on ordinary annuities; in the next example, we illustrate how money accumulates with an ordinary annuity.

Example 9.6

A college student is saving to buy a new car and deposits \$150 into a savings plan at the end of each month. Suppose that the plan pays interest monthly at an *APR* of 6 percent. Create a table that shows the monthly balance at the end of each month for a six-month period.

Solution

We create a table with columns to keep track of the interest accrued each month as well as the additional deposit each month. We observe that the monthly interest rate is $\frac{APR}{n} = \frac{0.06}{12} = 0.005$. The last column of the table keeps track of the new balance in the account each month and represents the sum: *previous balance + interest on previous balance + additional deposit.* The first column is the new balance from the previous month.

End of Month i	Previous Balance	Interest on Previous Balance	Additional Deposit	New Balance
1	0	0	\$150	\$150
2	\$150	150 × 0.005 = \$0.75	\$150	\$300.75
3	\$300.75	300.75 × 0.005 = \$1.50	\$150	\$452.25
4	\$452.25	452.25 × 0.005 = \$2.26	\$150	\$604.51
5	\$604.51	604.51 × 0.005 = \$3.02	\$150	\$757.53
6	\$757.53	757.53 × 0.005 = \$3.79	\$150	\$911.32

The table created in Example 9.6 helps us understand how interest and additional deposits accumulate when additional money is deposited into an account. We can also use a formula that gives the relationship between A, the amount in the account at a future date; PMT, the amount of the regular payment into the account; APR, the annual percentage interest rate; n, the number of payment and compounding periods per year (which are assumed to be the same with this formula); and t, the number of years. Here is the **accumulated savings formula**:

$$A = PMT \times \frac{\left[\left(1 + \frac{APR}{n}\right)^{nt} - 1\right]}{\left(\frac{APR}{n}\right)}$$

(Sometimes the amount A is called **future value** and denoted by FV.)

In the next example, we'll confirm that this formula gives us the same result we obtained using the table in Example 9.6 for month 6, and find what the account will be worth in the future under the same assumptions.

Example 9.7

Use the accumulated savings formula to find how much money the college student from Example 9.6 (who, at the end of each month, deposits \$150 into a savings plan that pays interest monthly at an APR of 6 percent) has in the account at the end of

a. Six months

b. One year

c. Four years

Solution

We use the formula

$$A = PMT \times \frac{\left[\left(1+\frac{APR}{n}\right)^{nt}-1\right]}{\left(\frac{APR}{n}\right)}$$

to find A, with $APR = 0.06$, $n = 12$, $PMT = \$150$, and different values of t for parts a, b, and c.

a. Here $t = \frac{1}{2}$ (years), so

$$A = \$150 \times \frac{\left[\left(1+\frac{0.06}{12}\right)^{(12)(1/2)}-1\right]}{\left(\frac{0.06}{12}\right)} = \$150 \times \frac{(1.005)^{6}-1}{0.005} \approx \$911.33$$

Note the difference of \$.01 with the value in the table in Example 9.6, which is due to rounding the calculations in the table to two decimals.

b. With $t = 1$ (year),

$$A = \$150 \times \frac{\left[\left(1+\frac{0.06}{12}\right)^{(12)(1)}-1\right]}{\left(\frac{0.06}{12}\right)} = \$150 \times \frac{(1.005)^{12}-1}{0.005} \approx \$1850.33$$

c. Substituting $t = 4$ gives

$$A = \$150 \times \frac{\left[\left(1+\frac{0.06}{12}\right)^{(12)(4)}-1\right]}{\left(\frac{0.06}{12}\right)} = \$150 \times \frac{(1.005)^{48}-1}{0.005} \approx \$8114.67$$

We can use the accumulated savings formula to help set a "savings goal" if we want to plan ahead and have money for something particular, as the next example illustrates.

Example 9.8

Suppose a new college graduate wants to accumulate \$30,000 for a deposit on a home in eight years by making regular end-of-the-month savings deposits. Assume an *APR* of 8 percent compounded monthly.

a. Find how much the graduate should deposit at the end of each month to reach the goal.

b. Find how much of this accumulated amount comes from the deposits and how much from the interest.

Solution

a. The goal is to accumulate A = \$30,000 in $t = 8$ years, with $APR = 0.08$ and $n = 12$. We want to find *PMT*, so we need to solve the following equation for *PMT*: $30{,}000 = PMT \times \dfrac{\left[\left(1+\frac{0.08}{12}\right)^{(12)(8)} - 1\right]}{\left(\frac{0.08}{12}\right)} \approx \text{PMT} \times \dfrac{(1.006666667)^{96} - 1}{0.006666667}$

so $PMT \approx 30{,}000 \times \dfrac{0.006666667}{(1.006666667)^{96} - 1} \approx 224.1003737.$

So the graduate needs to make monthly payments of \$224.11 to have \$30,000 at the end of eight years. Note that to make sure we will reach the goal, we need to round the payment up to the next cent, instead of rounding it to the nearest cent.

b. We assume the *APR* remains at 8 percent during the 8-year period. At the end of this time period, the graduate has deposited \$224.11 per month × 12 months per year × 8 yrs = \$21,514.56. So \$30,000 – \$21,514.56 = \$8,485.44 is the approximate amount of (compound) interest earned.

We now consider **installment loans** in which the borrower partially repays the loan with equal, regular payments, for example on a monthly basis, for a fixed amount of time. (Typical installment loans are college tuition loans, home mortgages, and car loans.) In the next example, we calculate how much of the principal (the amount borrowed) is paid off if we make regular monthly payments.

Example 9.9

A student wants to purchase a computer for $2000 and the store offers an annual interest rate of 18 percent if the student makes payments of $160.00 each month. Set up a table to determine how much the student will still owe after one year.

Solution

When the student makes the initial $160 payment at the end of the first month, the interest for that month is $2000 \times \frac{0.18}{12} = 30$. The interest is paid first, leaving $160 – $30 = $130 of the payment to go toward paying off the principal. At the end of the first month, the new principal is $2000 – $130 = $1870. The interest for the second month is then charged on the new principal and is $1870 \times \frac{0.18}{12}$ = $28.05. After the second month's interest is paid, at the end of the second month, 160 – 28.05 = $131.95 of the payment goes to principal. The results for the rest of the 12-month period are summarized in the table. Note that each month the interest is less and the payment toward the principal is greater. At each step, we round all amounts to the nearest cent.

End of Month *i*	Previous Principal	Payment	Interest Paid (on previous principal)	Payment Toward Principal	New Principal
1	$2,000.00	$160	$30.00	$130.00	$1,870.00
2	$1,870.00	$160	$28.05	$131.95	$1,738.05
3	$1,738.05	$160	$26.07	$133.93	$1,604.12
4	$1,604.12	$160	$24.06	$135.94	$1,468.18
5	$1,468.18	$160	$22.02	$137.98	$1,330.20
6	$1,330.20	$160	$19.95	$140.05	$1,190.15
7	$1,190.15	$160	$17.85	$142.15	$1,048.00
8	$1,048.00	$160	$15.72	$144.28	$903.72
9	$903.72	$160	$13.56	$146.44	$757.28
10	$757.28	$160	$11.36	$148.64	$608.64

End of Month i	Previous Principal	Payment	Interest Paid (on previous principal)	Payment Toward Principal	New Principal
11	\$608.64	\$160	\$9.13	\$150.87	\$457.77
12	\$457.77	\$160	\$6.87	\$153.13	\$304.64

At the end of the year, the principal remaining is \$304.64; one more full payment of \$160 plus a smaller payment will exhaust the loan.

If we wanted to pay off the loan in Example 9.9 in exactly 12 equal payments, it's clear that we would need to pay a bit more than \$160 each month. We can calculate this amount with the following loan payment formula:

$$PMT = p \times \frac{\left(\frac{APR}{n}\right)}{\left[1 - \left(1 + \frac{APR}{n}\right)^{-nt}\right]}$$

In this formula, *PMT* is the regular payment amount; *p* is the initial principal, that is, the amount borrowed; *APR* is the annual percentage rate; *n* is the number of regular payment periods per year; and *t* is the term, in years, of the loan.

Example 9.10

Consider the \$2000 loan to buy a computer at 18 percent *APR*.

a. Use the loan payment formula to determine the payment needed to pay off the loan for the computer in 12 regular monthly payments.

b. How would the payments change if we take two years to pay off the loan at the same interest rate?

c. How much interest do we pay on the total loan under the one-year payment plan and how much interest do we pay under the two-year payment plan?

Solution

For this example, $p = 2000$, $n = 12$, and $APR = 0.18$.

a. If we pay off the loan in 12 payments, $t = 1$, so

$$PMT = p \times \frac{\left(\frac{APR}{n}\right)}{\left[1 - \left(1 + \frac{APR}{n}\right)^{-nt}\right]} = 2000 \times \frac{\left(\frac{0.18}{12}\right)}{\left[1 - \left(1 + \frac{0.18}{12}\right)^{-(12)(1)}\right]} = \frac{2000 \times 0.015}{1 - (1.015)^{-12}} \approx 183.36$$

As expected, this payment of \$183.36 per month is greater than the \$160 payment in Example 9.9.

b. Now $t = 2$, so

$$PMT = p \times \frac{\left(\frac{APR}{n}\right)}{\left[1 - \left(1 + \frac{APR}{n}\right)^{-nt}\right]} = 2000 \times \frac{\left(\frac{0.18}{12}\right)}{\left[1 - \left(1 + \frac{0.18}{12}\right)^{-(12)(2)}\right]} = \frac{2000 \times 0.015}{1 - (1.015)^{-24}} \approx 99.85$$

If we pay off the loan in two years, our monthly payments will be \$99.85.

c. With the one-year payment plan, we pay a total of (12)(\$183.36) = \$2200.32, so \$200.32 is paid in interest. Under the two-year payment plan, we'd pay (24)(\$99.85) = \$2396.40, so \$396.40 is paid in interest.

Loans accumulated on credit cards differ from installment loans because the credit card company does not require the borrower to pay off the balance in a particular period of time. These types of loans are sometimes referred to as **open-end installment loans** because the borrower can make variable payments each month, with only a minimum payment required by the credit card company. Because credit card interest rates are typically high, if we make only the minimum payment, it takes a long time to pay off the loan, as we saw in Activity 4.1. We can use the loan payment formula to calculate the payments needed to pay off a credit card loan in a fixed amount of time.

Example 9.11

A student accumulated a balance of \$1500 on a credit card that has an *APR* of 20.99 percent. Assume that no additional purchases are made.

a. How much will the student need to pay each month to pay off the balance over a period of one and a half years?

b. Suppose the credit card company allows the student to make no payments for six months, but charges interest during that time. How much additional interest will the student owe at the end of the six-month period?

c. If after the six-month period of not paying anything to the credit card company, the student wants to pay off the accumulated balance (including the extra interest charged) in one year, how much will the student need to pay each month?

Solution

a. We use the loan payment formula with $p = 1500$, $APR = .2099$, $n = 12$, and $t = \frac{3}{2}$. Thus,

$$PMT = p \times \frac{\left(\frac{APR}{n}\right)}{\left[1 - \left(1 + \frac{APR}{n}\right)^{-nt}\right]} = 1500 \times \frac{\left(\frac{0.2099}{12}\right)}{\left[1 - \left(1 + \frac{0.2099}{12}\right)^{-(12)(3/2)}\right]} \approx \frac{1500 \times 0.017491667}{1 - (1.017491667)^{-18}}$$
$$\approx 97.8601$$

The student will have to make monthly payments of \$97.87 to pay off the loan in one and a half years.

b. The credit card company is earning interest from the student during the first six months; we use the compound interest formula introduced at the beginning of this topic, $A = \left(1 + \frac{APR}{n}\right)^{nt} \times p$, to find the amount of interest the credit card company has earned from the student at the end of six months with an interest rate of $r = 20.99$ percent. $A = \left(1 + \frac{APR}{n}\right)^{nt} \times p = \left(1 + \frac{0.2099}{12}\right)^{(12)(1/2)} \times 1500 \approx 1664.47$. The student will owe \$1664.47, which includes an additional \$164.47 in interest.

c. Now we need to calculate the payment needed to pay off the credit card balance of \$1664.47 in one year. So $p = 1664.47$ and we use the loan payment formula:

$$PMT = p \times \frac{\left(\frac{APR}{n}\right)}{\left[1 - \left(1 + \frac{APR}{n}\right)^{-nt}\right]} = 1664.47 \times \frac{\left(\frac{0.2099}{12}\right)}{\left[1 - \left(1 + \frac{0.2099}{12}\right)^{-(12)(1)}\right]} \approx 154.98$$

The student will need to pay \$154.98 per month.

Mortgages are installment loans that help people purchase homes. Mortgage interest rates are much lower than credit card rates and other loan interest rates because the home serves as a payment guarantee. Various mortgage options are available from different banks. These include the length of the loan, the down payment required, additional fees to obtain the loan, and whether the loan is a **fixed rate mortgage**, in which the interest rate doesn't change over the term of the loan or an **adjustable rate mortgage**, in which the interest rate changes as the prevailing rates change. (In Activity 9.1 we investigate some of these mortgage option scenarios.)

Summary

In this topic, we investigated simple and compound interest and looked at the effect of compounding using various compounding intervals and lengths of time over which we collect interest. We also determined how much money we should save each month to reach a particular savings goal. Finally, we investigated installment loans and credit card loans, analyzing the cost of borrowing money and evaluating various loan situations.

Explorations

1. Three students each have $1000 to invest from their summer jobs. Armen invests his money in an account that earns simple interest at an *APR* of 5 percent. Barok invests his money in an account that earns 4.9 percent interest per year compounded annually. Carrie invests her money in an account that earns 4.8 percent interest per year compounded monthly. Find how much money each student has in his or her account after

    a. 2 years

    b. 10 years

    c. 30 years

2. Here's an old story: A man walks into a New York City bank and asks for a $5000 loan, offering his Ferrari, worth $250,000 as collateral. He tells the loan officer that he needs the money for two weeks for an important venture. The loan officer, having the car as security and after checking references, gives the man the money he requested, with a signed agreement that he will pay the money plus $45 in interest when he returns in two weeks. The bank officer takes the car keys and the car is parked in the bank's underground lot. The man returns in exactly two weeks, pays the loan and interest, and reclaims his car. The bank officer asked him why he was willing to pay such a high interest rate. His reply: where else can I safely park my car for two weeks in New York City for only $45?

    a. What annual interest rate did the man pay?

    b. How much would the man need to repay at the end of two months if he borrowed $5000 with the same rules and same annual interest rate?

3. If a Virginia colonist had invested $50 in July of 1776 in an account with an *APR* of 5 percent, how much would that account be worth now if interest is compounded

    a. Annually?

    b. Quarterly?

    c. Daily?

4. Compare the annual percentage yield (*APY*) for three banks: bank 1 offers an *APR* of 3.8 percent compounded daily; bank 2 offers an *APR* of 4.1 percent compounded monthly; bank 3 offers an *APR* of 4.5 percent compounded quarterly.

5. Some banks advertise that interest is **compounded continuously**. We can think of this as compounding infinitely many times per year (rather than 12 times per year, as with monthly compounding, or 365 times per year, as with daily compounding). With continuous compounding, the compound interest formula has the following form: $A = p \times e^{(APR)(t)}$, where A is the amount in the account after t years, p is the starting principal, and APR is the annual percentage rate. The number e that appears in the formula is a special irrational number: $e \approx 2.71828$. (Note that many calculators have a special "e^x" key and Excel uses EXP(x) to evaluate e^x.) Find how much money a student would have in an account if he invested $1000 with interest compounded continuously, after 2 years, 10 years, and 30 years if the *APR* is

   a. 3 percent

   b. 5 percent

   c. 7 percent

6. A new parent wants to have $80,000 in a college savings account in 18 years. Although it's difficult to predict what the *APR* will be over 18 years, we can get a sense of how much the parent should save each month using various assumptions. Calculate the monthly savings needed if the money is invested in an account with monthly compounding and an *APR* of

   a. 4 percent

   b. 7 percent

   c. 9 percent

7. Set up a table like the one given in Example 9.9 and confirm that a monthly payment of $183.36 will result in the loan being paid off in 12 months, as we found using the formula in the solution to Example 9.10a.

8. Suppose that a friend who spent her junior year abroad has accumulated $2400 worth of debt on her credit card. The card has an *APR* of 19.6 percent. If she doesn't charge anything additional on the card, how much should she pay each month to pay off the card in

   a. 9 months?

   b. 15 months?

   c. 3 years?

9. You have $1500 in credit card debt on several cards, each of which has an outrageously high *APR*. You want to consolidate your debt and formulate a plan to pay it off in one year, so you are comparing two credit card offers. Card A offers an *APR* of 18.6 percent while card B offers no interest for the first six months then an *APR* of 20.5 percent after that.

   a. If you transfer the balance of $1500 to card A, find how much you will need to pay each month in order to pay off the whole balance in one year.

   b. Suppose you pay the amount computed in part a of this Exploration but under the scenario offered by card B. That is, you pay that amount each month, but for the first six months, no interest is being charged, so you are reducing your principal. How much would you need to pay each month for the final six months of the year to pay off the remaining balance on card B by the end of the year?

   c. How much interest would you pay under each of the plans?

10. Two car dealers each are offering a $7,000 loan. The first dealer offers an *APR* of 7.8 percent and the loan must be repaid in monthly payments over three years. The second dealer offers an *APR* of 8.4 percent, with monthly payments and a loan term of four years.

    a. Find the monthly payment under each scenario.

    b. Find the total amount of interest paid under each scenario.

11. Credit card companies often charge a higher rate of interest for cash advances than they do for purchases. Obtain detailed information from two credit card companies about their interest rates on purchases and cash advances and analyze these interest rates.

TOPIC 10

Introduction to Problem Solving

OBJECTIVES

After completing this topic, you will be able to

- » Identify a variety of problem-solving techniques
- » Recognize which problem-solving techniques can be used to solve a particular problem
- » Use a variety of problem-solving techniques to answer questions that concern personal financial matters
- » Recognize when to use various computational methods studied in previous topics

In this topic, we look at the problem-solving techniques used in Topics 1–9 and analyze how to apply them to solve other problems that relate to questions of personal finance.

In previous topics, we have used a variety of problem-solving techniques to develop a deeper understanding of the problems and issues investigated. Here is a summary of some of the techniques used:

1. Decide what information is relevant.
2. Represent the information in a different form.
   a. Make a table or chart.
   b. Draw a picture, graph, or diagram.
   c. Use an equation or formula.
3. Examine a simple case or try several special cases.
4. Break a problem into smaller problems or identify a sub-goal or sub-problem.
5. Work backward.
6. Look for a pattern.

In the following example, we look at previous topics to identify the problem-solving techniques used in those examples.

Example 10.1

Identify one problem-solving technique from the summary list above that was used in each of the following examples from previous topics. In each case, write a question related to the content of the example that could be answered using that technique.

a. Example 1.7, page 13

b. Example 2.4, page 39

c. Example 5.4a, page 91

d. Example 6.4, page 110

e. Example 6.7, page 112

f. Example 8.6b, page 143

Solution

a. In Example 1.7, we used problem-solving technique *2b. Draw a picture, graph, or diagram,* by creating a stemplot to represent the given data. We could use this technique to help answer the question: How selective are these colleges, and if I can only apply to five of these colleges, which five should I choose?

b. In Example 2.4, we used problem-solving techniques *2a. Make a table or chart* and *2b. Draw a picture, graph, or diagram,* by creating a table and a graph from the information given. We could use one of these problem-solving techniques to answer the question: How many minutes of walking should a person add to his or her daily activities to be able to consume an extra piece of fruit or candy bar without gaining weight?

c. In Example 5.4a, we decided which information line from the tax table instruction was relevant for each of the salaries given. The problem-solving technique represented was *1. Decide which information is relevant.* In this context, we could use this problem-solving technique to answer the question: How much federal tax corresponds to a specific income?

d. In Example 6.4, we used problem-solving technique *6. Look for a pattern* to help develop a formula that gave the family's debt as a function of time in months. We could use this formula to find the family's debt after eight months, for example.

e. Example 6.7 provided an example of problem-solving technique *3. Examine a simple case or try several special cases.* In this example, we looked at several special cases to answer the question: How does the number of presentations grow with the number of brands of chocolate in a taste test? We could use this problem-solving technique to answer the question: If we have *n* brands of chocolate and choose an order of presentation at random, what is the chance that a specific brand is chosen first?

f. In Example 8.6b, we identified variables that would be useful in ranking colleges. In this way, we broke the problem of rating colleges into several smaller problems, so we were using problem-solving technique *4. Break a problem into smaller problems or identify a sub-goal or sub-problem.* In this case, we can use this technique to answer the question: Which college is best for me?

The following example illustrates problem-solving technique 5. *Work backward.*

Example 10.2

Use problem-solving technique 5. *Work backward* to answer the following question. In Topic 4, Exploration 3, we learned that to calculate the body mass index (BMI) of a person we perform the following steps:

1. Multiply the person's weight by 0.4536 (this gives the weight in kilograms).
2. Divide the weight in kilograms by the square of the height in meters. Height in meters is obtained by multiplying the height in inches by 0.0254.

A person who is 68 inches tall wants to get his or her BMI within the normal range of 20 to 25. What weight range would give this BMI range?

Solution

We first note that for a person who is 68 inches tall, the two previous steps are

1. Multiply the person's weight by 0.4536.
2. Divide the number obtained by $(68 \times 0.0254)^2 \approx 2.9832$.

   So $BMI = \frac{weight \times (0.4536)}{2.9832}$.

To find the weight of a person with a BMI of 25, we start with BMI = 25 and work backward, reversing the order of the two steps:

1. Multiply the desired BMI of 25 by 2.9832: $25 \times 2.9832 \approx 74.58$.
2. Divide the number obtained in (1) by 0.4536: $\frac{74.58}{0.4536} \approx 164.4$.

So the person should weigh approximately 164.4 pounds to have a BMI of 25. Using the same procedure, we can find the weight that corresponds to a BMI of 20:

1. $20 \times 2.9832 \approx 59.664$
2. $\frac{59.664}{0.4536} \approx 131.5$

Because BMI increases as weight increases, looking at the two extreme values of BMI is sufficient. Thus a person who is 68 inches tall should weigh between 131.5 and 164.4 pounds to keep the BMI within the range 20 to 25.

As we have seen in previous topics, drawing a graph can help us identify trends and make estimates and predictions. In the next example, we use problem-solving technique *2b. Draw a picture, graph, or diagram* to make an "educated" guess.

Example 10.3

Suppose we own a home and use oil to heat the house. In May 2005, the oil company offers the opportunity to buy, at the current price, the oil needed for next winter. The following table gives the company's average monthly price per gallon of oil for the last three years. Use problem-solving technique *2b. Make a picture, graph, or diagram*, to decide whether or not we should buy the oil in advance.

Month	Jan 2002	Feb	Mar	Apr	May	Jun	Jul	Aug	Sep	Oct	Nov	Dec
Price	0.83	0.82	0.91	0.94	0.94	0.92	0.95	0.99	0.99	1.09	1.06	1.12
Month	Jan 2003	Feb	Mar	Apr	May	Jun	Jul	Aug	Sep	Oct	Nov	Dec
Price	1.16	1.41	1.52	1.26	1.17	1.21	1.14	1.16	1.13	1.16	1.19	1.25
Month	Jan 2004	Feb	Mar	Apr	May	Jun	Jul	Aug	Sep	Oct	Nov	Dec
Price	1.34	1.36	1.34	1.31	1.38	1.40	1.44	1.51	1.58	1.83	1.81	1.72
Month	Jan 2005	Feb	Mar	Apr	May							
Price	1.73	1.75	1.92	1.93	1.87							

Solution

We graph the function that gives the average monthly price per gallon, with time as the explanatory variable.

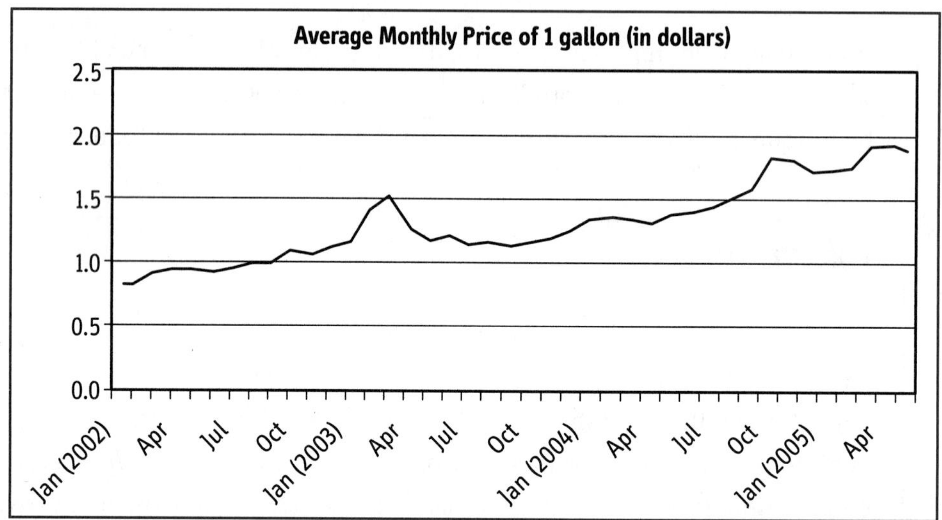

Looking at this graph, we see that the price of heating oil has had many small variations but generally has been increasing over the three-year period. Although the price decreased from April 2005 to May 2005, based on past behavior, we cannot expect the price to continue to decrease. During the past two years, the price was lower in May than in the next winter months (January through March). Thus, based on the information we have, it seems reasonable to buy heating oil in advance.

Example 10.4

Use the information given in Example 10.3 to estimate the price of heating oil the company might charge in December 2005. Explain how you made that estimate and what problem-solving techniques you used.

Solution

A reasonable way of predicting the price of oil would be to find the regression line for the given data. To do so, we make a scatterplot of the data, using a time scale from 0 to 40 on the horizontal axis (0 represents January 2002, 1 represents February 2002, and so on, ending with 40 representing May 2005).

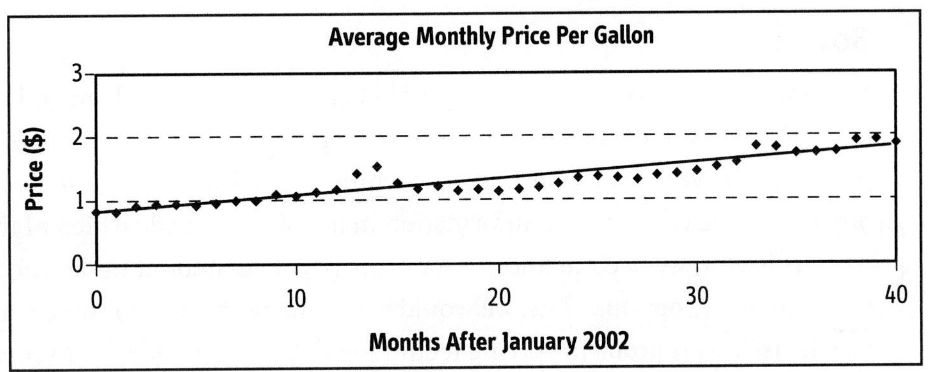

With the help of technology (see Activity 6.2), we obtain an equation for the regression line: $y = 0.0245x + 0.8151$. To estimate the price of heating oil in December 2005, we first note that December is the seventh month after May, so it corresponds to $x = 47$; then we calculate the corresponding value of y in the equation of the line: $y = (0.0245) \cdot (47) + 0.8151 \approx 1.967$, so we estimate the price of heating oil for December 2005 to be \$1.97 per gallon. We applied problem-solving techniques *2b. Draw a picture, graph, or diagram* and *2c.Use an equation or formula* to draw the scatterplot and use the equation of the regression line.

Often we need to use more than one technique to solve a problem. In the following example, we consider a problem whose solution might require combining several problem-solving techniques.

Example 10.5

Josh is planning to buy a car that costs \$15,000. He plans to make a down payment of \$2000 and to finance the rest with a loan to be paid in five years. Josh needs to decide whether he should get a loan through the car dealership, from the credit union, or if he should accept his uncle's offer of a loan to be paid at the end of five years with a simple interest of 4 percent per year. Among the three options, Josh wants to choose the loan that will require the smallest monthly payment. Identify which problem-solving technique(s) would help solve Josh's problem and describe how to use them.

Solution

We first need to find out more information about each type of loan and then select the relevant data using problem-solving technique *1. Decide what information is relevant.* We could also use technique *2a. Make a table or chart,* by organizing the relevant loan information in a table. To decide which of the three options is best, we need to find the monthly payment amount under each of the three financing options; thus, we would break the problem into three smaller problems. This is problem-solving technique *4. Break a problem into smaller problems or identify a sub-goal or sub-problem.*

In the next example, to help Josh solve his problem, we will proceed as described in the solution to Example 10.5. The first step is to give specific information about each of the three loans.

a. The car dealer has a special finance offer: \$500 discount on the price of the car and an annual percentage rate (*APR*) of 7.5 percent on a 5-year loan with equal monthly payments.

b. The credit union charges 6 percent interest per year (that is, $APR = 6$ percent) on a 5-year fixed-installment car loan with monthly payments.

Both car loans are **fixed-installment** (or **amortized**) **loans,** in which the number of payments is determined in advance and the same amount of money is paid each time. In both cases, the interest is compounded monthly and is charged on the principal (the amount of money still owed) at the end of the month. The **loan term** for both loans is 5 years, and the number of payments is $5 \times 12 = 60$. The **starting loan principal** (initial amount borrowed) is 12,500 if the loan is from the dealership and 13,000 otherwise.

c. If Josh decides to take a loan from his uncle, he will make a payment only at the end of 5 years, but he will have to set aside the same amount each month during the 5-year period of the loan. With simple interest, the interest is calculated on the original principal for each year during the term of the loan.

Example 10.6

Make a table summarizing the information collected that is relevant to Josh's car-financing problem described in Example 10.5.

Solution

The following table summarizes the information on the three possible loans for Josh:

Loan Source	Loan Principal	*APR*	Loan Term and Type
Uncle	$13,000	4%	5 years–simple interest
Car Dealer	$12,500	7.5%	60-month installment loan
Credit Union	$13,000	6%	60-month installment loan

To help Josh make his decision, we calculate the monthly amount he will need to spend for each of the three loans. We start with the simplest of the three, the uncle's offer.

Example 10.7

Suppose Josh takes a loan of $13,000 from his uncle under the conditions described previously. Josh wants to set aside the same amount of money every month for five years and pay off the loan at the end of the fifth year. What amount should Josh set aside every month?

Solution

To find the total debt after 5 years, we first find the interest. Recall from Topic 9 that the simple interest after 5 years at 4 percent per year on a principal of $13,000 is computed using the formula, $I = p \times r \times t$, where p is the principal, r is the annual interest rate expressed as a decimal, and t is the duration of the loan in years. So, in this case we have: $I = \$13{,}000 \times 0.04 \times 5 = \$2{,}600$. The total amount Josh needs to return at the end of the 5-year period is *Total amount* = *Principal* + *Interest* = $13,000 + $2,600 = $15,600. If Josh takes his uncle's loan, the amount he needs to set aside each month is $\frac{\$15600}{60} = \$260$.

To find Josh's monthly payment if he takes one of the installment loan options, we use the loan payment formula discussed in Topic 9 to find the payment *PMT* for a fixed-installment loan for a total of t years, with an initial principal p, and annual interest rate *APR*. The formula is

$$PMT = p \times \frac{\left(\frac{APR}{n}\right)}{1 - \left(1 + \frac{APR}{n}\right)^{-nt}}$$

Because we want to find the monthly payment, $n = 12$ and our formula is:

$$PMT = p \times \frac{\left(\frac{APR}{12}\right)}{1 - \left(1 + \frac{APR}{12}\right)^{-12t}}$$

In the next example, we use this formula to solve Josh's problem.

Example 10.8

Follow the steps outlined in Example 10.5 to find which of the three loan options would offer Josh the lowest monthly payment.

Solution

We collected the relevant information and organized it in a table (see Examples 10.5 and 10.6). To compare them, we need to find the monthly expense in each of the three cases. In Example 10.7, we found that for the loan from Josh's uncle, Josh's monthly expense would be $260.

Suppose Josh takes the car dealer's financing offer. In this case, the loan principal is \$12,500, the annual interest is 7.5 percent and the loan is for 5 years, so we set p = 12,500, APR = 0.075, and t = 5, in the formula for monthly payment:

$$PMT = p \times \frac{\left(\frac{APR}{12}\right)}{1-\left(1+\frac{APR}{12}\right)^{-12t}}$$

We obtain

$$PMT = 12{,}500 \times \frac{(0.00625)}{1-\left(1+\frac{0.075}{12}\right)^{-(12)(5)}} = 12{,}500 \times \frac{(0.00625)}{1-1.00625^{-60}} \approx 250.47$$

If Josh accepts the loan from the car dealer, his monthly payment will be \$250.47.

For the loan from the credit union, p = 13,000, APR = 0.06, and t = 5, so $PMT = 13{,}000 \times \frac{(0.005)}{1-1.005^{-60}} \approx 251.33$. If Josh takes the loan from the credit union, his monthly payment is \$251.33. Comparing these calculations, we see that the three options have similar monthly costs, but the offer from the car dealer is lowest. To minimize his monthly payments, Josh should choose to take the loan from the car dealer. (After all, if he takes this loan instead of the credit union one, he will save \$0.86 per month for a total savings of 60 × \$0.86 = \$51.60 over five years. He could buy an extra pizza each year!)

Credit cards provide an example of another type of installment loan. An open-end installment loan is a loan that requires periodic (usually monthly) payments, but the payments do not have to be of equal amounts. Most credit cards require a minimum payment to cover the interest on the balance owed and just a very small amount of the principal. The longer you take to pay credit card debt, the more interest you pay and the more profit the issuing bank makes. A method often used by banks to calculate finance charges on credit card accounts is the *average daily method.* With this method, the finance charge is the daily interest on the average daily balance on the credit card account for the billing cycle, multiplied by the number of days in the billing cycle. We illustrate this method in the following example.

Example 10.9

Tammy has a balance of $480 on her credit card on January 26. She charges a purchase of $41 on February 3 and a purchase of $63 on February 6 and pays $280 on February 14 (through an electronic check that is posted the same day). There are no other transactions on the account during this billing period, which starts on January 26 and ends on February 24. This credit card charges an interest rate of 9.49 percent per year on purchases and 21 percent on cash advances. Calculate the finance charges on Tammy's next credit card statement. Also indicate the problem-solving techniques used to solve this problem.

Solution

To answer this question, we consider two simpler steps (problem-solving technique 4. *Break a problem into smaller problems or identify a sub-goal or sub-problem*). We first find the average daily balance, and then we find the finance charge.

To find the average daily balance, we create a table (problem-solving technique 2a. *Make a table or chart*) that shows the date and amount of each transaction made during the billing period and the corresponding balance on each of those days.

Date	Transaction	Balance
1/26		$480
2/3	$41	$521
2/6	$63	$584
2/14	-$280	$304

The average daily balance is the sum of the daily balance for each day of the billing period divided by the number of days in that period. For 8 days (from January 26 through February 2), the daily balance was $480; for 3 days (February 3 through February 5), the daily balance was $521; for 8 days (February 6 to

February 13), the balance was \$584; and for the last 11 days of the billing period, the balance was \$304. We use a formula (problem-solving technique *2c. Use an equation or formula*) to find the average daily balance.

The sum of the daily balances is $(480 \times 8) + (521 \times 3) + (584 \times 8) + (304 \times 11) = 13{,}419$, and there are 30 days in this billing period, so the average daily balance is $\frac{\$13419}{30} = \$447.30$.

The daily interest rate is obtained by dividing the annual interest rate by 365, the number of days in a year: $\frac{0.0949}{365} = 0.00026$. The finance charge on Tammy's credit card account for this billing period is the daily interest on the average daily balance times the number of days in the billing period: $0.00026 \times \$447.30 \times 30 \approx \$3.49$.

Summary

In this topic, we identified six problem-solving techniques (see the list at the beginning of the topic). We looked at when these were used in previous topics and identified questions that we could answer using these problem-solving techniques. We analyzed problems in which finding the solution required us to use two or more problem-solving techniques. We also explored the average daily method used by some credit card companies to compute interest on credit card charges.

Explorations

1. For each of the problem-solving techniques listed at the beginning of Topic 10, find an Example (not mentioned in this topic), an Exploration, or an Activity in which you used the technique.

2. The following table gives the percentage of the U.S. population living below the poverty level for the period from 1990 to 2002. (Source: *World Almanac and Books of Facts 2004*, page 382.)

Year	Percentage of U.S. Population Living Below Poverty Level
1990	13.5
1991	14.2
1992	14.8
1993	15.1
1994	14.5
1995	13.8
1996	13.7
1997	13.3
1998	12.7
1999	11.8
2000	11.3
2001	11.7
2002	12.1

Do you predict that the poverty rate will increase or decrease in the next several years? Explain which problem-solving technique you used to answer the question.

3. Marian has received an inheritance of $50,000 and would like to invest it in a regular savings account or a money market savings account. The regular savings account pays an interest of 1.49 percent per year and is compounded quarterly. The money market savings account pays an interest of 1.485 percent and is compounded daily. Identify which technique(s) from the list given in this topic, would help Marian make a decision and describe how you would use the techniques. (You do not need to solve Marian's problem.)

4. Refer to Example 10.5. Josh wants to pay off the car loan in three years (instead of five years). Which of the three loans would give him the smallest monthly payment?

5. Identify which techniques were useful in answering the question in Exploration 4, and describe how you used them.

6. Laurie's credit card statement of September 5 showed a debt of $300. On September 15, she pays $50, and on September 20 and September 27, she charges purchases of $21.50 and $35, respectively. She makes no other payments or charges until the end of the billing period, September 6 through October 3 (both days included). Her credit card company charges interest monthly on the average daily balance for purchases at an annual rate of 14.2 percent.

   a. Identify which problem-solving techniques would be useful in answering the question: What will be the balance in Laurie's credit card account at the end of the billing period?

   b. Describe how you would use the techniques you identified.

7. Answer the question in Exploration 6 using the techniques you identified.

8. Refer to Exploration 6. Suppose in addition to the payments and purchases described, Laurie takes a cash advance of $100 on September 23. The credit card account charges interest of 21 percent on cash advances, also computed monthly but only on the daily average of cash advances.

   a. What will be the balance on Laurie's credit card account at the end of the billing period?

   b. Identify the problem-solving techniques used to answer the question.

TOPIC 11

Decision Making

OBJECTIVES

After completing this topic, you will be able to

- » Determine when a decision is a decision under certainty and when it involves probability considerations
- » Decide what criteria are important for a variety of decision-making situations
- » Devise a ranking system to rate alternative choices on different criteria
- » Apply two decision-making methods, the cutoff screening method and the weighted sum method, to help make decisions

How do you decide what compact disc player or car to buy? Is cost your only consideration? How do you "factor in" reliability or appearance? Should your class trip involve a visit to an amusement park, where inclement weather might close the park on the day of the trip, or should you schedule an indoor concert instead?

Some decisions require us to use information about which there is some uncertainty, such as weather conditions. Other decisions involve information that we already know or is readily available, such as the price per pound of produce. In Topic 21, we will investigate making decisions that involve uncertainty and probability considerations. In this topic, we look at decisions where the information impacting the decision is assumed to be known for certain or is able to be obtained. Such decisions are called **decisions under certainty** and include consumer purchase decisions such as buying a car.

Example 11.1

For each of the following decision-making situations, determine if it involves a decision under certainty or not. Give reasons for your answers, but do not actually solve the problems.

a. José is in fifth grade. His grandfather gave him $1000 to invest for college, so he will invest it for seven years. He has one opportunity to invest it at a yearly interest rate of 8 percent compounded annually and, of course, he will put the interest back into the account. He has another opportunity to invest his $1000 at a yearly interest rate of 7.5 percent, compounded daily with, again, the interest put back into the account. José must decide where to invest his money.

b. You are in charge of the hot-dog stand for home football games. Because home games are not played every week and because you cannot resell leftover food, you don't want to order too much. On the other hand, you want to service the customers who want refreshments. You need to decide how much food to order.

c. A prominent fast-food company has come up with a new kind of burger that they hope will be very successful. The company's marketing executives need to decide whether to test market the new burger in a limited area and then promote it nationally based on its success in the test market, or to promote it nationally without test marketing.

d. Amisha needs to buy a new pair of running shoes. She has found three pairs, each at a different store, that she likes equally well, and all are on sale. The first pair was originally $65. It is on a table marked 25 percent off, and she has a coupon for an additional 15 percent off of that sale price. The second pair originally cost $62 and is marked 1/3 off. The final pair was originally $69. It's marked 35 percent off and today, for one day only, everything at that store is marked down another 10 percent off of the sale price. Amisha must decide which pair of shoes to buy.

e. The manager of the sanitation crew in a large city must decide what routes to assign to the city's fleet of trash collection trucks in order to collect trash once per week at every home and business in the city.

Solution

The problems described in parts a and d of this example involve information that is known with certainty and so they involve decisions under certainty. In part a of this example, we can figure out exactly how much money José will have after seven years under each scenario because the interest rate is fixed. Amisha knows exactly how much each pair of shoes costs and can determine the final price for each. In part e of this example, if the manager assumes that the trucks are all operating so he knows how many trucks there are, it is a decision under certainty. Although this problem is a complicated one, the manager can find out where the streets and alleys and homes and businesses are and can (theoretically, at least) find the most efficient and economical routes to assign the city's trucks. In part b of this example, the weather is a factor that impacts upon sales and involves probability considerations, so the decision in part b of this example is not a decision under certainty. Forecasters can track weather patterns over the years and calculate the probabilities associated with various weather conditions on any particular day to help with this decision. Similarly, in part c of this example, how the new burger is received involves the uncertainty of consumer demand.

We will investigate some methods for making decisions under certainty. Many of these decisions require comparisons of cost, time, or quality of various alternative choices. The appropriate decision depends on the criteria, and we use various problem-solving techniques in our analysis.

Example 11.2

Assuming that cost is her only criterion (and she wants to minimize it), which pair of shoes should Amisha buy in Example 11.1d?

Solution

Because we are dealing with currency, we will round all computed amounts to the nearest cent. The first pair of shoes was \$65 but is marked 25 percent off. Thus she would pay 75 percent of \$65 or $0.75 \times \$65 = \$48.75$. With her 15 percent off coupon, she will pay 85 percent of \$48.75 for a final cost of $0.85 \times \$48.75 = \$41.44$.

The second pair is marked 1/3 off of \$62. Thus she will only pay 2/3 of \$62 for a final cost of $\frac{2}{3} \times \$62 = \$41.33$. The third pair was originally \$69 but is 35 percent off. Thus she will pay 65 percent of \$69 or $0.65 \times \$69 = \$44.85$. The additional 10 percent off gives the final cost of $\$44.85 - \$4.48 = \$40.37$. So assuming that minimizing cost is her only objective, Amisha should buy the third pair.

We will investigate how to incorporate criteria other than cost into the decision-making process. This process uses some ideas similar to those discussed when we considered ratings in Topic 8, but with a more personal slant.

Example 11.3

Name at least four characteristics or criteria you would want to evaluate or measure in some way before making a decision on each of the following "purchases."

a. A stereo set

b. What college to attend

c. A car

d. An exercise machine

e. A bicycle

Solution

The answers will vary because of individual values and tastes. Some possibilities are given.

a. For a stereo set: cost, reliability, sound quality, repair rate, availability of upgrades

b. For a college: location, size of school, academic reputation, programs available, and attractiveness of the campus

c. For a car: cost, size, gasoline consumption, availability, reliability, and location of dealer

d. For an exercise machine: cost, durability, type, size, and ease of use

e. For a bicycle: weight, comfort, looks, cost, and reliability

We now investigate how to incorporate multiple criteria into making a decision about which digital camera we should purchase.

Example 11.4

Suppose we want to purchase a digital camera and have four brands to consider. We collect information on the price and weight of each of these brands initially. Given the information in the following table, which brand might we choose to purchase and why?

Brand	Price	Weight
Kodak	$860	17 oz
Nikon	$887	14 oz
Olympus	$764	19 oz
Sony	$748	10 oz

Solution

If price and weight were the only criteria, assuming that we want a light-weight camera for ease of carrying, we would likely choose the Sony camera, which is the lightest and least expensive camera.

But what camera would we have decided to buy in Example 11.4 if the least expensive camera had not been the lightest? There are several methods that we can use when we are making a decision based on more than one criterion and the decision is not obvious. In the following examples, we discuss two of these: the cutoff screening method and the weighted sum method.

In the **cutoff screening method**, the decision maker predetermines a cutoff for each criterion. For example, a possible cutoff for the criterion of price is that the price of the camera be no more than $800. With a cutoff determined for each criterion, the decision maker then goes through the criteria one-by-one and eliminates any choices (brands of cameras in the camera example) that do not meet the required cutoff for that criterion. After all criteria have been checked, if one choice remains, then we have a decision. If more than one choice remains, the decision maker can see if he or she wants to make any of the cutoffs more restrictive or if there is another criterion to use to make a final decision. If all choices have been eliminated, then the decision maker must determine if any of the cutoffs can be relaxed or more possible choices included.

Example 11.5

Suppose we include three additional criteria on which to base our decision of which digital camera to buy. The four brands are rated also on print quality, next-shot delay, and flash range. Print quality was rated by comparing the output from the cameras, using a scale from 1 to 10, with 10 being excellent. The ratings and other information from the manufacturer appear in the following table.

Brand	Price	Weight	Print Quality Rating	Next-Shot Delay	Flash Range
Kodak	$860	17 oz	9	6 sec	10 ft
Nikon	$887	14 oz	10	8 sec	18 ft
Olympus	$764	19 oz	10	4 sec	10 ft
Sony	$748	10 oz	6	6 sec	8 ft

a. Using the cutoff screening method with the following cutoffs, decide which camera to buy. Suppose we want a camera that costs less than $900, weighs one pound or less, with a print quality of 9 or higher, a next-shot delay of no more than 6 seconds, and a flash range of at least 10 feet. Do any models meet all of these cutoffs? If no models meet the cutoffs, determine what we should do. If one or more models meet the cutoffs, what models are they?

b. Develop a table in which the dollar price of each camera is changed to a "price rating," with the lowest price camera receiving a "high" rating of 10. Set the price ratings of the other cameras relative to the least expensive one.

Solution

a. We will take the criteria in the order listed and eliminate from consideration models that don't meet the cutoffs. The maximum cost of $900 does not help us eliminate any models from consideration. Requiring a weight of one pound or less leads us to eliminate the Kodak and Olympus models from consideration. Having a print quality of 9 or better eliminates the Sony model. Now only the Nikon is under consideration. We check to see if that model meets all the remaining criteria, and we see that it does not. The next-shot delay of the Nikon is 8 seconds, so no model meets all of our criteria. We can decide to look at additional models or relax one of the cutoffs. One possibility is that we might relax the 6 or less second next-shot delay criterion and decide that the Nikon is our best choice.

b. The Sony camera with a price of $748 receives a price rating of 10. Because the camera with the next highest price is fairly close in price, we might assign it a rating of 9. Depending on how important the differences in prices among the models are, we might assign a price rating of 6 and 5, respectively, to the Kodak and Nikon models. Alternatively, we could assign one rating point per $20 (rounding) difference in price. Using that mechanism, we would assign the Kodak a price rating of 4 and the Nikon a price rating of 2. We will use the first approach and use the price ratings given in the following table.

Brand	Price Rating	Weight	Print Quality Rating	Next-Shot Delay	Flash Range
Kodak	6	17 oz	9	6 sec	10 ft
Nikon	5	14 oz	10	8 sec	18 ft
Olympus	9	19 oz	10	4 sec	10 ft
Sony	10	10 oz	6	6 sec	8 ft

In Example 11.5, we assigned a price rating so price is rated using a consistent system in which a higher rating indicates a preferred choice. There were two possible price-rating schemes described in the solution to the example, and there may be additional logical ways to rate the price. What is important is that the rating uses a consistent scale and that we use the same highest rating for all the criteria. We will also adopt a rating system for the weight, next-shot delay time, and flash range, so that a higher rating is the preferred choice and the rating system is consistent across the criteria with a highest rating of 10.

Example 11.6

The following table gives a rating system, based on 10 as the highest, most preferred rating, for the criteria of weight, next-shot delay, and flash range. Explain the logic of the system.

Brand	Price Rating	Weight Rating	Print Quality Rating	Next-Shot Delay Rating	Flash Range Rating
Kodak	6	7	9	9	6
Nikon	5	8	10	8	10
Olympus	9	6	10	10	6
Sony	10	10	6	9	5

Solution

The lightest camera (the Sony) gets a 10 in the weight rating column. We give the next lightest camera (the Nikon) an 8 because there is a jump in weight between the Sony and Nikon cameras. We reduce the rating approximately one point for each 2 ounces of additional weight. For the next-shot delay criterion, the Olympus, with a 4-second delay, is rated 10. The cameras with 6-second delays are rated 9 and the camera with an 8-second delay rates an 8, a one-point rating reduction for each additional 2 seconds of delay. The rating of 10 for flash range is assigned to the Nikon camera, which has the greatest flash range. The Kodak and Olympus cameras are given a flash range rating of 6 (a loss of one rating point per 2 feet of distance in flash range) and the Sony is assigned a 5.

By setting up the ratings within each criteria, we have set the stage for our second decision-making method. We now look at how to use the **weighted sum method** to determine the best camera for our use. Our goal is to get a single numerical rating for each of the cameras, so we can compare them. We want to factor into our numerical rating how important each criterion is to us, relative to the other criteria. To do this, we assign an "importance factor," called a **weight**, to each of the following criteria: price, weight rating, print quality, next-shot delay rating, and flash range rating, by giving each a weight between 1 and 10. A weight of 10 is assigned to the criterion that is most important. The weights for the other criteria will be chosen based on how important these criteria are to the decision maker, relative to the most important criterion.

Suppose that print quality is the most important criterion to us as decision makers. Thus, we assign a weight of 10 to print quality. The next most important criterion is price, and it's almost as important as print quality, so we assign it a weight of 9. Next-shot delay is next in importance and gets a weight of 8. Finally, weight rating and flash range are assigned weights of 5, because they are less important to us. (Note that these weights might differ for different decision makers.)

After assigning weights to each of the criteria, we compute a **weighted sum** for each model of camera. To compute the weighted sum for a particular camera model, we multiply the weight for each of the five criteria by the rating of that criterion for that camera. Then we add these terms to get a weighted sum for that model. Because higher ratings and weights indicate preferred alternatives, the preferred choice is the one with the highest weighted sum.

Example 11.7

Use the weights of 9, 5, 10, 8, and 5 for price, weight, print quality, next-shot delay, and flash range, respectively, to compute a weighted sum for each model of camera.

Solution

We compute the weighted sum for the Kodak model as follows:

$\text{Sum}_{\text{Kodak}}$ = (**wt for price rating**) ∗ (*price rating*) + (**wt for weight rating**) ∗ (*weight rating*) + (**wt for print quality rating**) ∗ (*print quality rating*) + (**wt for next-shot delay rating**) ∗ (*next-shot delay rating*) + (**wt for flash range rating**) ∗ (*flash range rating*) =

$\mathbf{9} * 6 + \mathbf{5} * 7 + \mathbf{10} * 9 + \mathbf{8} * 9 + \mathbf{5} * 6 = 281.$

The weighted sums for the other models are computed in a similar manner and are displayed in the table.

Brand	Price Rating	Weight Rating	Print Quality Rating	Next-Shot Delay Rating	Flash Range Rating	Weighted Sum
Weight for Weighted Sum	9	5	10	8	5	
Kodak	6	7	9	9	6	281
Nikon	5	8	10	8	10	299
Olympus	9	6	10	10	6	321
Sony	10	10	6	9	5	297

Using the weighted sum method, we choose the model with the highest weighted sum. For our choice of weights, this method would lead us to choose the Olympus camera.

We could use the weighted sum method to rate the four cameras we chose by putting them in order according to the weighted sum. So using this system, the Olympus is rated #1, followed by the Nikon, the Sony, and the Kodak, as #2, #3, and #4, respectively. In the next example, we see how the ratings might change using different weights for the weighted sum.

Example 11.8

Use the rankings of the cameras as determined for Example 11.7, but change the weights used for the criteria to: 9, 10, 7, 7, and 5 for price, weight, print quality, next-shot delay, and flash range, respectively. Compute the weighted sum for each camera with these weights and determine the ranking of the cameras from most preferred to least preferred.

Solution

We use the new weights to compute the weighted sums as follows:

$$\text{Sum}_{\text{Kodak}} = \mathbf{9} * 6 + \mathbf{10} * 7 + \mathbf{7} * 9 + \mathbf{7} * 9 + \mathbf{5} * 6 = 280$$

$$\text{Sum}_{\text{Nikon}} = \mathbf{9} * 5 + \mathbf{10} * 8 + \mathbf{7} * 10 + \mathbf{7} * 8 + \mathbf{5} * 10 = 301$$

$$\text{Sum}_{\text{Olympus}} = \mathbf{9} * 9 + \mathbf{10} * 6 + \mathbf{7} * 10 + \mathbf{7} * 10 + \mathbf{5} * 6 = 311$$

$$\text{Sum}_{\text{Sony}} = \mathbf{9} * 10 + \mathbf{10} * 10 + \mathbf{7} * 6 + \mathbf{7} * 9 + \mathbf{5} * 5 = 320$$

Using these weights, the Sony is rated #1, the Olympus is #2, followed by the Nikon and Kodak as #3 and #4, respectively.

Examples 11.7 and 11.8 show that using different weights can change the weighted sum and thus change a decision or ranking, which might then be turned into a "rating." When using the weighted sum method or when using a weighted sum to rank different choices, it is important to have a clear rationale for why we choose the weights we use. When considering rankings or ratings that have been set up by someone else, it is also important to understand how the rankings were determined.

Summary

In this topic, we studied two methods we can use to help us make decisions that involve "certain" information—that is, information that is known or that we assume to be known. These methods are the cutoff screening method and the weighted sum method. We discussed various criteria that might be important to consider for making different decisions, and we also investigated how to create consistent rankings of possible choices, relative to each criterion. Finally, we looked at the link between the weighted sum method and rankings or ratings of various alternatives.

Explorations

1. For each situation described below, determine if it involves a decision under certainty or not.

   a. Tyrell has a new job and has to decide between two medical insurance plans. If he chooses the first plan, the employer pays for the full cost of insurance, but Tyrell will have a copayment of $20 for every visit to the doctor's office and a copayment of $50 for each emergency room visit. If he chooses plan B, then he will need to pay $100 per month, but will have no copayment when he visits a doctor or the emergency room.

   b. Michelle wants to celebrate her son's tenth birthday and asked him if he preferred to go to an amusement park for a day and take one friend, or have a sleep-over party with six friends. Michelle's son does not have a special preference, so Michelle will choose the one that is least expensive.

2. Name four characteristics or criteria you would want to evaluate or measure in some way before making a decision to purchase each of the following:

   a. A bike helmet

   b. A fax machine

   c. A microwave oven

   d. A car

3. You are considering the purchase of a computer and have consulted several computer magazines that contain ratings and prices for the five models you are investigating. You decide to include price, speed, and expansion capabilities as three criteria on which you will judge the models under consideration. The ratings, with 10 being the highest for speed and expansion, and the prices are summarized in the following table.

Brand	Price	Speed Rating	Expansion Rating
Dell	$1921	8	10
NEC	$2020	8	8
Gateway Essential	$1528	6	6
Compaq	$2410	10	8
Sony	$2500	8	6

a. One model **dominates** a second model if it is better in all criteria than the second model. Do any of the models given in the table dominate another model? If so, which one dominates and which one is dominated? Explain how you would use this information.

b. Change each value in the price column to a relative rating based on a highest rating of 10.

c. Assign a weighting factor to each of the three criteria of price, speed, and expansion to use the weighted sum method for making a decision. Give a justification for your choices.

d. Use your weighting factors to find the weighted sum for each of the five models. Which model would you choose to buy, based on your weighted sum?

e. What other criteria might you want to include before you decide which computer to buy?

4. Use weights of 5, 10, and 7 for price, speed, and expansion, respectively, in Exploration 3 to find the weighted sum of each model and determine which model is preferred using the weighted sum method.

5. Consider the following table that lists the price for each of four models of clothes dryers and rates drying performance and ease of use based on a high score of 10.

Model	Kenmore	KitchenAid	Frigidaire	General Electric
Price	$580	$480	$310	$490
Performance Rating	10	8	9	8
Ease of Use Rating	8	8	6	9

a. If you were to use the cutoff screening method to make a decision about which dryer to choose, what cutoffs would you choose for each of the criteria?

b. Choose the preferred model based on the cutoffs you identified in part a of this exploration. Explain how you arrived at your choice.

c. Explain why it is not appropriate to use the price values as given in the table in a weighted sum.

d. Assign a weighting factor to each of the three criteria of price, performance, and ease of use, to use for the weighted sum method for making a decision. Give a justification for your choices.

e. Use your weighting factors to find the weighted sum for each of the four models. Which model would you choose to buy, based on your weighted sum?

6. Consider the values given in the following table for a selection of countries. (Sources: *The New York Times Almanac 2004,* pages 481–483 and *The World Almanac and Book of Facts 2004,* page 97.)

Country	Per Capita Total Spending on Health (in US dollars at ave. exchange rates)	Daily Calorie Supply, 1997	Infant Mortality Rate per 1000 Births
Afghanistan	8	1,523	162
Australia	1,698	3,001	6

Country	Per Capita Total Spending on Health (in US dollars at ave. exchange rates)	Daily Calorie Supply, 1997	Infant Mortality Rate per 1000 Births
China	45	2,844	37
France	2,057	3,551	5
Germany	2,422	3,330	5
India	23	2,415	65
Italy	1,498	3,504	5
Kenya	28	1,971	69
Mexico	311	3,137	28
Poland	246	3,344	9
South Korea	584	3,336	5
Spain	1,073	3,295	5
Turkey	150	3,568	40
United States	4,499	3,642	7

a. Assign weights to the three given social indicators (per capita total spending on health, daily calorie supply, and infant mortality rate) to specify their importance in contributing to a "social" index for a country.

b. Set up and describe a way to use the given social indicators and your weights from part a of this exploration to give a single "social" index for each of these countries. Justify your choices and explain what the rating shows.

c. What other variables might you include in such an index?

7. Consider the "social" index for countries presented in Exploration 6.

a. How sensitive is your "social" index to the choice of weights you assigned for each of the three social indicators? (Pick a different set of weights and compute the "social" index to help answer this question.)

b. How sensitive is your "social" index to the choice of relative ratings you assigned within each of the social indicators?

8. Suppose you are considering a job offer in each of four cities. The cities and the salary offered for each job are given in the following table. You don't want to make your decision based solely on salary, but want to consider the "liveability" of the city as well. Climate is one factor to consider.

City	Salary	Climate		
Seattle	$36,000			
Orlando	$31,000			
Chicago	$39,000			
Philadelphia	$34,000			

   a. Identify two other criteria on which you will base your decision and fill in the following table, assigning a relative rating to each city for each criterion.

   b. Use the weighted sum method to make a decision about which job to choose.

9. The entertainment industry uses rating schemes and critics' judgments to boost sales and create advertisements. For each of the following, decide on at least four criteria you would use to set up a rating scheme. Then pick three specific examples for each and rate them on the four criteria. Finally, use one of the methods discussed to decide which gets the highest overall rating.

   a. Newly released movies

   b. First-run television shows

   c. Music videos

   d. Compact disc recordings of music

   e. Fast-food restaurants

10. Suppose you used a scale of 1 to 100 to assign the weights to rank each of the criteria important to a decision (instead of the 1 to 10 system used previously). Do you think the results of the decision might change? Experiment with this scale using the data in Example 11.6 to help answer the question.

TOPIC 12

Inductive Reasoning

OBJECTIVES

After completing this topic, you will be able to

- » Distinguish between inductive and deductive reasoning
- » Recognize and use different forms of inductive reasoning
- » Decide whether or not a conclusion reached through inductive reasoning is valid
- » Identify assumptions made when using inductive reasoning and identify when a conclusion reached through valid inductive reasoning might be false

Reasoning is an essential activity of the human brain. We use reasoning to draw conclusions daily. It is important, however, to differentiate between valid and invalid reasoning. A conclusion is useful as part of human knowledge only when it has been obtained from valid reasoning. In this topic and the next, we discuss the two major forms of reasoning: inductive reasoning and deductive reasoning.

Inductive reasoning argues from particular cases to a general rule; **deductive** reasoning argues from general cases to particular cases. We use inductive reasoning when we draw a general conclusion from experiments or particular observations. The truth of a conclusion obtained through an inductive argument from valid premises is, at best, highly likely to be true, but not necessarily true. Deductive reasoning is used when conclusions are made through logical inference from the premises. A conclusion obtained through valid deductive reasoning from true premises is necessarily true.

Example 12.1

Decide whether each of the following situations describes inductive or deductive reasoning—that is, decide if the conclusion must follow from the premises or not, assuming the premises are true. In each case, decide if the conclusion is reasonable from the information given and state what assumptions are made to draw the conclusion.

a. My fourteen-year-old brother likes to play the guitar. My cousin, who is fifteen, also likes to play the guitar. Most teenagers like to play the guitar.

b. Thirty-four million Americans wear contact lenses and 85 percent of them choose soft contacts. Because 85 percent of 34 million is 28,900,000, we can conclude that almost 29 million Americans wear soft contact lenses.

c. A British study followed 9,000 women through pregnancy and after childbirth. Researchers recorded symptoms of depression at 18 and 32 weeks of pregnancy and at 8 weeks and 8 months after childbirth. These depression symptom scores were compared with depression scores for women at other times of their lives. The researchers concluded that symptoms of depression were not more common during pregnancy and after childbirth (postpartum) than at other times in a woman's life. (Source: *bmj.com,* http://bmj.bmjjournals.com/cgi/content/full/323/7307/257.)

Solution

a. This is an example of inductive reasoning. We drew a conclusion after observing two particular cases. This conclusion does not seem to be valid, however, because two teenagers from the same family can hardly be representative of all teenagers. The (faulty) assumption here is that these two cases are sufficient to make a general statement about the whole group.

b. This paragraph shows a case of deductive reasoning. From the fact that 85 percent of the 34 million Americans who wear contact lenses wear soft contact lenses, it follows that there are $0.85 \cdot 34{,}000{,}000 = 28{,}900{,}000$ Americans who wear soft contact lenses. In turn, this means that almost 29 million Americans wear soft lenses. The conclusion is not only reasonable but it is definitely true. We are assuming that the information given is true—that 85 percent of those who wear contact lenses wear soft lenses and that 34 million Americans wear contact lenses.

c. The researchers used inductive reasoning to conclude that there is no such a thing as "postpartum depression," but that depression occurs among women who have given birth recently just as frequently as among those who have not given birth recently. The conclusion is a reasonable one to make, assuming that the data collected are accurate, there are no biases in the selection of the 9,000 women, and that eight weeks after delivery is close enough to the delivery date so that any case of depression after delivery would still be present then. (Not everyone agrees with this last assumption. In fact, some experts say the number of women suffering from depression might be higher because some cases might have been missed in the study since the first measure was done after eight weeks.)

Forms of Inductive Reasoning

There are four main types of inductive reasoning: prediction, generalization, causal inference, and analogy. A **prediction** is a form of inductive argument that concludes with a claim about what will happen in the future, based on past or present observations. Financial analysts and weather forecasters often use this form of reasoning. The following, from the article, "A Plunge in Profits is Raising Risk for Stock Market and Economy," which appeared in *The New York Times* on July 29, 2001, is an example of a prediction.

> In profit reports released over the last two weeks, large publicly traded companies said that their earnings fell an average of 17 percent in the second quarter, compared with a year earlier.
>
> The rapid reversal has stunned Wall Street. Just seven months ago, analysts expected profits to rise 9 percent for the full year. They now forecast that earnings will drop more than 8 percent this year, or $30 billion, according to Thomson Financial/First Call, which surveys analysts' forecasts. The decline would be the first in annual profits since 1991.

As with any prediction, there is no absolute certainty that the earnings will drop more than 8 percent. But financial analysts are coming to this conclusion based on their experience and the fact that large publicly traded companies' earnings fell considerably compared with last year.

A second form of inductive reasoning is **generalization**. This occurs when a conclusion is drawn about a whole class or group based on the knowledge of some cases from that group. Studies that make conclusions from sampling methods use generalization. (Sampling is discussed in more detail in Topic 17.) A December 2005 survey by the National Highway Traffic Safety Administration (NHTSA) estimates that at any given time of the day, 6 percent of drivers are using a handheld phone. This means that approximately 974,000 vehicles on the road are driven by someone on a handheld phone. The survey also revealed that 10 percent of drivers in the 16- to 24-year-old-group used handheld phones while driving. (Source: *National Highway Traffic Safety Administration,* www.nhtsa.dot.gov.)

Here the conclusion that, at any given time, 6 percent of Americans are driving and at the same time using a handheld phone, is based on the observation of 43,000 drivers, which is a sample that does not include all American drivers. As with most cases of inductive reasoning, the conclusion is not necessarily true, but is highly probable, assuming the study was well designed.

Another form of inductive reasoning is **causal inference**, in which a conclusion is made about the cause of some situation when only the result is known. This form of reasoning is often used in daily life. The following example is excerpted from the story "Echo Guilt" from the book *Tell Me Everything and Other Stories,* by Joyce Hinnefeld.

> One day in the summer my husband, Jack, found a frog at the edge of our vegetable garden, in the back yard behind this house that we've rented, three hours north of New York City on a steep slope up from the Hudson River. Convinced that this was an animal that had somehow strayed too far from its native environment—surely the river—Jack teased the frog into a paper bag.

In this passage, Jack's conclusion that the frog has come from the river is very likely and perhaps the most reasonable conclusion to make. Of course, this is highly probable, but it is not necessarily the case. It could be, for example, that someone who lives nearby bought it at a pet store and somehow the frog escaped.

A fourth form of inductive reasoning occurs when we make a conclusion about something (events, people, objects) because of its similarity with other things. This is called reasoning by **analogy**. We use this form of reasoning when we interpret something unknown by relating it with something we know with similar characteristics. In July 2001, a local newspaper reported that people in Schuylkill County, Pennsylvania, called 911 to report that they saw a burning airplane crash into Peach Mountain. Later experts found that what the residents saw was most likely part of a meteoroid explosion that showered the region.

Those who thought they saw a burning airplane used analogy to arrive at that conclusion. They saw a burning ball of fire falling from the sky, similar to the ones they have probably seen in the movies many times that represent falling airplanes. Because of the similarity, they concluded it must have been a falling airplane. In this case the analogy did not work; that is, the conclusion was not correct.

Example 12.2

Each of the following scenarios contains inductive reasoning. Decide what type of inductive reasoning each uses and explain how you arrived at that decision.

a. Here is the article, "Study: More Parents Choose to Homeschool," published in *The Morning Call,* on August 3, 2001:

> ### Study: More Parents Choose to Homeschool
>
> Special to *The Morning Call*
>
> About 850,000 of the nation's 50 million children are being taught at home rather than in schools, mostly by parents who are well educated and live in cities, a new government study estimates.
>
> The report released this week by the Education Department, calculates that 1.7 percent of American children were home-schooled in 1999, resulting in a total estimate higher than in the past. "The number of parents taking direct responsibility for teaching their children through homeschooling is approaching a million, and we expect the next report on homeschooling will reflect growth in the population and new homeschooling opportunities," said Education Secretary Rod Paige.
>
> *Continues on next page*

Continued

Study: More Parents Choose to Homeschool

The new figures come from an Education Department telephone survey of 57,278 households conducted from January through May 1999. The new report says the number of homeschoolers could be as high as 992,000 or as low as 709,000. The 850,000 takes the average of the two.

It also paints a clear portrait of the average homeschooler, finding that they are more likely than other students to live with two or more siblings in a two-parent family, with only one parent working outside the home.

Parents of homeschoolers are, on average, better-educated than other parents—a greater percentage have college degrees—though their income is about the same. Like most parents the vast majority of those who homeschool their children earn less than $50,000, and many earn less than $25,000.

Most say they homeschool their children to give them a better education and not necessarily out of religious beliefs, although religion was second on a list of reasons.

b. From *Fear of Math, How to Get Over It and Get On with Your Life,* by Claudia Zaslavsky, Rutgers University Press, 1994:

> The economy of the United States has changed drastically within the past few decades. There has been enormous growth in communications and finance, while manufacturing jobs have gone abroad or have been restructured to require higher-level skills on the part of the workers. With advances in technology occurring frequently, people will have to learn to work smarter, rather than faster, than in the past. Increasingly the jobs in tomorrow's economy will require a knowledge of mathematics. New technologies will call for the ability to apply mathematics and science in practical ways, and rapid changes will demand that workers learn new skills throughout their lives. In other words, lifelong learning will be the pattern of the future.

c. From the article, "Left-Handers and the Debate on How They Came to Be That Way," *The New York Times*, May 16, 2000:

> His studies of handedness, he said, began with his interest in the fact that yeast, single-cell organisms, have two distinct sides. His theory about the right-handed gene and the absence of this gene in lefties, he (Dr. Klar) said, grew out of studies of mutant mice.
>
> In most mice, the heart is on the left side. But there is a mutant strain of mice that have hearts on the right. When the mutants mate, half the offspring have hearts on the left and half on the right.
>
> Last year, Dr. Klar said, geneticists identified a gene that is present in normal mice but totally missing in the mutants. "If it can work in mice," Dr. Klar asked, "why can't it work for handedness in people?"
>
> The beauty of the theory is that it explains a phenomenon that has long baffled geneticists: how can identical twins with exactly the same genetic makeup have different handed-ness, as 18 percent of them do? Dr. Klar's explanation is that these twins lack the right-handed gene, and each one has an equal chance of being right-handed or left-handed.

d. From the story, "The Speckled Band." *The Adventures of Sherlock Holmes*, by A. Conan Doyle:

> "You have come in by train this morning, I see."
>
> "You know me then?"
>
> "No, but I observe the second half of a return ticket in the palm of your left glove. You must have started early, and yet you had a good drive in a dog-cart, along heavy roads, before you reached the station."
>
> The lady gave a violent start, and stared in bewilderment at my companion.

> "There is no mystery, my dear madam" said he, smiling. "The left arm of your jacket is spattered with mud in no less than seven places. The marks are perfectly fresh. There is no vehicle save a dog-cart which throws up mud in that way, and then only when you sit on the left-hand side of the driver."

Solution

a. The conclusion about the number of children being homeschooled is obtained from a sample survey of 57,278 households. One type of induction used is generalization, which allows the surveyors to draw a conclusion about the total number of homeschooled children in the general population. The article also uses prediction because they "expect the next report on homeschooling will reflect growth in the population..."

b. The reasoning here is a prediction. Based on an analysis of the current job market, the author predicts that employees will need to have the ability to learn and apply mathematics concepts throughout their careers.

c. From the observation that the heart is on the right side in mice who are missing a certain gene, Dr. Klar concludes that there is a specific gene present in all right-handed persons. He is using analogy to draw this conclusion.

d. Here Sherlock Holmes uses induction in the form of causal inference to conclude that his visitor has traveled by train and has arrived recently. He arrives at this conclusion after observing a return train ticket in the lady's hand and fresh mud stains on her clothes.

Inductive reasoning leads to conclusions that are likely to be true, but might not be true. In Example 12.3, we will consider why specific conclusions might not be true.

Example 12.3

For each of the scenarios described in Example 12.2, identify what premises are used and explain why it is possible that the conclusion might be false.

Solution

a. The premises used here are that the households surveyed are a good sample of the general population and therefore the information obtained from them can be generalized. The conclusion could be false if the households surveyed contained, for example, a disproportionately large number of households with college educated adults. This might result in too high an estimate of home-schooled children.

b. One premise is that the author's analysis is accurate and another is that advances in technology will continue. The prediction might not be true if the author's analysis of the present situation is inaccurate or if, in the future, the technology situation changes.

c. One premise is that there is a gene in humans similar to the one found in the normal mice. As with any analogy, the conclusion might be false. Although there may be some similarities between two objects or situations, it does not necessarily follow that they are identical in all aspects. In this case, there are similarities in the basic biology of mice and humans, but we know there are also many differences. For this reason, the researcher can be sure of his theory only if he identifies the human gene he suspects exists.

d. Two premises are that the ticket is current and that the mud was not planted as a diversion. This conclusion could be false if, for example, the lady's ticket was an old one that she just found in her purse or the mud stains were perhaps purposely made to give the impression she had just traveled by dog-cart.

In the previous example, we saw that even with good inductive reasoning the conclusion may not be true. Bad inductive reasoning often leads to false conclusions. A generalization based on two specific people from the same family, as in Example 12.1a, does not result in a valid conclusion. In the next example, we will look at other cases of inductive reasoning in which the conclusion is not based on sound reasoning.

Example 12.4

For each of the following scenarios, describe the fallacies in the reasoning.

a. Since it has rained for two full days, I predict that tomorrow will be a beautiful, sunny day.

b. Smoking can't be as bad as they say it is. My aunt Margaret, who smokes a pack of cigarettes a day, is 80 years old and still in good health.

c. As reported in *The New York Times*, July 29, 2001, the number of head injuries among bicycle riders has increased 10 percent since 1991, while the use of helmets has increased considerably; therefore, I will not use my helmet anymore because it seems that using helmets increases the chances of getting a head injury.

Solution

a. This seems more like a wish than a prediction. A prediction must be based on some relevant evidence or study. To predict whether it will be sunny or not, we must analyze the atmospheric conditions. It is quite possible to have several consecutive rainy days, so the fact that it has already rained for two days is not a reason to think that it will be sunny tomorrow.

b. The reasoning is invalid in this case because we are using one individual case to draw a general conclusion about the effects of smoking in the human body. In this case, we do know that this conclusion is wrong because there is much scientific evidence, based on valid inductive reasoning, of the health risks of smoking.

c. This is a misuse of the causal inference form of reasoning. Although the two facts are occurring together, there is no reason to believe that one is causing the other. (According to the newspaper article, some experts think people are assuming riskier behaviors because of the sense of security a helmet provides.)

Summary

In this topic, we discussed the differences between inductive and deductive reasoning and noted that correct deductive reasoning always leads to a true conclusion, but a conclusion reached through inductive reasoning is not necessarily true. We also analyzed different forms of inductive reasoning—prediction, generalization, causal inference, and analogy—and investigated fallacies in inductive reasoning.

Explorations

1. The following passage is taken from *L is for Lawless* (Henry Holt and Company, Inc., New York, 2005, p. 237), one of the books in a mystery series by Sue Grafton. In this scene, two characters are discussing something that had happened years ago. A man had pulled a bank robbery and was arrested some time later, but without the cash and the jewels he stole. In this passage, identify where deductive reasoning is used and where inductive reasoning is used and explain how they are different.

   "Unless he had time to go to some other town and come back," I said. "It's like saying you always find something the last place you look. I mean, it's self-evident. Once you find what you are looking for, you don't look any place else. The last you saw him, he had the sacks full of cash. By the time he was arrested, they were gone.

   Therefore, the money had to have been hidden some time in that period. By the way, you never said how long it was."

   "Half a day."

   "So he probably didn't have time to get far."

   "Yeah, that's true..."

2. Consider each of the following reasoning scenarios. For each, identify if inductive or deductive reasoning is used, determine if the conclusion is reasonable from the information given, and state what assumptions are made to draw the conclusion.

   a. Arrest rates in large cities in the U.S. for certain offenses such as disorderly conduct, drunkenness, and vagrancy have been declining. This shows that America's large cities are becoming more peaceful.

   b. An advertisement for a particular brand of fruit bar claims that mothers will feel good about feeding it to their children because it is made from "real fruit juice."

   c. I go to the store with $5.00 in my wallet. I want to buy 7 pounds of bananas that cost $0.49 per pound. I will have enough money also to buy a small candy bar.

   d. The majority of Americans prefer football over other sports. This conclusion was based on a survey taken by middle school students of their peers. Their survey showed that approximately 65 percent preferred football as their favorite sport.

   e. At 9 A.M. on the morning of the New Hampshire Republican primary, reporters predicted that Bush would beat the other candidates in that primary, based on up-to-the-minute returns.

   f. My airplane from San Francisco to Allentown, Pennsylvania, stops in Chicago. I have stopped in Illinois.

3. Each of the following scenarios contains inductive reasoning. Decide what type of inductive reasoning each contains.

   a. "The men, Dorothy thought, were about as old as Uncle Henry, for two of them had beards." (Source: *The Wonderful Wizard of Oz,* by L. Frank Baum, Books of Wonder, 1987, p. 21.)

   b. "From the limited polls that have been taken on this issue (stem cell research), we know that Americans wanted Mr. Bush to advance research with astonishing potential and to reassure them that, in so doing, the United States would not cross some terrible line into unthinkable evil." (Source: "Bush's Gift to America's Extremists," *The New York Times*, August 19, 2001.)

c. "On Friday, Cisco shares closed at $19.06, less than one-fourth their March 2000 high of $80.06. But because profit forecasts have plunged, the company's shares, by some estimates, trade somewhere between 47 times estimated 2002 earnings and 73 times. As a result, Cisco and some other technology stocks could fall further, pessimists among the analysts warn." (Source: "A Plunge in Profits Is Raising Risk for Stock Market and Economy," *The New York Times*, July 29, 2001.)

d. ""Ayi!" she gasped. On the side of a ground swell lay Jello, his body torn in bloody shreds, his face contorted. Beside him lay her backpack!

Instantly she knew what had happened; Amaroq had turned on him. Once Kapugen had told her that some wolves had tolerated a lone wolf until the day he stole meat from the pups. With that, the leader gave a signal and his pack turned, struck, and tore the lone wolf to pieces. "There is no room in the wolf society for an animal who cannot contribute," he had said.

Jello had been so cowed he was useless. And now he was dead." (Source: *Julie of the Wolves,* by Jean Craighead George, Harper & Row, Publishers, Inc., 1972, p. 121.)

4. For each of the scenarios detailed in Exploration 3, describe what premises are used and why the conclusion might be false.
5. Find newspaper or magazine articles that illustrate the four types of inductive reasoning: prediction, generalization, causal inference, and analogy. Explain how each article exhibits the particular type of reasoning.

TOPIC 13

Deductive Reasoning

OBJECTIVES

After completing this topic, you will be able to

- Relate logical statements given in words to symbolic logical statements, and use truth tables to assess logical statements
- Identify logical statements and formulate their negations
- Recognize and use the three different types of compound logical statements
- Formulate and use the contrapositive and converse of an if-then statement
- Form simple deductive arguments and analyze correct and incorrect deductive reasoning

Deductive reasoning is the form of reasoning that we use to derive logical consequences from given true statements. We often use deductive reasoning when we want to prove a point, whether it is a mathematical theorem, a legal argument, or a scientific theory. Deductive reasoning always leads to true statements, provided the premises are true.

The following excerpts from an address to the House of Representatives by Representative Osborne give examples of logical statements and deductive reasoning (Source: *Congressional Records website*, www.congress.gov, May 1, 2003):

> Mr. Speaker, last weekend, the National Football League draft was conducted. Over 200 players were selected in the draft. Each player eventually will be represented by an agent. The difficult thing is that many of these people who call themselves agents have no special qualifications.

> We find that many of them have no legal training, no expertise in writing contracts, some misrepresent themselves, some offer illegal inducements, particularly to undergraduates, such as cars, cash, clothes, and sometimes even drugs, to get young people to commit to a contract while they still have eligibility, which makes them ineligible, of course. A few even have criminal records. Most of them will tell a player that they will get them drafted higher. ....
>
> Currently, Mr. Speaker, there are only 15 States that have tough laws regulating actions by sports agents. There are 17 States, including my home State of Nebraska, that have no laws at all regulating sports agents, and then there are 18 States remaining that have some laws. It is kind of a hodgepodge, a patchwork; and there is no consistency and no teeth in the regulations. So the majority of young people coming out of college really are not protected by any laws that would govern sports agents.

In these comments, the representative uses deductive reasoning when he draws the conclusion that young players are not protected by laws that would govern sports agents. His conclusion follows from current data on existing state regulations.

It is important to be able to tell whether or not a specific form of deductive reasoning is valid. When the reasoning is valid, then the truth of the premises necessarily implies the truth of the conclusion. The validity of reasoning is governed by the rules of logic. We will explore the kinds of statements used in reasoning and the basic logic principles that lead to valid reasoning.

A **statement** is a sentence that is either true or false, but not both. We explore this definition of a statement in the first example.

Example 13.1

For each of the following sentences, decide whether it is a statement or not, and explain why you gave the answer you did.

a. "I am the man who accompanied Jacqueline Kennedy to Paris, and I have enjoyed it." (Source: a quote from President John F. Kennedy.)

b. The population of Pennsylvania is 12,281,054, and the population of the U.S. is 150,287,967.

c. "If automakers can produce greater fuel efficiency across the Atlantic, why can't they do it here?" (Source: "The Low Cost of Lowering Auto Emissions," *The New York Times,* August 1, 2001.)

d. "Studies of embryonic stem cells in mice have been conducted for two decades, but in 1998, Dr. James A. Thomson, a developmental biologist at the University of Wisconsin, shook up the stem cell world by reporting that he had isolated human embryonic stem cells." (Source: "Patent Laws May Determine Shape of Stem Cell Research," *The New York Times,* August 17, 2001.)

e. Please do not go out in this storm.

f. "Currently, Mr. Speaker, there are only 15 states that have tough laws regulating actions by sports agents." (Source: an address by Representative Osborne quoted previously in this topic.)

Solution

The sentences in parts a, b, d, and f are all statements because they are true or false. Note that we might or might not know whether they are true or false, but we know that one or the other must be the case. The sentence in part c of this example contains a question, which is neither true nor false, so it is not a statement. The sentence in part e of this example is not a statement because it is a request, which is neither true nor false.

Negation of a Statement

The **negation** of a statement is the statement obtained by negating the original statement. If we represent the original statement by P, its negation is **not P**. If P is true, then "not P" is false; and if P is false, then "not P" is true; that is, P and "not P" have opposite truth values. We can summarize the relationship between a statement P and its negation using a **truth table**. A truth table gives all possible truth values for the statements under consideration. Here is a truth table for a statement P and its negation:

P	not P
T	F
F	T

The table shows that P can either be true (T) or false (F). If P is true, then "not P" is false. If P is false, then "not P" is true, by the definition of "not P."

We must be careful about the language we use when forming the negation of a statement, as the next example illustrates.

Example 13.2

Based on the results of a study published in a scientific journal, we can make the following statement: "Using appropriate technology to control greenhouse gases in four highly polluted major cities—Sao Paulo, Brazil; Mexico City; Santiago, Chile; and New York City—would save 64,000 lives over the next 20 years." Find the negation of this statement and say whether it is true or false.

Solution

The negation of this statement is, "Using appropriate technology to control greenhouse gases in four highly polluted major cities—Sao Paulo, Brazil; Mexico City; Santiago, Chile; and New York City—would not save 64,000 lives over the next 20 years." This new statement is true if the original statement is false, and it is false if the original statement is true. (We could also negate the original statement by saying, "It is not the case that using appropriate technology to control greenhouse gases in four highly polluted major cities—Sao Paulo, Brazil; Mexico City; Santiago, Chile; and New York City—would save 64,000 lives over the next 20 years," but that is an awkward statement and would not be a particularly useful negation.)

Compound Statements

Simple statements are often combined into more complex statements, called **compound** statements. To form a compound statement, we connect simple statements through **logical connectors**. The truth value of a compound statement depends on the truth value of each of the components and on the connector used. The three connectors that are used most often are **and**, **or**, and **if-then**. If we represent the original statements by P and Q, then using each of these connectors, we obtain three compound statements: **P and Q**; **P or Q**; and **if P then Q**.

A statement of the form **P and Q** is called the **conjunction** of P and Q. For example, the statement (from an address to the Senate by Senator Grassley, *Congressional Records website,* www.congress.gov, July 26, 2001):

> These two efforts will provide complete elimination of the marriage penalty for low- and many middle-income working families **and** will also benefit married couples with higher incomes.

is the conjunction of the statement P, "These two efforts will provide complete elimination of the marriage penalty for low- and many middle-income working families," and the statement Q, "These two efforts will benefit married couples with higher incomes."

The conjunction of "P and Q" is true when both of the statements, P and Q, are true, and it is false if either one of P or Q is false or both P and Q are false. This relationship is summarized in the truth table given here. Note that we need to include four rows in the table to cover all possible combinations of truth values for both P and Q.

P	Q	P and Q
T	T	T
T	F	F
F	T	F
F	F	F

The conjunction can also be expressed by connecting words other than the word *and,* such as the words *while, but,* and *yet.* For example, the statement "P and Q" given previously could be expressed as:

> These two efforts will provide complete elimination of the marriage penalty for low- and many middle-income working families **but** will also benefit married couples with higher incomes.

A compound statement of the form **P or Q** is called a **disjunction**. It is true when *at least* one of P or Q is true, that is, when P is true or Q is true or both are true. The following statement is an example of a disjunction:

> This evening, I will study for the history test **or** I will complete the math project.

The compound statement is composed of the statement P, "This evening, I will study for the history test," and the statement Q, "This evening, I will complete the math project." The compound statement will be true in the case where the speaker only studies for the history exam that evening, or the speaker only completes the math project, or the speaker does both—studies history and completes the math project—that evening. The truth table for the disjunction "P or Q" covers all possible combinations of truth values for the two simpler statements P and Q.

P	Q	P or Q
T	T	T
T	F	T
F	T	T
F	F	F

Note that the word *or* in English may also mean that only one and not both of the two connected statements is true. In this case, we call it an *exclusive or* as opposed to the *inclusive or* of the disjunction "P or Q" defined previously. Whether the *or* is inclusive or exclusive in a specific statement is often determined by the context but sometimes needs to be clarified. In standard logic, we assume *or* means "inclusive or" unless otherwise stated.

A compound statement of the form **if P, then Q** is called a **conditional** statement. In this situation, the statement P is called the **antecedent** and Q is called the **consequent** of the conditional statement. The conditional statement "if P, then Q" is true when Q is true (and P is either true or false), or when P is false (and Q is either true or false). That is, it is true in all cases, except the case when the antecedent (P) is true and the consequent (Q) is false. The following statement (also from Senator Grassley's address mentioned previously) is an example of a conditional statement:

> **If** the first $6,000 of a single individual is taxed at 10 percent, **then** the first $12,000 of a married couple filing jointly will be taxed at 10 percent.

Here, the statement P is "The first $6,000 of a single individual is taxed at 10 percent," and the statement Q is "The first $12,000 of a married individual filing jointly will be taxed at 10 percent." The only case in which this statement would be false is when the first $6,000 of a single individual is taxed at 10 percent, and the first $12,000 of a married couple filing jointly is taxed at a rate other than 10 percent. Note that if it is not the case that the first $6,000 of a single individual is taxed at 10 percent, then the statement is true whether or not the first $12,000 of a married couple filing jointly is taxed at 10 percent.

The truth table for "if P, then Q" is as follows:

P	Q	if P, then Q
T	T	T
T	F	F
F	T	T
F	F	T

There are other ways of expressing the conditional. Sometimes the word *then* is omitted, resulting in a statement of the form "if P, Q." For example, the previous conditional statement can be expressed as:

> If the first $6,000 of a single individual is taxed at 10 percent, the first $12,000 of a married couple filing jointly will be taxed at 10 percent.

Other ways of expressing the same statement are to say "Q, if P" or "Q whenever P," as the following rewordings of the previous example show.

> The first $12,000 of a married couple filing jointly will be taxed at 10 percent, if the first $6,000 of a single individual is taxed at 10 percent.
>
> The first $12,000 of a married couple filing jointly will be taxed at 10 percent whenever the first $6,000 of a single individual is taxed at 10 percent.

The following example will help us identify different types of simple and compound statements.

Example 13.3

For each of the following statements, decide whether the statement is simple or compound. If the statement is compound, decide whether it is of the form "P and Q," "P or Q," "not P," or "if P, then Q," and give P and Q. (These statements are all taken from two speeches published in *Vital Speeches of the Day*, "Breaking the Deadlock in U.S. Trade Policy," by Murray Weidenbaum, June 2001, and "The Intellectual Climate of the U.S," by William Brody, July 2001.)

a. "...globalization is responsible for abuses of labor rights and of the environment and it reduces the sovereignty of individual nations."

b. "Eventually, of course, some deals will have to be made if any legislative action is to occur at all."

c. "For a variety of reasons, including disagreements between the developed and developing nations, the World Trade Organization meetings in Seattle concluded in failure."

d. "If these issues are not managed intelligently and creatively, the domestic consensus in favor of open markets may ultimately erode."

e. "The people hurt by globalization are being ignored while the winners are enjoying all the benefits."

f. "We may have a faculty member teaching microeconomics at Harvard in the fall, in Singapore in the winter, and at Hopkins in the summer, or we may have faculty members doing collaborative research across institutions."

Solution

a. This statement is a compound statement of the form "P and Q," which is a conjunction. Here P is "Globalization is responsible for abuses of labor rights and of the environment," and Q is "It [globalization] reduces the sovereignty of individual nations."

b. This is a conditional statement, that is, a compound statement of the type "if P, then Q." In this case, the statement P (the antecedent) is "Any legislative action is to occur," and Q (the consequent) is "Some deals will have to be made."

c. This statement is a simple statement.

d. This statement is a compound statement of the form "if P, then Q." P, the antecedent, is "These issues are not managed intelligently and creatively," and Q, the consequent, is "The domestic consensus in favor of open markets may ultimately erode."

e. This is a compound statement. It is a conjunction that can be expressed in the form "P and Q," where P is the statement, "The people hurt by globalization are being ignored," and Q is the statement, "The winners are enjoying all the benefits."

f. This is a disjunction, that is, a compound statement of the form "P or Q." Here P is "We may have a faculty member teaching microeconomics at Harvard in the fall, in Singapore in the winter, and at Hopkins in the summer"; Q is "We may have faculty members doing collaborative research across institutions."

Negation of Compound Statements

The negation of any statement P can be expressed as "It is not the case that P." Usually, however, this is not a useful form of the negation. For example, the negation of the statement (from Senator Grassley's address mentioned earlier):

> These two efforts will provide complete elimination of the marriage penalty for low- and many middle-income working families **and** will also benefit married couples with higher incomes.

could be stated in a quite non-useful form as:

> It is not the case that these two efforts will provide complete elimination of the marriage penalty for low- and many middle-income working families and will also benefit married couples with higher incomes.

A more useful form of the statement's negation is:

> These two efforts will **not** provide complete elimination of the marriage penalty for low- and many middle-income working families **or** will **not** benefit married couples with higher incomes.

Note that this last statement is of the form "(not P) or (not Q)."

The disjunction "(not P) or (not Q)" is the logical **equivalent** to the statement "not (P and Q)," and therefore it is the negation of "P and Q" because the statements "(not P) or (not Q)" and "not (P and Q)" are both true or they are both false. So,

> **"not (P and Q)" is equivalent to "(not P) or (not Q)"**

This is the case because the statement "(not P) or (not Q)" is false whenever the statement "P and Q" is true. To see this, note that the disjunction is false only in the case when both of the statements in the conjunction are false; that is, "(not P) or (not Q)" is false only when both "not P" and "not Q" are false, which is when both P and Q are true. This is precisely the only case when the conjunction "P and Q" is true. A truth table summarizes this discussion. Note that the last two columns have exactly the same truth values, which means the statements are equivalent. To find the truth values for the last column, we use the "not P" column and the "not Q" column. The *or* statement is false only when both "not P" and "not Q" are false, which is when P and Q are both true.

P	Q	not P	not Q	P and Q	not (P and Q)	(not P) or (not Q)
T	T	F	F	T	F	F
T	F	F	T	F	T	T
F	T	T	F	F	T	T
F	F	T	T	F	T	T

In a similar manner, we can see that the negation of a disjunction "P or Q" is the conjunction "(not P) and (not Q)." That is,

"not (P or Q)" is equivalent to "(not P) and (not Q)"

P	Q	not P	not Q	P or Q	not (P or Q)	(not P) and (not Q)
T	T	F	F	T	F	F
T	F	F	T	T	F	F
F	T	T	F	T	F	F
F	F	T	T	F	T	T

The negation of a conditional statement "if P, then Q" is a statement logically equivalent to "not (if P, then Q)." That is, a statement that is true exactly when "if P, then Q" is false. This is the case when P is true and Q is false. The statement "P and (not Q)" is true precisely when P is true and (not Q) is true, that is when P is true and Q is false. We can then say that

"not (if P, then Q)" is equivalent to "P and (not Q)"

A truth table confirms this relationship.

P	Q	not Q	if P, then Q	not (if P, then Q)	P and (not Q)
T	T	F	T	F	F
T	F	T	F	T	T
F	T	F	T	F	F
F	F	T	T	F	F

In the following example, we use the equivalent statements shown in the truth tables to compose useful negations of various compound statements.

Example 13.4

Give a useful negation of each of the following statements (from Example 13.3).

a. "...globalization is responsible for abuses of labor rights and of the environment and it reduces the sovereignty of individual nations."

b. "Eventually, of course, some deals will have to be made if any legislative action is to occur at all."

c. "...the World Trade Organization meetings in Seattle concluded in failure."

d. "If these issues are not managed intelligently and creatively, the domestic consensus in favor of open markets may ultimately erode."

e. "The people hurt by globalization are being ignored while the winners are enjoying all the benefits."

f. "We may have a faculty member teaching microeconomics at Harvard in the fall, in Singapore in the winter, and at Hopkins in the summer, or we may have faculty members doing collaborative research across institutions."

Solution

a. Because this is a compound statement of the form "P and Q," its negation is of the form "(not P) or (not Q)." So a useful negation of the given statement is "Globalization is not responsible for abuses of labor rights and of the environment, or it does not reduce the sovereignty of individual nations."

b. The negation of the conditional "if P, then Q" is a statement of the form "P and (not Q)." So, a useful negation of the given statement is "Some legislative action will occur and (but) no deals will be made."

c. This is a simple statement. A useful negation is "The World Trade Organization meetings in Seattle did not conclude in failure" or "The World Trade Organization meetings in Seattle were successful."

d. A useful negation of this conditional statement is "These issues are not managed intelligently and creatively and the domestic consensus in favor of open markets may not erode."

e. The negation of this statement is of the form "(not P) or (not Q)." It can be expressed as "The people hurt by globalization are not being ignored or the winners are not enjoying all the benefits."

f. This is a disjunction of the form "P or Q," so its negation is the conjunction "(not P) and (not Q)." A useful negation of the given statement is "We may not have a faculty member teaching microeconomics at Harvard in the fall, in Singapore in the winter, and at Hopkins in the summer, and we may not have faculty members doing collaborative research across institutions."

Contrapositive and Converse of a Conditional Statement

Two other conditional forms are related to the conditional "if P, then Q." These are the **contrapositive** and the **converse** of the conditional "if P, then Q." The **contrapositive** of "if P, then Q" is the conditional statement **if (not Q), then (not P)**. Note that to state the contrapositive of the statement "if P, then Q," we switch the positions of P and Q and negate both. The contrapositive of a conditional statement is equivalent to the original conditional. (You are asked to use a truth table to verify this in Activity 13.1.) This means that the contrapositive is true when the original statement is true, and it is false when the original statement is false. For example, consider again the statement:

> **If** the first $6,000 of a single individual is taxed at 10 percent, **then** the first $12,000 of a married couple filing jointly will be taxed at 10 percent.

Its contrapositive is

> **If** the first $12,000 of a married couple filing jointly is **not** taxed at 10 percent, **then** the first $6,000 of a single individual is **not** taxed at 10 percent.

The **converse** of the conditional statement "if P, then Q," is the conditional statement **if Q, then P**; that is, the converse is the conditional statement that interchanges the consequent and antecedent of the original statement. The converse is not equivalent to the original statement. It is possible for the converse to be false when the original statement is true or the converse might be true and the original statement might be false. It is also possible that a statement and its converse might both be true or might both be false. Consider once more the statement:

> **If** the first $6,000 of a single individual is taxed at 10 percent, **then** the first $12,000 of a married couple filing jointly will be taxed at 10 percent.

Its converse is

> **If** the first $12,000 of a married couple filing jointly is taxed at 10 percent, **then** the first $6,000 of a single individual is taxed at 10 percent.

Example 13.5

Consider the conditional statement, "If the Democratic candidate from New York is elected to the House of Representatives, then the Democrats will have a majority in the House."

a. Give the contrapositive of the conditional statement.

b. Give the converse of the conditional statement.

Solution

a. We first identify P and Q in the given conditional statement; P is the statement, "the Democratic candidate from New York is elected to the House of Representatives," and Q is the statement, "the Democrats have a majority in the House." The contrapositive is "if not Q, then not P," which is, "If the Democrats do not have a majority in the House of Representatives, then the Democratic candidate from New York was not elected to the House."

b. The converse of the given statement is "If the Democrats have a majority in the House of Representatives, then the Democratic candidate from New York was elected."

Quantified Statements

We often use statements that assert a truth about some or all elements of a set. These statements are called **quantified statements**, and contain words such as *all, every, no, there is*, or *there exists* called **quantifiers**. These are examples of quantified statements:

"All citizens can vote."

"Some dogs are dangerous."

"No candidate is sufficiently ahead in the polls."

"There is a country that has a name starting with the letter *C*."

Quantified statements can be grouped into two general classes: those equivalent to statements containing the **universal quantifier**, *all,* and those equivalent to statements containing the **existential quantifier**, *there exists.* For example, the statement, "Some dogs are dangerous" contains an existential quantifier since it can be rephrased as "There exists a (or there exists at least one) dog that is dangerous."

Sometimes we will need to reword statements to identify the quantifier. For example, the statement, "No candidate is sufficiently ahead in the polls" contains a universal quantifier because it is equivalent to "All candidates are not sufficiently ahead in the polls." We look at additional examples of quantifiers in the next example.

Example 13.6

For each of the following statements, identify the type of quantifier it contains, and write an equivalent statement using one of the quantifiers *all*, or *there exists.*

a. "Every team here is exceptional."

b. "…some of the material was accessed in 2004."

c. "No known human has ever received an injection of embryonic stem cells…"

d. "...not all of the files from the computer have been examined."

e. "No one is missed more than the Americans." (This refers to the fact that the American baseball team was eliminated before getting to Greece for the 2004 Olympic games.)

(Statements in part a and e of this example are from the article, "Olympic Baseball Pallid Without U.S-Cuba Clash," *The Providence Sunday Journal,* August 13, 2004; statements in parts b and d of this example are from the article, "Seized Terror-Target Files Were Accessed in Spring," *The Wall Street Journal,* August 13, 2004; and the statement in part c of this example is from the article, "Trace of Human Stem Cells Put in Unborn Mice Brains," *The New York Times,* December 13, 2005.)

Solution

a. This statement is equivalent to the statement, "All teams here are exceptional." It involves a universal quantifier.

b. This statement is equivalent to "There exists some material that was accessed in 2004." It involves an existential quantifier.

c. The statement is equivalent to "Every human has not received an injection of embryonic stem cells..." It contains a universal quantifier.

d. This statement is equivalent to the statement, "Some files from the computer have not been examined," or "There exists a file that has not been examined." It involves the existential quantifier.

e. An equivalent statement is "Every non-American team is not missed as much as the Americans," or "All non-American teams are not missed as much as the Americans." This statement involves a universal quantifier.

Special care is needed when stating the negation of a quantified statement because the type of quantifier used in the statement is different from the type of quantifier used in its negation. For example, the negation of "Some dogs are dangerous" is "All dogs are not dangerous," or equivalently, "No dog is dangerous." (Note that the statement, "Some dogs are not dangerous" is **not** a negation of "Some dogs are dangerous," because both statements are true.)

The general form of the negation of a statement involving the universal quantifier, such as "all A's are B," is "some A's are not B" (or equivalently, "there exists an A that is not B"), because if it is not true that all A's are B, then there must be at least one A that is not B. Similarly, the general form of the negation of a statement involving the existential quantifier, such as "some A's are B," is "all A's are not B." The next example illustrates how to formulate useful negations of quantified statements.

Example 13.7

Write a useful negation of each of the statements in Example 13.6.

Solution

a. The given statement ("Every team here is exceptional") contains a universal quantifier, so its negation contains an existential quantifier. The negation is "There is a team here that is not exceptional," or "Some teams here are not exceptional."

b. Because this statement ("...some of the material was accessed in 2004.") uses the existential quantifier, its negation will have the universal quantifier: "All of the material was not accessed in 2004," which is best expressed as, "None of the material was accessed in 2004."

c. The negation of the statement, "No known human has ever received an injection of embryonic stem cells..." involves an existential quantifier. It can be expressed as, "Some humans have received an injection of embryonic cells..."

d. The original statement, "...not all of the files from the computer have been examined" is equivalent to "some files from the computer have not been examined." Its negation is "All files from the computer have been examined."

e. The negation of the statement, "No one is missed more than the Americans" involves an existential quantifier, "There is a non-American team that is missed as much as the Americans."

Deductive Arguments

We conclude this topic with a brief discussion of what makes a deductive argument a good deductive argument. A deductive argument consists of premises (or hypotheses or assumptions) and conclusions that follow logically from those premises. There are two key elements for a good deductive argument: (1) the premises are true, and (2) the reasoning is valid. When these two elements are present, the conclusion is unquestionably true. The two forms of valid deductive reasoning are **direct reasoning** or **Modus Ponens** and **indirect reasoning** or **Modus Tollens**.

Direct reasoning or **Modus Ponens**, also known as the **Law of Detachment**, is stated as follows:

If the statement "if P, then Q" is true, and P is also true, then Q must be true.

An example of a good deductive argument using Modus Ponens is the following: "If two quantities x and y are related through an equation of the form $y = kx$ (where k is a constant), then y is directly proportional to x. The number of calories c that a 150-pound person uses when walking for m minutes is $c = 5.4m$. Then, the number of calories is directly proportional to the number of minutes the person walks."

Indirect reasoning or **Modus Tollens** is stated as follows:

If the statement "if P, then Q" is true and Q is false, then P must be false.

An example of Modus Tollens or indirect deductive argument is the following: "If the response variable is directly proportional to the explanatory variable, then the response variable is a linear function of the explanatory variable. The relative energy released by an earthquake is not a linear function of the earthquake's magnitude in the Richter scale. Conclusion: The relative energy released by an earthquake is not directly proportional to its magnitude in the Richter scale." We look at additional samples of valid and invalid reasoning in the next example.

Example 13.8

For each of the following arguments, decide whether the reasoning is valid or not. If it is valid, decide which form of deductive reasoning is used. If it is not valid, explain why not.

a. If the Flyers win today's game, then they advance to the Stanley cup playoffs. They did not advance to the Stanley cup playoffs. Therefore, they did not win today's game.

b. If the price of electricity goes up, our family will pay over $80 a month in electricity. The price of electricity is the same as last month. Then our family will not pay more than $80 this month.

c. When you are caught driving over the speed limit you get a ticket. You were given a ticket this morning. So, you must have been driving over the speed limit.

d. If the interest rates go down, then more houses are sold. The interest rates have gone down. Then the number of houses sold has increased.

Solution

a. Let P be the statement, "The flyers win today's game," and let Q be the statement, "They advance to the Stanly cup playoffs." The given argument has the form: "if P, then Q"; "not Q." Then, "not P." Assuming the statements "if P, then Q" and "not Q" are true, this is a valid argument. The form of reasoning used is the Modus Tollens or indirect reasoning.

b. This is not a valid argument. Let P be the statement, "The price of electricity goes up," and let Q be the statement, "Our family will pay over $80 a month in electricity." The given argument has the form: "if P, then Q." "not P." Then "not Q." This reasoning is not correct because assuming that the statements "if P, then Q" and "not P" are true, does not mean that "not Q" must be true. The truth of the statement "if P, then Q" guarantees that when P is true, then Q must be true, but says nothing about the truth of Q when P is false. In the case of the given argument, P is false ("not P" is true), so we cannot conclude that Q is false (or "not Q" is true). Note that the electricity bill may be larger this month if, for example, the air conditioner was used more often.

c. This is not a valid argument. Let P be the statement, "You are caught driving over the speed limit" and Q be the statement, "You get a ticket." The given argument has the form: "if P, then Q"; "Q." Therefore "P." Even when the statement, "if P, then Q" is true, and Q is true, the truth of P does not follow.

(Remember that the conditional is true when the consequent is true, whether or not the antecedent is true.) In this case, you might have been given a ticket for driving with bad tires, for example, or for not stopping at a stop sign.

d. This is a valid argument. Let P be the statement, "The interest rates go down," and let Q be the statement, "More houses are sold." The given reasoning has the form: "if P, then Q"; "P." Therefore "Q." The reasoning form is Modus Ponens or direct reasoning.

Summary

In this topic, we investigated statements used in deductive reasoning and considered three types of compound statements: conjunctions, disjunctions, and conditional statements. We explored how to negate simple and compound statements and considered how to formulate the contrapositive and the converse of a conditional statement. We also studied quantified statements, using the universal and existential quantifiers, and we practiced valid deductive reasoning. We used truth tables to help understand compound statements and logical reasoning.

Explorations

1. Identify which of the following are statements, and for those that are statements, say whether they are true or false.

   a. Be quiet.

   b. George Bush was president during the years 2000 and 2004.

   c. There are more than 2000 students currently enrolled at Pennsylvania State University.

   d. All college students attend parties at least once a week.

   e. Would you like to go to the movies?

2. For each of the following, determine if the expression is a statement or not, and explain why you gave the answer you did.

    a. "The human fascination with whales has led to a new counterweight to the pro-whaling forces—the hundreds of companies running whale-watching operations in 87 countries, including those seeking an end to the ban on commercial hunting." (Source: "Save the Whales! Then What?" *The New York Times,* August 17, 2004.)

    b. "His conducting has gained technical assurance over the years." (Source: "Hearing Echoes of Yesterday," *The New York Times,* August 17, 2004.)

    c. "A musical by Shostakovich? The colossal and inscrutable 20th century composer who has come to epitomize the tragic plight of the artist compelled to play a public role in a totalitarian state?" (Source: "Hearing Echoes of Yesterday," *The New York Times,* August 17, 2004.)

    d. "Conformity is the jailer of freedom and the enemy of growth." (Source: quote from President John F. Kennedy.)

3. For each of the following statements, decide whether or not the statement is simple or compound. If the statement is compound, decide whether it is of the form "P and Q," "P or Q," or "if P, then Q," and identify each of the statements P and Q.

    a. My GPA is over 3.0.

    b. Some courses at Kansas State University meet four times a week, while other courses meet three times a week.

    c. If the homework is not handed in on time, I will lose points on the grade.

    d. All students take Writing 100 or mathematics in their first term at college.

    e. She likes to watch soccer and she plays the clarinet.

    f. He will go to the play or he will come to the party.

    g. If Huntington-Hill is used to apportion representatives, then Delaware gets 1 seat and Pennsylvania gets 19 seats.

    h. If the graph of a company's profit over the years 1996 through 2004 is increasing and concave downward, then the rate at which the company's profit increases is decreasing.

4. Give a useful negation of each of the statements in Exploration 2. Make sure your statement is clear and understandable.

5. The truth table given in the text showed that the statement, "not (P or Q)" is logically equivalent to "(not P) and (not Q)." Explain why these two statements are equivalent.

6. Use truth tables to show that the negation of "if P, then Q" is not equivalent to "if P, then (not Q)." Explain how your truth table shows these statements are not equivalent.

7. For each of the if-then statements (conditionals) given next, state the converse and the contrapositive.

   a. If he is guilty, he will be convicted.

   b. If he is convicted, then he must be guilty.

   c. If I study for the exam, then I will get a good grade.

   d. If you want your clothes to be really clean, then use TIDE.

   e. If I run this red light, I will get to my class on time.

8. In each of the following situations, if the hypotheses allow a valid reasoning process, state the conclusion and describe why it is a valid conclusion. If the statements do not fit any valid reasoning process, write "no valid conclusion" and explain why there is no valid conclusion.

   a. In order to drink legally in Pennsylvania, you must be 21 years of age or older. You are not yet 21.

   b. If you use Brand H laundry detergent, your clothes will be "whiter than white." You do not use Brand H laundry detergent.

   c. If you study for at least two hours you will pass the test. You fell asleep and did not study at all.

   d. If a student plays football at this school he cannot play soccer. John does not play soccer.

e. If I can save enough money to afford the trip, I will go to Aruba on spring break. I did not go to Aruba on spring break.

f. If a student plays football at this school, he cannot play soccer. Eric plays football.

9. For each of the following quantified statements, (i) decide whether the statement involves a universal or an existential quantifier, (ii) write the statement using *all* or *there exists,* and (iii) decide whether the statement is true or false (give a reason for your answer).

   a. There is a country that has a name starting with the letter *C.*

   b. In the U.S., every citizen votes in the presidential elections.

   c. No European country has a larger population than the U.S.

   d. Some college students have full-time jobs outside the college.

10. Write the negation of each of the quantified statements in Exploration 9.

11. Explain, using the context of the statement, why the contrapositive statement given in the solution of Example 13.5a must have the same truth value as the original statement, but the converse given in the solution of Example 13.5b might not have the same truth value as the original statement.

TOPIC 14

Apportionment

OBJECTIVES

After completing this topic, you will be able to

- Use and understand the terminology of apportionment
- Recognize several methods of apportionment
- Use several methods of apportionment to decide how many representatives each subgroup of a larger group should have, given a fixed total number of representatives
- Calculate geometric means

The number of representatives from each state in the House of Representatives is proportional to the state's population. This number is updated every ten years when the national census takes place. Throughout the years, several different methods, called methods of apportionment, have been used or proposed to decide how many representatives each state should have. Apportionment methods are useful in any situation in which a small number of delegates or representatives are to be chosen to represent a larger population that consists of separate groups, in a manner consistent with the size of these groups. For example, we could use an apportionment method to decide how many student council members each dorm would elect, in such a way that the number of members is proportional to the size of the dorm. In this topic, we will discuss several apportionment methods.

The American congress consists of the Senate and the House of Representatives. The Senate is composed of two senators from each state. The House, on the other hand, has representatives from all states, but the number of representatives per state varies according to the state's population. This structure was established by the United States Constitution, which includes the following paragraph from Article I, Section 2.

> Representatives and direct taxes shall be apportioned among the several states which may be included within this Union, according to their respective numbers, which shall be determined by adding to the whole number of free persons, including those bound to service for a term of years, and excluding Indians not taxed, three fifth of all other persons. The actual enumeration shall be made within three years after the first meeting of the Congress of the United States, and within every subsequent term of ten years, in such manner as they shall by law direct. The number of representatives shall not exceed one for every thirty thousand, but each state shall have at least one representative...

Since 1790 a national census has been conducted every ten years to determine each state's population. The actual population considered has changed with time. Originally some sectors of the population were not counted, but since the 14th Amendment in 1868 the total emancipated slave population was included, and in 1940 it was determined that no Native American would be classified as "not taxed."

Apportionment is the process of distributing, according to some plan, the number of seats to which each state is entitled in the U.S. House of Representatives, in proportion to that state's population based on the ten-year census. It is also used for other processes of distribution proportional to population. The result of such a process is also called apportionment.

The total number of seats in the House is determined by Congress and has changed from the 65 set by the constitution from 1787 until the first enumeration in 1790, to the current number of 435. It is not possible to have an exact distribution of seats as the following example shows.

Example 14.1

According to the 2000 census, the population of New York state is 19,004,973 and the total population of the U.S. is 281,424,177. Assuming that each state is allotted a number of seats in the House of Representatives in exactly equal proportion to its population, how many of the total 435 seats would correspond to New York? Explain why this is not a practical answer.

Solution

Because the total number of seats is 435, and the total population is 281,424,177, each seat corresponds to $\frac{281{,}424{,}177}{435} \approx 646{,}952.131$ people. To find the number of representatives for New York, we then divide the state's population by the number of people per seat: $\frac{19{,}004{,}973}{646{,}952.131} \approx 29.376$.

The number of seats that correspond to New York is 29.376. Because the number of seats per state has to be a whole number, the number of seats for New York cannot be given by this answer. We will need to round this number up or down to obtain either 29 or 30 seats.

The fact that the number of seats allotted to each state must be an integer number makes apportionment a difficult task. Congress has used various methods of apportionment at different times, each method designed to correct problems that the previously used method had, but introducing a new bias. Mathematically it is not possible to find a "perfect" method, so the choice of method is a combination of mathematics and politics. We will describe the five methods used since 1850, after we introduce some appropriate terminology.

The **standard divisor** is the nation's total population divided by the total number of seats: $\textit{standard divisor} = \frac{\textit{total population}}{\textit{total number of seats}}$. It can be interpreted as the average number of people represented by any one House member. This is the number we calculated in Example 14.1 for the population given by the 2000 census and for the current total number of seats of 435.

A state's **standard quota** (also called the **exact quota**) is the number obtained by dividing the state's population by the standard divisor:

$$\textit{standard quota} = \frac{\textit{state's population}}{\textit{standard divisor}}$$

This is the fraction of the total number of seats that the state is entitled to have. If the standard quota for each state were always an integer, then this would be the allotted number of seats for each state and a fair apportionment would be guaranteed. Because the standard quota is generally a noninteger, it is natural to select the integer just below this number or the integer just above it. However, ordinary round-off does not quite work as the following example shows.

Example 14.2

Suppose the states of Delaware, New Jersey, New York, and Pennsylvania want to form a Middle Atlantic States Council with 15 members, in such a way that representation is proportional to each state's population and each state has at least one representative. The following table gives the population for each of these four states, from data collected in the 2000 census.

State	Population
Delaware	785,068
New Jersey	8,424,354
New York	19,004,973
Pennsylvania	12,300,670

a. Find the standard divisor and the standard quota for each of the given states.

b. Round off each standard quota using the usual round off method. Determine if these quotas can be used to apportion seats for the 15-member council, so that each state has at least one representative.

Solution

a. The total population to be represented on this Council is the sum of the populations of the four states: 785,068 + 8,424,354 + 19,004,973 + 12,300,670 = 40,515,065. Using this number as the total population and 15 as the total number of seats, we compute the standard divisor:

$$\textit{standard divisor} = \frac{40{,}515{,}065}{15} \approx 2{,}701{,}004.333$$

Now we compute each state's standard quota

$$\text{Delaware's } \textit{standard quota} = \frac{785{,}068}{2{,}701{,}004.333} \approx 0.2907$$

$$\text{New Jersey's } \textit{standard quota} = \frac{8{,}424{,}354}{2{,}701{,}004.333} \approx 3.1190$$

$$\text{New York's } \textit{standard quota} = \frac{19{,}004{,}973}{2{,}701{,}004.333} \approx 7.0363$$

$$\text{Pennsylvania's } \textit{standard quota} = \frac{12{,}300{,}670}{2{,}701{,}004.333} \approx 4.5541$$

b. Rounding off to the nearest integer, we get 0, 3, 7, and 5. Because every state must have at least one representative, we cannot assign 0 to Delaware, so we assign 1 seat to Delaware. The following table summarizes our result:

State	Population	Standard Quota	Rounded-Off Quota	Apportionment
Delaware	785,068	0.2907	0	1
New Jersey	8,424,354	3.1190	3	3
New York	19,004,973	7.03626	7	7
Pennsylvania	12,300,670	4.5541	5	5
Total			**15**	**16**

This method would result in one more seat than the 15 seats the council will have. We cannot use this method to apportion the 15 seats in a manner proportional to the population of each state and under the condition that all four states are represented.

Quota Methods

The methods of apportionment used by Congress since 1850 can be classified into two groups: **quota methods** and **divisor methods**. Quota methods work with the standard divisor and round off the standard quotas, according to some rules, while divisor methods work with divisors other than the standard divisor.

Quota methods consist of assigning seats in such a way that the final number corresponding to any given state is either the first integer below its standard quota, called the **lower quota**, or the first integer above its standard quota, called the **upper quota**. Thus if a state's standard quota is, for example, 6.356, then a quota method will assign to this state either 6 or 7 seats. Because the total number of seats is fixed, any apportionment method must allow for some states to receive their lower quota, while other states will receive their upper quota as the final number of seats. Two quota methods are described next.

Hamilton's Method

Hamilton's method consists of assigning the lower quota (except when it is 0) to each state at first and then assigning any remaining seats to the states whose standard quotas have the largest fractional part. Thus to implement this method we follow these steps:

1. Find the standard divisor.
2. Find each state's standard quota.
3. Assign to each state the integer part of its standard quota (this is the lower quota), unless this is 0. If the integer part of the standard quota for any state is 0, assign 1 seat to that state.
4. If there are seats remaining, then, eliminating any state that had 0 as its lower quota, find the state that has the largest fractional part and assign one more seat to this state. If there are still remaining seats, assign the next seat to the state with the next largest fractional part. Continue in this manner until all seats are assigned.

Example 14.3

Suppose that the Middle Atlantic States Council discussed in Example 14.2 will now have 19 members. Use Hamilton's method to apportion the number of seats for each of the four states.

Solution

We will follow the four steps described previously.

1. The standard divisor is the total population of the four states divided by the total number of seats:

$$\textit{standard divisor} = \frac{40{,}515{,}065}{19} \approx 2{,}132{,}371.842$$

2. To find each state's standard quota, we divide the state's population by the standard divisor and obtain these numbers (rounded to 4 decimals):

$$\text{Delaware's } \textit{standard quota} = \frac{785{,}068}{2{,}132{,}371.842} \approx 0.3682$$

$$\text{New Jersey's } \textit{standard quota} = \frac{8{,}424{,}354}{2{,}132{,}371.842} \approx 3.9507$$

$$\text{New York's } \textit{standard quota} = \frac{19{,}004{,}973}{2{,}132{,}371.842} \approx 8.9126$$

$$\text{Pennsylvania's } \textit{standard quota} = \frac{12{,}300{,}670}{2{,}132{,}371.842} \approx 5.7685$$

3. Choosing the lower quota when the lower quota is not 0, and 1 if the lower quota is 0, we first assign Delaware 1, New Jersey 3, New York 8, and Pennsylvania 5.

4. The total number of seats assigned so far is 1 + 3 + 8 + 5 = 17. We have two remaining seats, so we look at the fractional parts of those states with nonzero lower quotas. These fractional parts are as follows: New Jersey 0.9507, New York 0.9126, and Pennsylvania 0.7685. Thus New Jersey, with the largest fractional part, gets the first remaining seat. Because there is another seat remaining, we look for the state with second largest fractional part. The second remaining seat goes to New York.

We summarize the results in the following table.

State	Standard Quota	Integer Part or 1	Fractional Part	Apportionment by Hamilton's Method
Delaware	0.3682	1		1
New Jersey	3.9507	3	0.9507	4
New York	8.9126	8	0.9126	9
Pennsylvania	5.7685	5	0.7685	5
Total		**17**		**19**

Lowndes' Method

The only difference between Lowndes' method and Hamilton's method is that Lowndes' method uses the **relative fractional part** to decide which states get the extra seats. The relative fractional part of a number is the ratio of its fractional part divided by its integer part. For example, the relative fractional part of 3.79 is $\frac{0.79}{3} \approx 0.2633$. To use Lowndes' method, we follow steps 1 through 3 of Hamilton's method and then replace step 4 with the following:

4. If there are any remaining seats, compute the relative fractional parts of the standard quotas (with the exception of those that had a lower quota of 0) and assign one of the remaining seats to the state with the largest relative fractional part. If seats still remain, we assign the next seat to the state with the second largest relative fractional part. We continue in this manner until all seats are assigned.

Example 14.4

Use Lowndes' method to apportion the number of representatives in the 19-member council discussed in Example 14.3.

Solution

We have already computed each state's standard quota and their integer and fractional parts (see the table at the end of Example 14.3). Now we calculate each state's relative fractional part for all states other than Delaware, which has the lower quota of 0 and automatically gets 1 representative. The relative fractional parts are as follows:

New Jersey's *relative fractional part* $= \frac{0.9507}{3} \approx 0.3169$

New York's *relative fractional part* $= \frac{0.9126}{8} \approx 0.1141$

Pennsylvania's *relative fractional part* $= \frac{0.7685}{5} \approx 0.1537$

The state with the largest relative fractional part is New Jersey, and Pennsylvania has the second largest relative fractional part, so they each get one additional representative. The apportionment is given in the last column of the following table:

State	Standard Quota	Integer Part or 1	Fractional Part	Relative Fractional Part	Apportionment by Lowndes' Method
Delaware	0.3682	1			1
New Jersey	3.9507	3	0.9507	0.3169	4
New York	8.9126	8	0.9126	0.1141	8
Pennsylvania	5.7685	5	0.7685	0.1537	6

The use of the relative fractional part instead of the fractional part gives smaller states a slight advantage over larger states because if two states have equal fractional parts, the relative fractional part of the state with the smaller standard quota will be larger. (Consider two states with standard quotas of 3.6781 and 7.6781. Their relative fractional parts are $\frac{0.6781}{3} \approx 0.2260$ and $\frac{0.6781}{7} \approx 0.0969$, respectively.)

Divisor Methods

Divisor methods use **modified divisors** instead of the standard divisor. The quotient of a state's population divided by a modified divisor is called the state's **modified quota**. Each divisor method uses a different procedure to round a state's modified quota, and the modified divisor is chosen in such a way that when the modified quotas are computed and rounded off to an integer value, the total number of resulting seats is exactly the number of seats available.

We describe four divisor methods: Jefferson's method, Adam's method, Webster's method, and the Huntington-Hill method.

Jefferson's Method

Jefferson's method is also known as Vinton's method, or the method of greatest divisors. The method consists of finding a (modified) divisor D so that when each state's modified quota is *rounded down* and all are added, then the total number is exactly the number of seats to be allotted. For example, if a state's population is 12,005,678 and the modified divisor is $D = 1{,}789{,}000$, then this state's modified quota is $\frac{12{,}005{,}678}{1{,}789{,}000} \approx 6.71$, so the number of seats allotted to this state would be 6, which is 6.71 rounded down.

Because we do not know in advance which divisor D will work, we might have to try a few values of D until we find the right one. We illustrate this method with an example.

Example 14.5

Use Jefferson's method to apportion the 19 seats of the Middle Atlantic Council discussed in Examples 14.3 and 14.4.

Solution

In Example 14.3, we computed each state's standard quota using the standard divisor $= \frac{40{,}515{,}065}{19} \approx 2{,}132{,}371.842$ and obtained the values in the following table.

State	Population	Standard Quota
Delaware	785,068	0.3682
New Jersey	8,424,354	3.9507
New York	19,004,973	8.9126
Pennsylvania	12,300,670	5.7685
Total	**40,515,065**	

If we round down each standard quota, with the exception of Delaware's, which we need to set as 1, we have Delaware 1, New Jersey 3, New York 8, and Pennsylvania 5. This gives us a total of 17 while the total number of seats is 19. This tells us that the standard divisor cannot be the modified divisor, and it gives us an idea of which divisors to try. Because using the standard divisor leads to too few seats allotted, we need to try a divisor that will give larger modified quotas. The population of each state does not change with the choice of divisor, so the way to get larger modified quotas is to divide by a smaller number. Therefore, we should try a number less than the standard divisor of 2,132,371.842 as our first guess for modified divisor. Let's try $D = 2{,}000{,}000$. Then the modified quotas for each state are as follows:

Delaware's *modified quota* $= \frac{785{,}068}{2{,}000{,}000} \approx 0.393$

New Jersey's *modified quota* $= \frac{8{,}424{,}354}{2{,}000{,}000} \approx 4.212$

New York's *modified quota* ≈ 9.502

Pennsylvania's *modified quota* ≈ 6.150

Rounding down the modified quotas (with the exception of Delaware that has to get at least one seat), we get 1, 4, 9, and 6. These add up to 20, which is too large. Now we know that the divisor must be greater than 2,000,000 and less than 2,132,371.842. If we try D = 2,100,000, the modified quotas are 0.374, 4.012, 9.050, and 5.857, respectively. Rounding down all but Delaware's, we obtain 1, 4, 9, and 5. Because their sum is 19, we have found the modified divisor $D = 2{,}100{,}000$ that allows us to allot 1 seat to Delaware, 4 to New Jersey, 9 to New York, and 5 to Pennsylvania.

The following table summarizes our work:

State	Std. Quota (D=2,132,371.842)	Std. Quota Rounded	Modified Quota (D= 2,000,000)	Modified Quota Rounded	Modified Quota (D= 2,100,000)	Number of Seats by Jefferson's Method
Delaware	0.3682	1	0.393	1	0.374	1
New Jersey	3.9507	3	4.212	4	4.012	4
New York	8.9126	8	9.502	9	9.050	9
Pennsylvania	5.7685	5	6.150	6	5.857	5
Total		**17**		**20**		**19**

As discussed in Example 14.5, the modified divisor that results when applying Jefferson's method is a number less than the standard divisor; this smaller modified divisor produces greater modified quotas. This explains the alternative name for this method, **method of greatest divisors**.

Adam's Method

Similar to Jefferson's method, Adam's method uses the same idea of finding a modified divisor that would lead to the total number of seats. The only difference between these two methods is that in Adam's method the modified quotas are *rounded up*. This means that the modified divisor will be greater than the standard divisor and will produce smaller modified quotas. Adam's method is also called the **method of smallest divisors**.

Webster's Method

Webster's method is another divisor method, also called the **method of major fractions**. In this method, modified quotas are rounded off to the nearest integer in the standard way. (That is, if its fractional part is 0.5 or larger, the modified quota is rounded up; otherwise it is rounded down.) The divisor D is chosen so that the modified quotas, when rounded to the nearest integer, add up to the total number of seats.

Huntington-Hill Method

The Huntington-Hill method, also called the **method of equal proportions**, is the method currently used by Congress. In this method, the modified quotas are rounded to an integer value using the **geometric mean**. The geometric mean of two numbers a and b is the square root of their product,

$$\text{geometric mean of } a \text{ and } b = \sqrt{a * b}$$

If q is a modified quota, and q is between the integers n and $n + 1$, then the Huntington-Hill method uses the geometric mean of n and $n + 1$ as the cutoff point to decide whether q will be rounded up to $n + 1$ or rounded down to n. That is, if $q \leq \sqrt{n * (n + 1)}$, then q is rounded down to n, and if $q > \sqrt{n * (n + 1)}$, then q is rounded up to $n + 1$.

Example 14.6

Suppose the Huntington-Hill method is used to apportion the 19 seats of the council discussed in Example 14.5. Will the modified divisor $D = 2{,}100{,}000$ result in a correct apportionment of the 19 seats? If the answer is no, would a correct divisor be greater than or less than D?

Solution

In Example 14.5, we computed the modified quotas corresponding to the divisor $D = 2{,}100{,}000$, which are as follows: Delaware, 0.374, New Jersey, 4.012, New York, 9.050, and Pennsylvania, 5.857. Because a state cannot have 0 seats, Delaware will get 1 seat (independently of the geometric mean). To see how many seats New Jersey will get, we compute the geometric mean of 4 and 5 since 4.012 is between 4 and 5. This geometric mean is $\sqrt{4 * 5} \approx 4.472$. Because New Jersey's modified quota is 4.012, and $4.012 < 4.472$, the number of seats for New Jersey is rounded down to 4. Similarly, we compute the corresponding geometric means for the remaining states, obtaining $\sqrt{9 * 10} \approx 9.487$ for New York and $\sqrt{5 * 6} \approx 5.477$

for Pennsylvania; we compare each of the geometric means with the modified quota of that state. Because the modified quota for New York is 9.050, and $9.050 < 9.487$, New York gets 9 seats. Because Pennsylvania's modified quota is 5.857, and $5.857 > 5.477$, Pennsylvania will get 6 seats. The total number of seats, using this method and the data given, is $1 + 4 + 9 + 6 = 20$. So the modified divisor D = 2,100,000 does not give an apportionment totaling 19 seats. To get smaller modified quotas, we need to use a divisor that is greater than the standard divisor. We explore this method further in Activity 14.2.

Summary

In this topic, we analyzed six methods of apportionment. Two of the methods, Hamilton's and Lowndes', are quota methods that use the standard divisor to obtain the standard quotas. The other four methods discussed use modified divisors and are Jefferson's method, Adam's method, Webster's method, and the Huntington-Hill method.

Explorations

1. Consider Lowndes' method for apportioning representatives.
   a. Describe this method.
   b. Does this method favor larger or smaller states?
   c. Explain in detail your answer to part b of this exploration and how it works mathematically.
2. For each of the following divisor methods, indicate whether the divisor is greater than or less than the standard divisor and explain why.
   a. Adam's method
   b. Jefferson's method
   c. Webster's method
   d. Huntington-Hill method

of the three rates of increase given is $\frac{5+2+3}{3} \approx 3.33$. We assume that, in general, Sam's expenses have increased by approximately 3.33 percent. Because $35{,}000 + 0.0333 * 35{,}000 + 6{,}000 = 42{,}165.50$, Sam should try to negotiate a salary of no less than $42,165.50.

b. We used problem-solving technique 8. *Generalize* when we estimated the effect of inflation overall, based on knowing the rate of inflation on three items.

Example 15.4

Suppose you win $20,000,000 in the lottery. You will receive your prize in 20 yearly installments of $1,000,000 each and will pay a 35 percent tax per year, so your net income will be $650,000 each year. Because of inflation, the value of the $650,000 will be, in today's dollars, less each year.

a. Assuming inflation will be 2 percent per year during the next 20 years, find the value in today's dollars of $650,000 after 20 years. Identify the problem-solving techniques you used to answer this question, and explain how you used them.

b. Use problem-solving technique 9. *Look at a related problem (sometimes simpler) or look for analogies,* to estimate the value, in today's dollars, of the total amount received after 20 years.

Solution

a. If v_1 is the value in today's dollars of the $650,000 installment after one year, and inflation is 2 percent, then $v_1 + 0.02v_1 = 650{,}000$. So, $v_1 \cdot (1 + 0.02) = 650{,}000$ or $v_1 \cdot (1.02) = 650{,}000$. Solving for v_1 we see that the value in today's dollars of the $650,000 installment after one year is $v_1 = \frac{650{,}000}{1.02}$. The following year the value in today's dollars will be

$$v_2 = \frac{v_1}{1.02} = \frac{\left(\frac{650{,}000}{1.02}\right)}{1.02} = \frac{650{,}000}{(1.02)^2}$$

At the end of the third year, the value in today's dollars of $650,000 will be

$$\frac{\left(\frac{650{,}000}{(1.02)^2}\right)}{1.02} = \frac{650{,}000}{(1.02)^3}$$

We now see a pattern: to find the value in today's dollars of $650,000 after 20 years, we calculate $\frac{650{,}000}{(1.02)^{20}} \approx 437{,}431.37$ and find that the value of the installment, in today's dollars, will be approximately $437,431.37. We used problem-solving technique 3. *Examine a simple case or try several special cases,* when we looked for the value after one, two, and three years. These special cases helped us see a pattern, so we also used problem-solving technique 6. *Look for a pattern.*

b. To find the value, in today's dollars, of the total amount received after 20 years we would need to find the value, in today's dollars, of $650,000 for each of the next 20 years, and then add those values. We will instead solve a simpler similar problem: assume that the installment received each year is always equivalent to $540,000 (this is roughly the average between the current value and the value after 20 years). To solve this simpler problem, we need find only the total of 20 payments of $540,000 each, which is 20 * ($540,000) = $10,800,000.

In the following example, we identify problem-solving techniques used in examples from previous topics.

Example 15.5

Identify the problem-solving techniques (from the list given at the beginning of this topic) that were used in each of the following examples from previous topics.

a. Example 7.4, page 125

b. Example 9.11b, page 164

c. Example 11.5a, page 190

d. Example 14.6, page 251

Solution

a. In Example 7.4, we made a table, so we used problem-solving technique *2a. Make a table or chart.* The table contained special values, which shows we also used problem-solving technique *3. Examine a simple case or try several special cases.* We used the table to look for a pattern, using problem-solving technique *6. Look for a pattern.* The pattern we found allowed us to generalize the results to a general formula. This last step used problem-solving technique *8. Generalize.*

b. In Example 9.11b, we found the additional interest accumulated for a credit card account in six months by first solving the related problem: "Find the total debt after that period." For this, we used problem-solving technique *9. Look at a related problem (sometimes simpler) or look for analogies.* To solve the related problem, we used problem-solving technique *2c. Use an equation or formula.*

c. In Example 11.5a, we used the cutoff screening method to decide which of four digital cameras we would buy. After considering which information was relevant (problem-solving technique 1) and making a table (technique 2a), we used indirect reasoning (technique 10). For example, we used indirect reasoning when we eliminated the Sony model because its print quality was 6 and we wanted a camera with a print quality 9 or better. The reasoning is this: if we choose a camera, then the camera must have print quality of 9 or better. Sony does not have a print quality of 9 or better, so we do not choose it.

d. Example 14.6 uses the Huntington-Hill method of apportionment. In this method we used technique *7. Guess (estimate) and check.* We guessed a modified divisor and then checked if the apportionment it gave was correct for the total number of seats available.

Most "real-life" problems require the use of several problem-solving techniques to solve them. The following example shows a situation in which we can use several of the techniques listed at the beginning of this topic.

Example 15.6

Suppose you are a homeowner and need to decide which of three options for purchasing house heating oil would be best for you. The company offers two plans: (1) the "Buy-now Plan" at $1.49 per gallon, in which you pay for oil up front for the amount you estimate you will use; (2) the "Economy Plan" at $1.59 maximum per gallon, in which you pay monthly payments spread out over 12 months for the estimated amount of oil you will need (same as last year's usage). According to the company, this plan is best for the consumer. If the market price drops, your price will also drop, and if the market price rises, your price will not go above $1.59. You may choose not to select one of the two plans so another option is (3) no special plan, so you pay for the amount of oil used as it is delivered and pay the price at the time of delivery. Explain which problem-solving techniques, from the list given in this topic, would help you decide which plan to use and give the order in which you would use those techniques.

Solution

First, we need to decide what information is relevant; some of the relevant information is given here and other information we would need to collect. Relevant information includes the description of the company's different plans, the records of amount of oil delivered to the household, and when it was delivered throughout the last year. (Because this will be the estimated amount of oil used in the first two plans, we need to know the total, but we may also need to know more about the pattern of oil expenditure in this household.) We also need information on oil prices throughout the last two years (to make an educated guess of what prices we should expect).

Next we divide the problem into two different sub-problems: (1) an analysis of the pattern of oil consumption throughout the year (or last two years, if possible); (2) an analysis of the pattern of oil prices for the last one or two years. To perform each of these two pattern analyses, we use tables and graphs. We also use generalization when, based on the pattern for the last one or two years, we assume (predict) that the pattern will continue.

We look for analogies by looking for similarities between last year's and the previous year's pattern. Finally, we examine three different cases (one for each of the company's plans).

In summary, by proceeding in the way described, we would be using the following techniques from the list: *1. Decide what information is relevant*; *2. Represent the information in a different form*; *3. Examine a simple case or try several special cases*; *8. Generalize*; *4. Break a problem into smaller problems or identify a sub-goal or sub-problems*; and *9. Look for a related problem or look for analogies.*

Summary

In this topic, we reviewed the list of problem-solving techniques discussed in Topic 10 and discussed these new ones: *7. Guess (estimate) and check*; *8. Generalize*; *9. Look at a related problem (sometimes simpler) or look for analogies*; and *10. Use indirect reasoning.* We solved some simple problems and identified the problem-solving techniques used, and we developed a strategy to solve a more involved problem, identifying the problem-solving techniques we could use to solve it.

Explorations

1. For each of the problem-solving techniques 7 through 10 in the list given in this topic, find an example of an activity you did in this course that used the technique.
2. For each of the following explorations, state all the techniques you used or you would use to solve it. Explain briefly how you used or would use each technique.
   a. Topic 5, Exploration 4, page 99
   b. Topic 8, Exploration 9, page 148
   c. Topic 11, Exploration 7, page 199
   d. Topic 14, Exploration 4g, page 254
3. On April 13, 2005, the current U.S. public debt (which includes savings bonds, treasury notes, and government securities) was $7,792,607,796,216. The government is paying interest on this debt at various interest rates. Suppose the average interest rate on the different types of securities is 3 percent per year, compounded monthly, and assume that the government does not

sell any securities for the rest of the year. How much should the government pay each month from May to December so the debt is less than 7 trillion dollars? Identify the problem-solving techniques you used. (Source: *Bureau of the Public Debt,* www.publicdebt.treas.gov.)

4. Tim is planning to open a savings account to deposit $7,000 he collected from his summer job. He finds two banks that offer interest of 2.4 percent per year. At the first bank interest is compounded quarterly, and at the second bank interest is compounded daily. Tim will not make any withdrawals during the next two years. By how much will the amounts in the two accounts differ at the end of two years? Identify the problem-solving techniques you used to answer the question.

5. Refer to Exploration 3 of Topic 10. Solve Marian's problem. Did you use any of the additional problem-solving techniques introduced in this topic?

6. Suppose you open a savings account at an interest rate of 1.5 percent per year, compounded monthly. You make an initial deposit of $1,000 and, starting one month after you open the account, you deposit $100 each month and do not make any withdrawals. Identify which problem-solving techniques you would use to answer the question: How long would it take for the account to reach $3,000? Explain how you would use those techniques.

7. Use the techniques identified in Exploration 6 of this topic to answer the question raised there.

8. Suppose you won $20,000,000 in the lottery. You may choose to receive the prize in a lump sum now and pay 50 percent of it in taxes or in 20 yearly installments of $1,000,000 paying 35 percent tax per year. If you choose to receive the prize in a lump sum, you will invest it in an account and expect a return of 5 percent per year. You will make yearly withdrawals to cover all your expenses (living expenses and donations to charity and family and friends). If you decide to receive the prize in installments, you think that after covering all your expenses, you will have a reasonable amount to invest and expect a return of 4.5 percent per year. Explain which problem-solving techniques, from the list given in this topic, would help you decide which of the two options would give you the largest amount of savings at the end of 20 years, and list the order in which you would use those techniques.

TOPIC 16

Averages and Five-Number Summary

OBJECTIVES

After completing this topic, you will be able to

- » Compute the mean (or average), median, and mode of a data set
- » Use measures of center to analyze and compare data sets
- » Find and interpret numerical measures of spread including the range and quartiles
- » Use the five-number summary and a boxplot to analyze data

Suppose you scored 82 points on the first assignment and 75 points on the second assignment of the same course. If both were graded on a scale of 100 points, on which of the two assignments did you do better? Because your grade for the first assignment is higher, you would probably say you did better on the first one. But did you really do better on the first assignment? Suppose that the average grade on the first assignment was 85 points and the average grade on the second one was 67, would this change your answer? Suppose you learned that half of the students in the class scored above 80 on the first assignment, while on the second assignment only one student scored above 69. Wouldn't you think that the second assignment was much harder than the first and that you did pretty well on it? In this topic, we discuss measures of center and spread that can be used to compare different sets of data.

Pictures of a data distribution give us a quick view of the distribution, but we often need to use a single number to summarize a collection of data or to compare two or more collections of data. For example, in Topic 1, Exploration 6, we looked at the average verbal SAT test scores for high school seniors for each state in the U.S. We were interested in using a *single* verbal SAT number (the average) to associate with each of the states.

The three commonly used measures of center are mean, median, and mode. The **mean** is the arithmetic average, commonly called the **average**, and is found by adding all the data values and then dividing that sum by the *number* of data values. If we have n data values, denoted $x_1, x_2, \ldots, x_n$, then the mean is: $\frac{x_1 + x_2 + \ldots + x_n}{n}$. The **median** is the middle observation in an ordered list of data. The median is found by first putting the data values in numerical order. If there is an odd number n of data values, the median is the middle data value, the data value in *position* $\frac{n+1}{2}$ in the ordered list. If there is an even number n of data values, the median is the mean of the two middle values, that is, the arithmetic average of the data values in *positions* $\frac{n}{2}$ and $\frac{n}{2} + 1$. The **mode** is the most frequently occurring data value; that is, it is the data value with the highest frequency. If every data value has the same frequency, then a data set has no mode. If there are two values of the variable that occur the same number of times in the data set and more frequently than the other data values, then a distribution is **bimodal**. The mode is appropriate for categorical data as well as for numerical data, while the mean and median can only be calculated for numerical data.

More important than knowing how to compute these measures of center is understanding how to use and interpret them by knowing their properties. In the following example, we compute measures of center and analyze how they are affected when one data value is changed.

Example 16.1

The following table gives the number of hazardous waste sites on the U.S. National Priority List that are in each of 15 centrally located states, as of September 20, 2002. (Source: *Environmental Protection Agency,* www.epa.gov/superfund/sites/npl/npl.html.)

State	Number of Sites
Colorado	15
Idaho	6
Illinois	39
Indiana	28
Iowa	13
Kansas	10
Minnesota	24
Missouri	23
Nebraska	10
North Dakota	0
Ohio	29
Oklahoma	10
South Dakota	2
Utah	16
Wisconsin	39

a. Find the mean, median, and mode of the number of sites and explain what these measures of center show about this data set.

b. Suppose the number of sites in Wisconsin was incorrectly recorded as 239 (instead of 39). How do the mean, median, and mode of the data set change?

c. Find how many of the observed data values (called **observations**) fall above the mean and median in each of parts a and b of this example and comment on these results.

Solution

a. To compute the mean, we add all the data values and divide by the number of data values, which is 15. The mean is

$$\frac{\textit{sum of \# of sites}}{15} = \frac{15+6+39+28+13+10+24+23+10+0+29+10+2+16+39}{15} = \frac{264}{15} = 17.6$$

To compute the median, we need to order the data. The following table shows the data values in increasing order. Because there are an odd number of data values (15), the median is the middle value, that is, the value in position $\frac{n+1}{2} = \frac{16}{2} = 8$. Counting eight data values from the smallest data value shows that the median is 15.

State	Number of Sites
North Dakota	0
South Dakota	2
Idaho	6
Kansas	10
Nebraska	10
Oklahoma	10
Iowa	13
Colorado	15
Utah	16
Missouri	23
Minnesota	24
Indiana	28
Ohio	29
Illinois	39
Wisconsin	39

Because the data value 10 occurs three times and is the most frequently occurring data value, it is the mode of this data set. The median, 15, is a bit smaller than the mean, 17.6, but fairly close to it.

b. If the largest data value were 239 instead of 39 the median and mode would still be 15 and 10, respectively; however, the mean $= \frac{\textit{sum of \# of sites}}{15}$ $= \frac{464}{15} \approx 30.93$. Here the median and mode are unchanged, but the mean is greatly affected by the one (incorrect) value.

c. In part a of this example, seven data values lie below the median and seven lie above it. Nine of the data values lie below the mean and six lie above it. In part b of this example, it is still true that seven data values lie on either side of the median, but only two data values (39 and 239) lie above the mean of 30.93; thirteen lie below the mean.

In the hypothetical situation created in Example 16.1b, 239 would be an **outlier**, which is a data value that is outside the general pattern of the data. It is important to identify outliers and why they occur, if possible. The mean is more affected by outliers than are the other measures of center, as Example 16.1 shows.

If the distribution of a set of data is **symmetric**, that is, a histogram or a stemplot of the data looks the same to the left and to the right of the "center," then the mean and median will be close together. If the distribution of a data set is **skewed**, that is, if on the histogram of the data, one tail end is longer than the other, then the mean will be farther out on the long tail than will be the median. (Note that in Example 16.1b, with the value 239 in the data set, the data values are skewed toward larger values and the mean is considerably greater than the median.) Here are two histograms, one of a hypothetical set of freshman SAT scores that is approximately symmetric and the other of another hypothetical set of freshman SAT scores that is skewed to the right, or positively skewed.

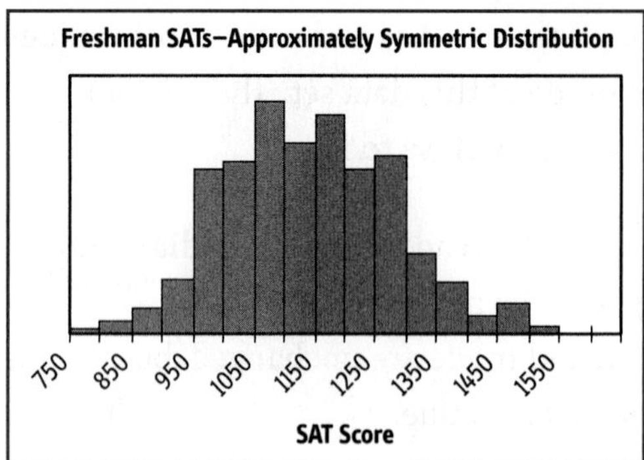

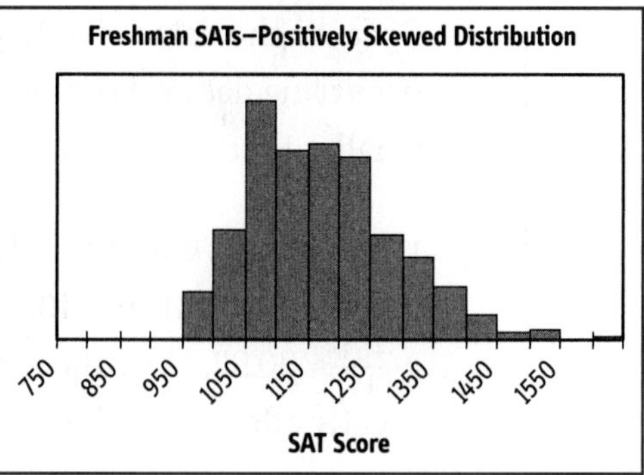

Measures of center give us a single number (or category in the case of categorical data and the mode) to summarize a data set. But we often need more information about a data set than a single value can convey. In the following examples, we look at the relationship between the mean, median, and mode and the stemplot.

Example 16.2

Create a stemplot of the hazardous waste-site data given in Example 16.1. Explain where the mean, median, and mode fall on this plot.

Solution

Because the data values go from 0 to 39, the digit in the tens place will be the stem and the digit in the units place will be the leaf. The stemplot appears next.

Number of Waste Sites (stem = tens digit; leaf = units digit)

Stem	Leaves
0	0 2 6
1	0 0 0 3 5 6
2	3 4 8 9
3	9 9

If we picture a smooth curve drawn along the right edge of the leaves, we can see that the curve has a peak in the 10s and appears to be fairly symmetric. Both the mean and the median fall on the 10s stem. From the stemplot, we can see that the mode is 10. (It is the value that occurs three times.) We can also see that the data in the stemplot are in numerical order; it is straightforward to find the median of such an ordered data set.

Example 16.3

The following table shows the 2003 annual salary, rounded to the nearest dollar, for each of the U.S. state governors. (Source: *The World Almanac and Book of Facts 2004,* p. 69.)

State	Salary ($)	State	Salary ($)	State	Salary ($)
Alabama	96,361	Louisiana	95,000	Ohio	122,800
Alaska	85,776	Maine	70,000	Oklahoma	110,299
Arizona	95,000	Maryland	135,000	Oregon	93,600
Arkansas	75,296	Massachusetts	135,000	Pennsylvania	142,142
California	175,000	Michigan	177,000	Rhode Island	105,000
Colorado	90,000	Minnesota	114,506	South Carolina	106,178
Connecticut	150,000	Mississippi	122,160	South Dakota	98,250
Delaware	114,000	Missouri	120,087	Tennessee	85,000
Florida	124,575	Montana	93,089	Texas	115,345
Georgia	127,303	Nebraska	85,000	Utah	100,600
Hawaii	94,780	Nevada	117,000	Vermont	127,456
Idaho	98,500	New Hampshire	100,690	Virginia	124,855
Illinois	150,691	New Jersey	157,000	Washington	142,286
Indiana	95,000	New Mexico	110,000	West Virginia	90,000
Iowa	104,795	New York	179,000	Wisconsin	131,768
Kansas	98,331	North Carolina	118,430	Wyoming	105,000
Kentucky	107,130	North Dakota	87,216		

a. Find the mean, median, and mode of these data.

b. Create a stemplot of the data.

c. Could you obtain the mean, median, and mode from the stemplot created? Explain why or why not.

Solution

a. The mean is the sum of all the salaries divided by 50, which is \$113,985.90. In order to find the median, we need to sort the data in order. The next table shows the data sorted by increasing salary.

State	Salary (\$)	State	Salary (\$)	State	Salary (\$)
Maine	70,000	Idaho	98,500	Ohio	122,800
Arkansas	75,296	Utah	100,600	Florida	124,575
Nebraska	85,000	New Hampshire	100,690	Virginia	124,855
Tennessee	85,000	Iowa	104,795	Georgia	127,303
Alaska	85,776	Rhode Island	105,000	Vermont	127,456
North Dakota	87,216	Wyoming	105,000	Wisconsin	131,768
Colorado	90,000	South Carolina	106,178	Maryland	135,000
West Virginia	90,000	Kentucky	107,130	Massachusetts	135,000
Montana	93,089	New Mexico	110,000	Pennsylvania	142,142
Oregon	93,600	Oklahoma	110,299	Washington	142,286
Hawaii	94,780	Delaware	114,000	Connecticut	150,000
Arizona	95,000	Minnesota	114,506	Illinois	150,691
Indiana	95,000	Texas	115,345	New Jersey	157,000
Louisiana	95,000	Nevada	117,000	California	175,000
Alabama	96,361	North Carolina	118,430	Michigan	177,000
South Dakota	98,250	Missouri	120,087	New York	179,000
Kansas	98,331	Mississippi	122,160		

From the table, we can see that the salaries range from \$70,000 to \$179,000. Because there are 50 data values, the median is the arithmetic average of the data values in positions $\frac{50}{2} = 25$ and $\frac{50}{2} + 1 = 26$. Thus, the median is $\frac{\$107{,}130 + \$110{,}000}{2} = \$108{,}565$. The data value \$95,000 occurs three times, and no other data value occurs that often or more often, so the mode is \$95,000.

b. We have several choices of how to split the data into a stem portion and a leaf portion. It looks as though the stemplot will have enough stems, but not too many if we create the stemplot with the digit in the ten thousands place as the stem and then use the digit in the thousands place as the leaf. To do this, we choose to truncate each salary and look at salary in thousands. By truncating the data, we simplify the stemplot and still retain approximate values of the data. The resulting stemplot is shown here. (Note that we no longer have the exact values of the data in the stemplot.)

U.S. State Governors' Salaries (stem = ten thousands place; leaf = thousands place)

Stem	Leaves
7	0 5
8	5 5 5 7
9	0 0 3 3 4 5 5 5 6 8 8 8
10	0 0 4 5 5 6 7
11	0 0 4 4 5 7 8
12	0 2 2 4 4 7 7
13	1 5 5
14	2 2
15	0 0 7
16	
17	5 7 9

c. We could not get the exact mean, median, or mode using the data in this form because we do not have the exact data values but only approximations.

A histogram of the data on governors' salaries (see Example 16.3), which appears next, shows approximately the same shape that the stemplot shows. Both pictures of the data show a longer tail toward the larger values; the distribution is skewed to the right. We expect a distribution that is skewed to the right to have a larger mean than median because the mean is "pulled up" by the larger data values on the right tail. This is the case with our data. (Recall that the mean of this data set is $113,985.90, which is greater than the median of $108,565.)

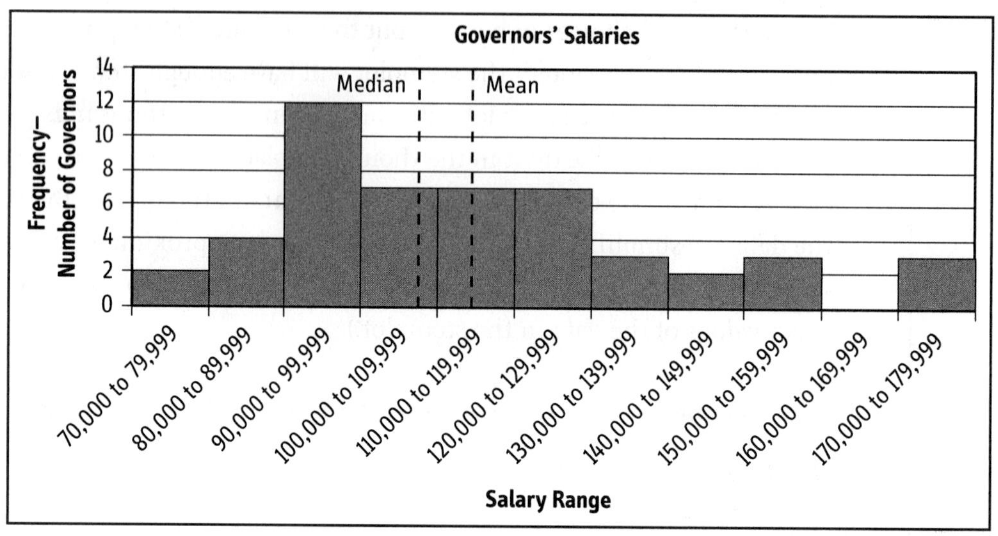

We can picture the median of a data set as the value on the horizontal axis for which half of the total area of the histogram lies to the left of the value and half lies to the right of the value.

The mean can be visualized by thinking about a playground ride. Picture the histogram balanced on a seesaw. The mean is the position of the center of the seesaw along the horizontal axis of the histogram when it is perfectly balanced.

Because a measure of center alone does not give enough information, we often need an additional numerical measure that gives us an idea of how spread out the data are. We can get a visual measure of the spread of a set of data by looking at a graph of the data, but there are three commonly used **numerical measures of the spread** of a data set: the range, the interquartile range, and the standard deviation. We will look at the first two of these in this topic and the third in Topic 17. The **range** of a data set is the difference between the maximum (the largest) data value and the minimum (the smallest) data value. The **interquartile range** is the length of the middle half of the data and is found by first identifying the first and third quartiles of the data.

To find the first quartile and the third quartile, we order the data values in increasing order, find the median, and divide the data into a lower half and an upper half. If n is odd, the median is not in either the lower or the upper half. The **first quartile, Q_1**, is the median of the lower half of the data values. The **third quartile, Q_3**, is the median of the upper half of the data values. So finding the first and third quartiles involves finding two additional medians. The interquartile range, denoted **IQR**, is the difference $Q_3 - Q_1$.

We can list the first and third quartiles, together with the median and the minimum and maximum data values, to give the **five-number summary** of a set of data. The five-number summary gives us a quick view of both the center and the spread of a data set and "divides" the data set into four parts; approximately one-fourth of the data values are between the minimum and Q_1, approximately one-fourth of the data values are between Q_1 and the median, approximately one-fourth of the data values are between the median and Q_3, and approximately one-fourth are between Q_3 and the maximum. So the quartiles are appropriately named because they divide the data, roughly, into quarters. We present the five-number summary of a data set as the ordered list of the following five values: minimum data value, Q_1, median, Q_3, maximum data value.

Example 16.4

a. Find the five-number summary, the range, and the interquartile range for the hazardous waste-site data from Example 16.1, and explain what the five-number summary shows about the data set.

b. Repeat part a of this example for the governors' salaries data set from Example 16.3.

Solution

a. The hazardous waste-site data set contains 15 observations. Thus, as we noted before, the data value in position 8 in the ordered list is the median. To get Q_1, we need to find the median of the data values in positions 1 through 7. The median of these seven data values is the data value in position 4; so $Q_1 = 10$. Similarly, Q_3 is the data value in position 12. Thus, $Q_3 = 28$. The five-number summary for this data set is: 0, 10, 15, 28, 39. The range is $39 - 0 = 39$, and the interquartile range is $28 - 10 = 18$. The five-number summary shows that there are roughly 10 units between the minimum and Q_1 and approximately the same between Q_3 and the maximum. The distance from Q_1 to the median is 5, which is less than the distance of 13 from the median to Q_3. These distances imply that the data are slightly skewed to the right.

b. The governors' salaries data set has 50 observations. The median is the average of the data values in positions 25 and 26 in the ordered list; thus, the median is \$108,565. The first quartile is the median of data values less than the median, so it is the median of the data values in positions 1 through 25. Thus Q_1 is the data value in position 13, and Q_1 = \$95,000. Similarly, Q_3 is the median of the data values greater than the median, so it is the median of data values in positions 26 through 50. Therefore, Q_3 is the data value in position 38, and Q_3 = \$127,303. The five-number summary for this data set is: \$70,000, \$95,000, \$108,565, \$127,303, \$179,000. The range is \$179,000 – \$70,000 = \$109,000, while the interquartile range is \$127,303 – \$95,000 = \$32,303. The five-number summary shows that there is a gap of \$25,000 between the minimum and the first quartile and an even greater gap (over \$50,000) between the third quartile and the maximum. Similarly, there is a larger gap between the median and the third quartile than there is between the first quartile and the median. This is consistent with a data set that is skewed to the right.

We can sketch a graph of the five-number summary of a data set, called a **boxplot**, by creating a horizontal (or vertical) number line that spans an interval just a bit larger than the interval from the minimum to the maximum of the data set. We put a short mark perpendicular to the number line and slightly above it (or to the right of it, if we use a vertical number line) where each of the numbers in the five-number summary is located. We then draw a box from the first quartile to the third quartile and draw lines to the minimum and maximum as shown in the boxplot of the hazardous waste-site data pictured here.

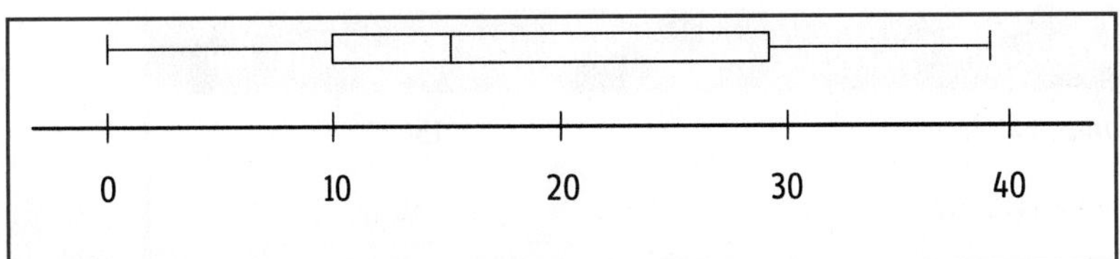

Number of Hazardous Waste Sites in 15 Centrally Located States

We can use two boxplots, sketched next to one another, or one above the other, to compare two data sets.

Example 16.5

Brownies and ice cream bars are two popular desserts. Sketch side-by-side boxplots of the following two data sets that give the calorie content of different brands of brownies and ice cream bars. What do the boxplots show? (Source: *The Corinne T. Netzer Encyclopedia of Food Values.*)

Brownie, Prepared According to Package Directions	Calories in One Piece
Pillsbury caramel fudge chunk	170
Betty Crocker caramel swirl	120
Betty Crocker German chocolate	160
Nestle double chocolate chip	150
Pepperidge Farm hot fudge	400
Duncan Hines milk chocolate	160
Betty Crocker microwave frosted	180
Robin Hood/Gold Medal pouch mix	100
Duncan Hines peanut butter	150
Pillsbury triple, chunky	170

Ice Cream Bar, One Piece	Calories per Bar
Good Humor Fat Frog	154
Good Humor Halo Bar	230
Heath	170
Haagen-Dazs, caramel almond	230
Good Humor chip candy crunch	255
Nestle premium milk choc w/almonds	230
Haagen-Dazs vanilla w/dk choc coating	390
Nestle crunch vanilla w/wh choc coating	350
Klondike Krispy	290
Oh, Henry vanilla with choc coating	320

Solution

We order the data in each of the sets to find the five-number summary.

Brownie	100	120	150	150	160	160	170	170	180	400

Ice Cream	154	170	230	230	230	255	290	320	350	390

For the calories-of-brownies data, the five-number summary is: 100, 150, 160, 170, 400.

For the calories-of-ice cream bars data, the five-number summary is: 154, 230, 242.5, 320, 390.

The following two boxplots show that the brownie calories have a larger range but a much smaller interquartile range. The middle half of the brownie-calorie data lies below the first quartile of the ice cream bar–calorie data set. The minimum value of the ice cream data lies at approximately the same point as the first

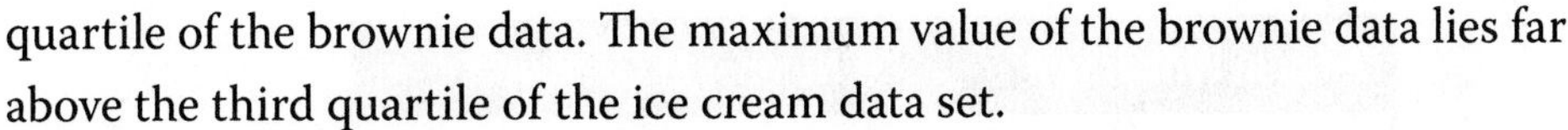
quartile of the brownie data. The maximum value of the brownie data lies far above the third quartile of the ice cream data set.

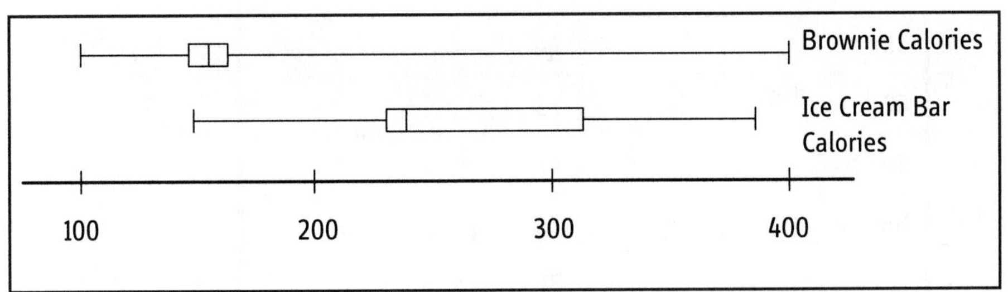

Summary

In this topic, we used data sets on number of waste sites by state, state governors' salaries, and calorie content in popular brands of brownies and ice cream, to explore measures of center and spread. We discussed the concepts of mean, median, mode, quartiles, range, and interquartile range. We also calculated the five-number summary and graphed the boxplot for data sets.

Explorations

1. Create a set of exam scores (for a 100-point exam) for a hypothetical class of ten students in which
   a. The mean, median, and mode are all the same.
   b. The mean is at least ten points greater than the median.
   c. The median is at least ten points greater than the mean.
2. A particular ATM machine dispenses $20 bills as requested by customers in amounts up to a maximum of $400. The following table shows the frequency of transactions in which customers withdrew specific amounts during a particular month. (Note that this particular machine has a $60 "quick cash" option.)

Amount Withdrawn	Number of Customers
$20	32
40	44
60	93
80	43
100	56
120	31
140	33
160	25
180	16
200	45
220	23
240	18
260	21
280	11
300	27
320	14
340	13
360	10
380	7
400	18

a. How many transactions were there in which customers withdrew cash at this ATM machine during the month?

b. Find the mode for the amount of cash withdrawn from this particular ATM machine during the month.

c. Find the median amount of cash withdrawn from this ATM machine during the month.

d. Find the mean amount of cash withdrawn from this ATM machine during the month.

e. Create an appropriate graph for this data set and explain what it shows.

3. The following table gives estimates of the percent change in population for each U.S. state over the years 1990 to 2000. (This data set was used in Example 1.8.)

State	Percent Change	State	Percent Change
Alabama	10.1	Montana	12.9
Alaska	14	Nebraska	8.4
Arizona	40	Nevada	66.3
Arkansas	13.7	New Hampshire	11.4
California	13.8	New Jersey	8.9
Colorado	30.6	New Mexico	20.1
Connecticut	3.6	New York	5.5
Delaware	17.6	North Carolina	21.4
Florida	23.5	North Dakota	0.5
Georgia	26.4	Ohio	4.7
Hawaii	9.3	Oklahoma	9.7
Idaho	28.5	Oregon	20.4
Illinois	8.6	Pennsylvania	3.4
Indiana	9.7	Rhode Island	4.5
Iowa	5.4	South Carolina	15.1
Kansas	8.5	South Dakota	8.5
Kentucky	9.7	Tennessee	16.7
Louisiana	5.9	Texas	22.8
Maine	3.8	Utah	29.6
Maryland	10.8	Vermont	8.2
Mass.	5.5	Virginia	14.4

State	Percent Change	State	Percent Change
Michigan	6.9	Washington	21.1
Minnesota	12.4	West Virginia	0.8
Mississippi	10.5	Wisconsin	9.6
Missouri	9.3	Wyoming	8.9

a. Find the five-number summary for this data set and identify which states fall in the lowest quartile and which states fall in the highest quartile.

b. Explain what the five-number summary shows about this data set.

c. Find the average of the percent change for all the states. Is this the same as the overall percent change for the total U.S.? Explain why or why not.

d. Suppose you include the District of Columbia, which had a percent population change over the time period 1990 to 2000 of –5.7. How does the five-number summary change? Find the new five-number summary and compare it to the five-number summary for the 50 states.

4. An article about children and earning power, entitled "Where Wealth Begets Children," appeared in *The New York Times* on October 24, 1999. The article was accompanied by the following table, which gives median yearly earnings for families with mothers ages 40 to 44.

Manhattan Neighborhoods	1 Child	2 Children	3 or More Children
Chinatown	$36,520	$19,357	$19,000
Washington Heights	37,000	28,085	35,800
Morningside Heights	19,924	30,240	39,372
East Harlem	48,000	27,500	22,488
Central Harlem	52,000	42,148	43,732
Chelsea	48,750	24,200	14,000
Greenwich Village	60,300	44,500	80,150

Manhattan Neighborhoods	1 Child	2 Children	3 or More Children
Upper West Side	98,650	122,000	100,400
Upper East Side	120,000	142,000	302,975

a. Describe what trends this table shows and what might explain these trends.

b. Explain why the median income was given (instead of a different "representative" value for earning power).

c. What additional information might help you understand the differences?

d. What kind of graph could you use to present the information in the table? Create a graph to present these data.

5. A pamphlet published by the Commonwealth of Pennsylvania, "Use Water Wisely," contains the following information: "Be aware of how much water you use! Awareness is the first step in conservation. The following table indicates how much water the average person uses each day." Explain how this "average" might have been obtained.

Use	Gallons Per Day
Toilet	19
Bathing and Hygiene	15
Laundry	8
Kitchen	7
Housekeeping	1
TOTAL	**50**

6. A recent newspaper article reported on the consumption of alcohol by college students. The report indicated that the average amount of alcohol consumed per college student has decreased over the last ten years. The report went

on to discuss the rise among college students of "binge drinking," which is defined as consuming large amounts of alcohol at one sitting.

a. Explain how both of these observations can be true.

b. What data would you want to collect, in addition to the average amount of alcohol consumed per college student, to understand how to address the problem?

7. The sale of special, simpler Internet names has become a big business as unique Internet names become more scarce. The website http://GreatDomains.com is a retailer of domain names and websites, and on July 26, 2004, it offered 21 sites for sale in the "Food and Beverage" category. The median price for the 21 sites in this category was $7500, but the mean cost was more than double that, at $15,071. Explain how such a median and mean are possible.

8. Next you will find a table containing the number of hazardous waste sites for a group of 12 New England and mid-Atlantic states. (Source: *Environmental Protection Agency,* www.epa.gov/superfund/sites/npl/npl.html.)

State	Number of Sites
Connecticut	15
Delaware	16
Maine	13
Maryland	18
Massachusetts	31
New Hampshire	18
New Jersey	113
New York	90
Pennsylvania	94
Rhode Island	12
Vermont	9
Virginia	30

a. Find the five-number summary for this data set and sketch a boxplot.

b. Explain what your five-number summary and boxplot show about this data set.

9. The following excerpt is from an AP newspaper article dated April 30, 2001, entitled "Study: 'Safe' Levels of Lead Still Harm IQ" (http://archives.cnn.com/2001/fyi/teachers.ednews/04/30/lead.iq.ap/). It describes the results of a study of the relationship between blood lead concentration in children and IQ scores. In particular, the second paragraph describes differences in mean IQ test scores for two groups of children.

a. Describe how the children were grouped into the two groups.

b. Is the description of the mean given in the article correct? Explain your answer.

BALTIMORE (AP) -- Children exposed to lead at levels now considered safe scored substantially lower on intelligence tests, according to researchers who suggest one in every 30 children in the United States suffers harmful effects from the metal.

Children with a lead concentration of less than 10 micrograms per deciliter of blood scored an average of 11.1 points lower than the mean on the Stanford-Binet IQ test, the researchers found. The mean is the intermediate value between the lowest and highest scores.

"There is no safe level of blood lead," said Dr. Bruce Lanphear, lead author of the lead study presented Monday at the Pediatric Academic Societies annual meeting.

Children are most commonly exposed to lead by inhaling lead-paint dust or eating paint flakes. Lead-based paint was widely used in homes throughout the 1950s and 1960s until it was banned in 1978.

At high levels, lead can cause kidney damage, seizures, coma and death.

10. A study to determine how fast cars travel on Main Street, which has a speed limit of 25 mph, tracked cars traveling at the following speeds: 24, 20, 32, 25, 52, 35, 28, 26, 29, 30.

    a. What measure should be used to identify a "typical" speed on this street?

    b. Explain why you made the choice you did in part a of this exploration.

11. The following graph, from *The Morning Call*, July 31, 2004, shows the 30-year average rainfall for the months January through July, compared to the 2004 monthly precipitation in Allentown, Pennsylvania, for those same months.

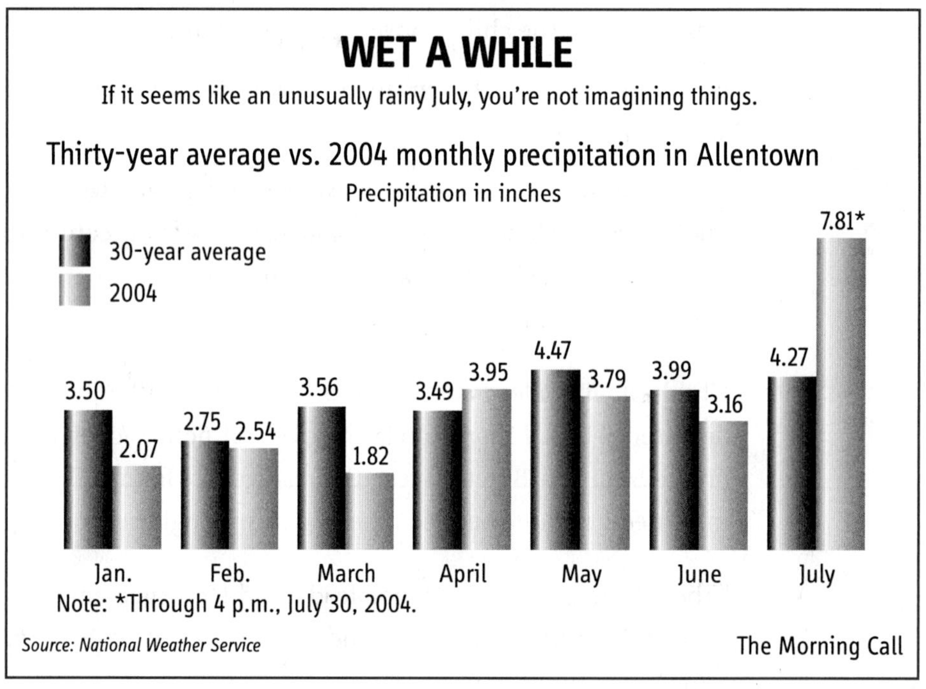

    a. Describe what the graph shows.

    b. What can you say about rainfall in Allentown in 2004?

TOPIC 17

Standard Deviation, z-Score, and Normal Distributions

OBJECTIVES

After completing this topic, you will be able to

- Compute and interpret the standard deviation of a set of data
- Use the standard deviation to compare the distributions of several data sets
- Compute and interpret the z-score of a data value and use z-scores to compare data from different data sets
- Visualize data that are normally distributed and use the empirical rule to understand a normal distribution

In Topic 16 we looked at two measures of the spread of a data set: the range and the interquartile range. The range as a measure of spread uses only the maximum and minimum values in the data set. The interquartile range (the third quartile minus the first quartile) uses more information about the data set; it measures the width of the middle 50 percent of the data and is resistant to the effects of outliers. But the variability of a data set involves the entire collection of data because each data value contributes to the variability of the whole. A third measure of the spread or variability in a data set, which depends on each data value in the set, is the standard deviation.

Suppose we take three samples of seven cars traveling along a residential stretch of Main Street that has a posted speed limit of 25 miles per hour and record the speed of each of the cars. Here are three hypothetical sets of seven sample speeds. (Note that in each sample we have mean = median = 25 and range = 40.)

Sample 1: 5, 25, 25, 25, 25, 25, 45

Sample 2: 5, 15, 20, 25, 30, 35, 45

Sample 3: 5, 5, 5, 25, 45, 45, 45

Although the mean, median, and mode are the same in all three samples, there is a real difference in the variability of the data in each of these samples. For example, in Sample 1, five of the cars were traveling at the posted speed limit and only one was traveling over the speed limit; in Sample 3, three cars were traveling 20 miles over the speed limit. We would like to be able to compare the three samples using a measure that incorporates each data value in the sample. In the context of the setting of this example, how much each individual car's speed deviates from the mean of 25 is important information. (You will be asked to compute the standard deviation of each of these samples in Exploration 1.) The **standard deviation** is a measure of the "typical" deviation from the mean for the values in the data set.

The standard deviation of a data set can be obtained by performing a few computations as described in the following paragraph. We'll illustrate this process using the calories-of-brownies data set introduced in Example 16.5 and reproduced here.

Brownie, Prepared According to Package Directions	Calories in One Piece
Pillsbury caramel fudge chunk	170
Betty Crocker caramel swirl	120
Betty Crocker German chocolate	160
Nestle double chocolate chip	150
Pepperidge Farm hot fudge	400
Duncan Hines milk chocolate	160

Brownie, Prepared According to Package Directions	Calories in One Piece
Betty Crocker microwave frosted	180
Robin Hood/Gold Medal pouch mix	100
Duncan Hines peanut butter	150
Pillsbury triple, chunky	170

Here are the steps needed to compute the standard deviation of a data set:

1. First compute the mean. (The mean of the brownies data set is 176.)
2. Next calculate the **deviation** of each data value from the mean; that is, subtract the mean from each data value. (These values are shown in the second column of the following table. Note that some of these deviations are positive and some are negative because some of the data values are larger than the mean and some are smaller. In fact, if we add all these deviations, their sum should be 0 or very close to 0, allowing for possible round-off errors.)
3. Now square these deviations from the mean, as shown in the third column of the table. (All the values in this column are positive because they are squares.)
4. Next, sum the squares of the deviations; this gives us the total sum of the squared deviations.
5. Divide the sum obtained in step 4 by one less than the number of data values.
6. Finally take the non-negative square root of the quotient found in step 5 to obtain the standard deviation.

Calories per Serving—Brownies	*(Calories - Mean)*	*(Calories - Mean)*^{2}
170	-6	36
120	-56	3,136
160	-16	256
150	-26	676
400	224	50,176

Calories per Serving–Brownies	(Calories - Mean)	(Calories - Mean)^2
160	-16	256
180	4	16
100	-76	5,776
150	-26	676
170	-6	36
Sum of Columns	**0**	**61,040**

The sum of the squares of the deviations is 61,040; we divide this sum by 9 (one less than the number of data values) to get approximately 6782.22. The standard deviation of this data set is the non-negative square root of 6782.22, which is approximately 82.35. (Note that to get an average of the squared deviations, it might seem reasonable to divide by the number of data values, because we divide the sum of the data values by the number of data values to get the mean. There are important statistical reasons why we divide by one less than the number of data values when computing the standard deviation of a sample.)

When computing the standard deviation, we first look at how far above or below the mean each data value falls by computing each data value's deviation from the mean. We then square these deviations and obtain an "average" squared deviation. By taking the non-negative square root of that "average" squared deviation, we obtain a measure of a "typical" deviation from the mean for the data set. (Recall that the radical sign $\sqrt{x}$ indicates we want the non-negative number or expression that when squared gives x. It is defined for $x \geq 0$.)

Example 17.1

Find the standard deviation of the calories-of-ice-cream-bars data set given in Example 16.5 (and repeated here) and compare with the standard deviation of the brownies data set obtained previously.

Ice Cream Bar, One Piece	Calories per Bar
Good Humor Fat Frog	154
Good Humor Halo Bar	230
Heath	170
Haagen-Dazs, caramel almond	230
Good Humor chip candy crunch	255
Nestle premium milk choc w/almond	230
Haagen-Dazs vanilla w/dk choc coating	390
Nestle Crunch vanilla w/wh choc coating	350
Klondike Krispy	290
Oh, Henry vanilla with choc coating	320

Solution

We again use a table to organize the work required to calculate the standard deviation. The first column of the table contains the original data. We set up the other columns as in the previous example.

Looking at the data set before we do the computations, we can predict the standard deviation of the ice cream bar data to be less than 82.35 (the standard deviation of the brownie-calorie data set) because the ice cream bar–calorie data values appear to deviate less from the "center" or mean of their data set. The mean of this data set is 261.9, which is larger than the mean of the brownie-calorie data set.

Calories per Serving–Ice Cream Bar	(Calories - Mean)	$(\text{Calories - Mean})^2$
154	-107.9	11,642.41
230	-31.9	1,017.61
170	-91.9	8,445.61
230	-31.9	1,017.61
255	-6.9	47.61

Calories per Serving–Ice Cream Bar	(Calories - Mean)	(Calories - Mean)^2
230	-31.9	1,017.61
390	128.1	16,409.61
350	88.1	7,761.61
290	28.1	789.61
320	58.1	3,375.61
Sum of Columns	**0**	**51,524.9**

The sum of the squares is 51,524.9, and 51,524.9 divided by 9 is approximately 5724.99. The standard deviation is the square root of 5724.99; $\sqrt{5724.99} \approx 75.66$, which is smaller than the standard deviation of the brownie-calorie data set.

The standard deviation, like the mean, is greatly influenced by outliers, as the next example shows.

Example 17.2

Suppose the calorie content of the Pepperidge Farms hot fudge brownie is only 200 calories instead of 400; recompute the mean and the standard deviation of the calories-of-brownies data set and compare with the previous results.

Solution

We will again set up a table to see how the mean and standard deviation are affected by not having the one large data value in the data set. The mean of the new data set is 156 (20 less than the mean of the data set without the adjusted outlier), so we see that the mean is significantly affected by this change.

Calories per Serving–Brownies	(Calories - Mean)	(Calories - Mean)^2
170	14	196
120	-36	1,296

Calories per Serving—Brownies	(Calories - Mean)	(Calories - Mean)^2
160	4	16
150	-6	36
200	44	1,936
160	4	16
180	24	576
100	-56	3,136
150	-6	36
170	14	196
Sum of Columns	0	7,440

The standard deviation is the square root of $\frac{7440}{9}$, which is $\sqrt{826.67} \approx 28.75$. The standard deviation was reduced drastically (from 82.35 to 28.75) by a change in one extreme data value.

A greater variability in a sample of data is indicated by a larger value of the standard deviation of the sample, while a smaller value of the standard deviation indicates less variability. The standard deviation of a particular data set is estimated to be large or small by comparing it to the standard deviation of another set of data.

The standard deviation is fairly tedious to compute using a table, as we did for the previous example. Organizing the computation of the standard deviation with a table is a good way to get a sense of what the standard deviation is; however, it is impractical to use except for small examples. Most calculators and computers have a built-in function that can be used to compute the standard deviation.

The median, five-number summary, range, and interquartile range are useful to describe a data set that contains outliers or a data set that is skewed. The mean and standard deviation are more frequently used for roughly symmetric data sets that have no outliers. We will look at one particular type of distribution that fits the description of "symmetric without outliers"; this type of data set occurs frequently with measurements and test scores and other naturally occurring phenomena and is called a **normal distribution**.

Consider the following frequency histogram of Verbal Scholastic Aptitude Test scores for a recent freshman class at a small college. The data appear to be fairly symmetric, with the mean and median occurring at approximately the center peak. This peak is at approximately 550.

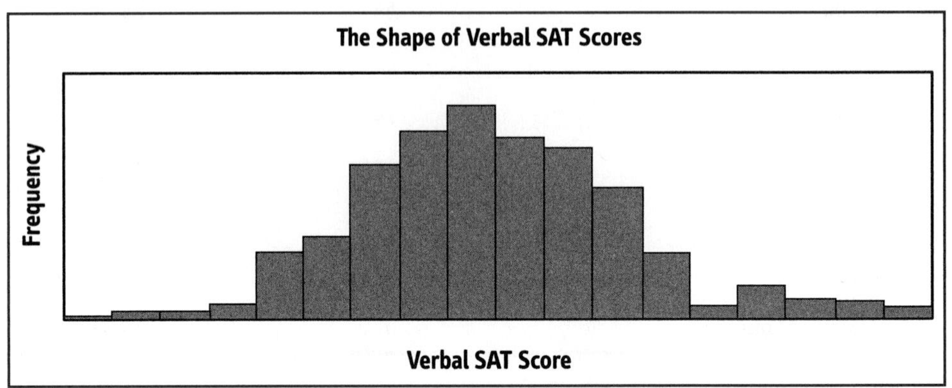

If we were to sketch in a smooth curve, tracing along the tops of the frequency bars of the histogram of the data set, it would roughly resemble a bell-shaped curve, as shown here. This type of curve is called a **normal curve**. Data that are normally distributed have a single peak and follow roughly a bell shape; a normal curve is sometimes called a **bell curve** because of the bell shape. The curve sketched here is the "idealized" version of the curve of a data set that follows a normal distribution exactly and allows us to visualize quickly the shape of a data set known to be approximately normal.

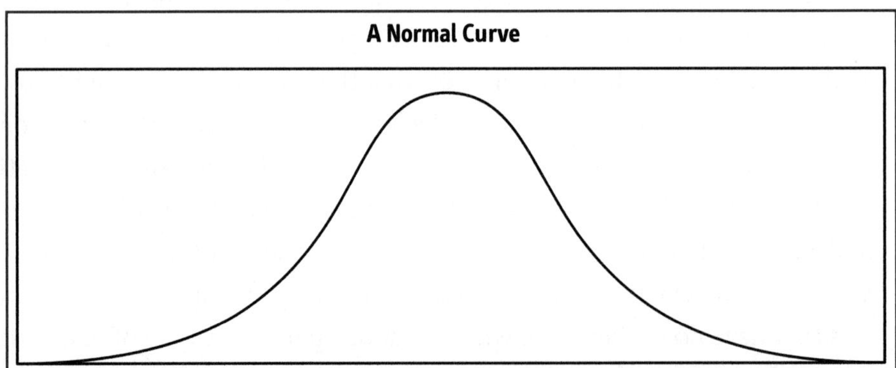

A data set that is normal or approximately normal is characterized by its mean and standard deviation. On the sketch of any normal curve, the mean is the point along the horizontal axis at which the peak occurs. In general, the larger the standard deviation, the more spread out the curve is and the less high the peak is. The following graph shows two normal curves with the same mean. The curve with the higher peak has a smaller standard deviation; the data values with a resulting histogram that gives rise to the taller and narrower normal curve are less spread out along the horizontal axis than those values leading to the shorter curve.

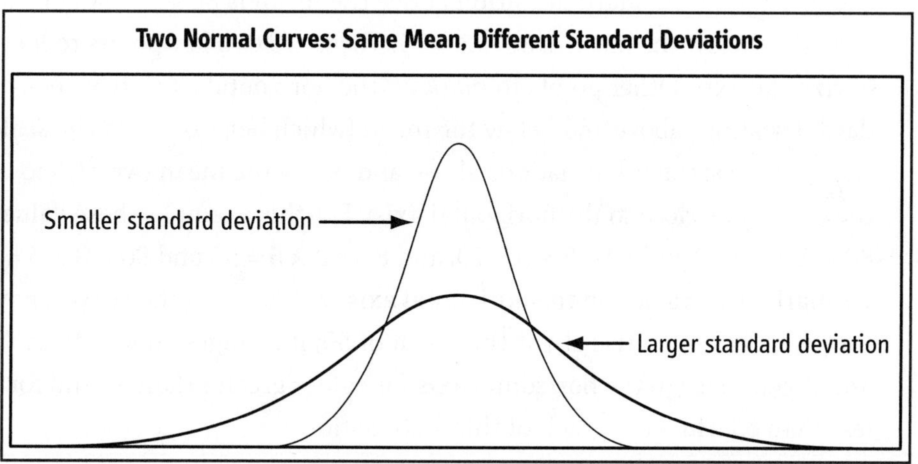

If we trace a normal curve, beginning at the left end of the graph and proceeding to the right, the curve is concave upward initially. The point where the curve changes from concave upward to concave downward lies at the mean minus the standard deviation along the horizontal axis. If we continue tracing along the curve, the curve is concave downward (around the peak) until we reach the mean plus the standard deviation. The curve changes to concave upward at that point. We also see that the curve is symmetric with respect to the mean and that at the left and right ends, it gets closer and closer to the horizontal axis. Although the curve continues indefinitely in both directions, values where the curve gets close to the horizontal axis are *mean* – 3 × (*standard deviation*) and *mean* + 3 × (*standard deviation*).

We can sketch a normal curve by marking several key points along the horizontal measurement axis. We'll describe how with an example. Suppose we know that grades in a large freshman course are approximately normally distributed, with a mean of 80 and a standard deviation of 5. We want to sketch this distribution of grades. First, we mark the mean of 80 on the horizontal axis, where the highest peak of the curve occurs. Then we identify the numbers that represent grades that are one standard deviation above the mean and one standard deviation below the mean because these points are where the normal curve changes concavity. One standard deviation above the mean is $80 + 5 = 85$; one standard deviation below the mean is $80 - 5 = 75$. These are two key points to locate on the horizontal axis. Other points to mark on the horizontal axis are values two standard deviations above and below the mean (which help us set up the scale), and values three standard deviations above and below the mean (which show us where the curve gets close to the horizontal axis). For this example, these values are $80 + 2 \times 5 = 90$ and $80 - 2 \times 5 = 70$, and $80 + 3 \times 5 = 95$ and $80 - 3 \times 5 = 65$. After we mark these values on the horizontal axis, we sketch in the curve, showing that the graph is symmetric about the mean of 80; it changes concavity at 75 and 85; and it gets close to the horizontal axis for values greater than 95 and for values less than 65. Here is a graph of this distribution.

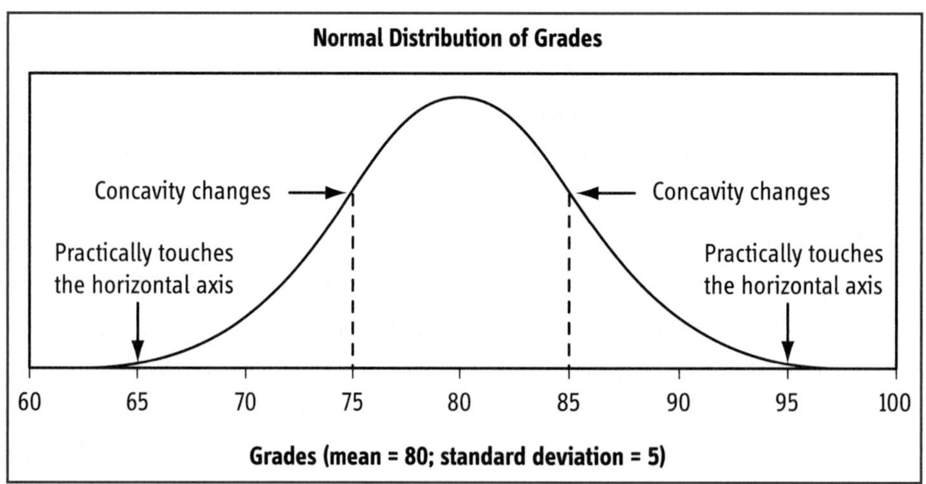

The basic characteristics of a normal curve (bell-shaped curve) are

» The peak occurs above the value on the horizontal axis that corresponds to the mean of the data set.

» The curve is symmetric with respect to the mean.

» The curve changes concavity above the two points on the horizontal axis located one standard deviation from the mean.

» The curve practically touches the horizontal axis at the two points on the horizontal axis that are located three standard deviations from the mean.

Example 17.3

A particular test is used to determine IQ scores (for Intelligence Quotient), and among a certain group of middle school students, IQ scores are known to have an approximately normal distribution with a mean of 100 and a standard deviation of 15.

a. Find the two values that represent IQ scores that lie one standard deviation above and one standard deviation below the mean. Also find values that represent scores that fall two and three standard deviations above the mean and those that fall two and three standard deviations below the mean.

b. Mark the mean and the six values you found in part a of this example on a horizontal axis.

c. Sketch the graph of the normal curve that represents IQ scores for the group of middle school students.

Solution

a. One standard deviation above the mean is the IQ value of $100 + 15 = 115$. One standard deviation below the mean is the value $100 - 15 = 85$. The value $100 + 2 \times 15 = 130$ is two standard deviations above the mean, and $100 + 3 \times 15 = 145$ is three standard deviations above the mean. The value $100 - 2 \times 15 = 70$

is two standard deviations below the mean, while 100 – 3 × 15 = 55 is three standard deviations below the mean.

b. We mark the horizontal axis, with the mean (100) in the center as shown here.

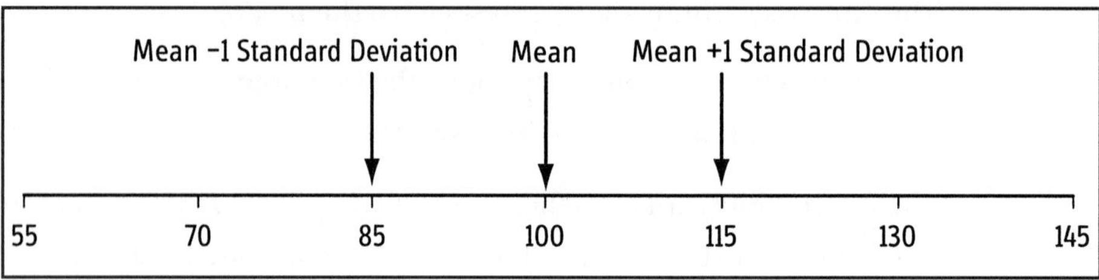

c. Tracing from left to right, the curve is very close to the horizontal axis for an IQ score of 55. The curve increases and changes from concave upward to concave downward above the IQ value of 85 (the mean minus the standard deviation). The curve reaches its peak above 100, and changes from concave downward to concave upward above the IQ value of 115 (the mean plus the standard deviation). The curve decreases and gets very close to the measurement axis for IQ values greater than 145.

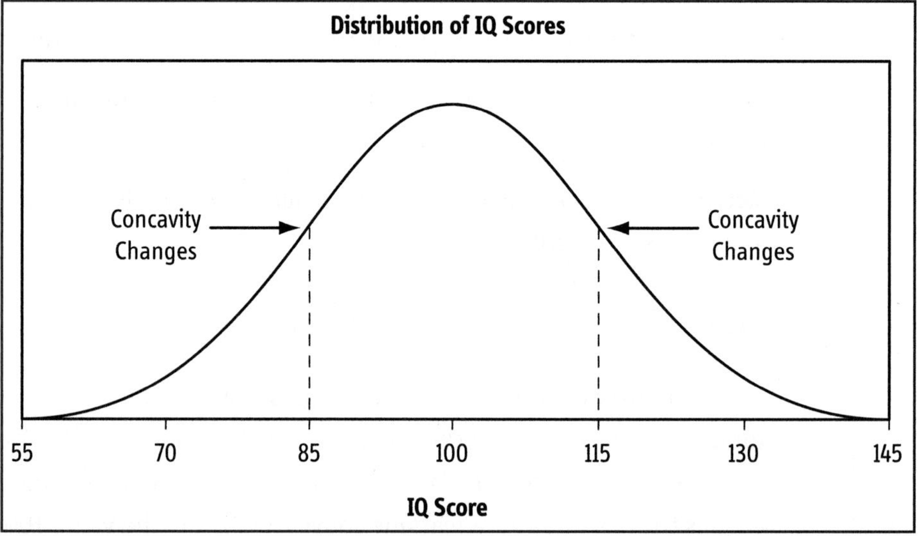

There is a useful **empirical rule** that applies to any data set that follows an approximately normal distribution. Here the phrase *empirical rule* refers to a rule that is derived from practical experience. The rule says that if the observations in a data set can be approximated by a normal curve, then approximately 68 percent of the data values are within one standard deviation of the mean; approximately 95 percent of the data values are within two standard deviations of the mean; and approximately 99.7 percent of the data values are within three standard deviations of the mean. For the normal curve shown in Example 17.3 and representing IQ values of a group of middle school students, with a mean of 100 and a standard deviation of 15, this rule says that approximately 68 percent of the data values fall between 85 and 115; approximately 95 percent of the IQ values fall between 70 and 130; and approximately 99.7 percent (that is, almost all) of the data values fall between 55 and 145.

In a histogram, the area of a frequency bar corresponds to the proportion of observations that fall into that category; we can then interpret the area under the normal curve between the relevant measurement values as the proportion of observations that fall between those values. For example, it is reasonable that on the graph with a mean of 100 and a standard deviation of 15, 95 percent of the area under the normal curve (and hence 95 percent of the observations) falls between 70 and 130.

Example 17.4

Suppose the verbal SAT scores for freshmen at a certain college follow an approximately normal distribution with a mean of 550 and standard deviation of 79.

a. Use the empirical rule to find an interval of values within which approximately 68 percent of the freshmen verbal SAT scores fall.

b. Use the empirical rule to find an interval of values within which approximately 95 percent of the freshmen verbal SAT scores fall.

c. Use the empirical rule to find an interval of values within which approximately 99.7 percent of the freshmen verbal SAT scores fall.

d. If there are 482 students in the freshman class, approximately how many will fall into each of the intervals computed in parts a, b, and c of this example?

Solution

a. The empirical rule says that we need to find the interval that covers all values within one standard deviation of the mean, either one standard deviation greater than the mean or one standard deviation less than the mean. Thus we need to calculate the *mean* – 1 × (*standard deviation*) and *mean* + 1 × (*standard deviation*). This is 550 – 1 × 79 = 471 and 550 + 1 × 79 = 629. So approximately 68 percent of the freshman verbal SAT scores will fall between 471 and 629.

b. For the "95 percent rule," we need to calculate the interval that covers all values within two standard deviations of the mean; that is, we calculate the *mean* – 2 × (*standard deviation*) and *mean* + 2 × (*standard deviation*). This is 550 – 2 × 79 = 392 and 550 + 2 × 79 = 708. We can say that approximately 95 percent of the freshmen have verbal SAT scores between 392 and 708.

c. To calculate the interval into which approximately 99.7 percent of freshmen verbal SAT scores fall, we calculate 550 – 3 × 79 = 313 and 550 + 3 × 79 = 787. So 99.7 percent of the freshman at this college have verbal SAT scores between 313 and 787.

d. We calculate 68 percent of 482 to get (0.68) × 482 = 327.76, or approximately 328 of the freshman have a verbal SAT score between 471 and 629. Ninety-five percent of 482 is 457.9, so approximately 458 freshmen have a verbal SAT score between 392 and 708. This also means that approximately 24 freshmen have a verbal SAT score that falls outside of this range, either greater than 708 or less than 392. The "99.7 percent rule" says that 480.55 or approximately 481 freshmen have a verbal SAT score between 313 and 787. Thus, one student would be expected to have a score outside of this range.

Sometimes we want to compare values from different data sets. One way to do this is to **standardize** the values; that is, evaluate how many standard deviations each observation lies away from the mean of its data set. We can then compare the standardized values and assess which observation is more extreme, taking into account the mean and standard deviation of the data set from which the observation was obtained. The standardized value of an observation is called a **z-score**. The z-score tells us how many standard deviations the observed data value lies away from the mean. To evaluate the z-score for an observation from a data set with a known mean and standard deviation, we subtract the mean from the observation and then divide this difference by the standard deviation. Thus we compute:

$$z = \frac{observation - mean}{standard\ deviation}$$

For example, suppose that verbal SAT scores of a particular college's freshman class are normally distributed with a mean of 550 and a standard deviation of 50. A student from this class who scored 620 on the verbal SAT test has a z-score of $z = \frac{620 - 550}{50} = 1.4$. This means that the student's verbal SAT score lies 1.4 standard deviations above the mean of the distribution. It tells us where this student's score falls relative to the mean and standard deviation of the scores of all students in the class. A student from this class who scored 520 on the verbal SAT test would have a z-score of $z = \frac{520 - 550}{50} = -0.6$.

Note that the standard deviation of any data set is 0 (if all the data values are the same) or a positive number. So if an observation is greater than the mean, its z-score is positive because, in this case, *observation* – *mean* is positive. If an observation is less than the mean, its z-value is negative because, in this case, *observation* – *mean* is negative.

Example 17.5

The distribution of IQ scores for a group of middle school students, as described in Example 17.3, has a mean of 100 and a standard deviation of 15. Assume that the distribution of IQ scores for college students is also approximately normally distributed and has a mean of 115 with a standard deviation of 18. A middle school student's IQ score is reported to be 134, while her older brother, who is a college student, has an IQ score of 144. How do these scores compare?

Solution

It's clear that the older brother's score is higher. But because the two scores come from different distributions, we can compare them by using the standardized scores. The middle school student's standardized score is $z = \frac{134 - 100}{15} \approx 2.27$, while the older brother's score is $z = \frac{144 - 115}{18} \approx 1.61$. The middle school student's IQ score lies 2.27 standard deviations above the mean of her distribution while her older brother's IQ score lies 1.61 standard deviations above the mean of his distribution. With respect to their own age groups, the middle school student has the higher IQ score.

Summary

In this topic, we looked at how to define and interpret a measure of the spread of a data set, the standard deviation. We saw that the standard deviation can be changed significantly by altering one extreme data value. We investigated characteristics of a normal distribution, including the empirical rule. Finally, we looked at z-scores as a way to compare values that come from different data sets.

Explorations

1. Consider again the three samples of the speeds of seven cars given at the beginning of this topic: Sample 1: 5, 25, 25, 25, 25, 25, 45; Sample 2: 5, 15, 20, 25, 30, 35, 45; and Sample 3: 5, 5, 5, 25, 45, 45, 45.

   a. Look at the data in each sample and without doing any calculations, rank the samples in order from smallest to largest standard deviation. Give reasons for your ranking.

   b. For each of the samples, create a table like the ones in Examples 17.1 and 17.2; for each sample, compute the deviation of each data value from the mean, the squared deviations, and finally the standard deviation of the data set.

   c. Explain what the three standard deviations tell you about the data sets.

2. The following tables provide information about the top women-owned businesses in the U.S. The first table gives year 2000 revenue in millions of dollars and number of employees for Pennsylvania women-owned companies with

year 2000 revenues of $70 million or higher. The second table gives the same information for Michigan women-owned companies with year 2000 revenues of $70 million or higher.

Pennsylvania Companies	2000 Revenue ($ million)	Number of Employees
84 Lumber	2000	4400
Charming Shoppes	1600	12000
Rodale	500	1300
Mothers Work	366.3	3800
Harmelin Media	200	105
McGettigan Partners	175	400
Wetherill Associates	135	500
Hanna Holdings	73.7	1284

Michigan Companies	2000 Revenue ($ million)	Number of Employees
Ilitch Holdings	800	700
Plastech Engineered Products	420	3500
Jerome-Duncan Ford	350	300
Elder Ford	287.6	108
Patsy Lou Williamson Auto. Gp.	221	240
Mexican Industries	174	1463
Rush Trucking	153	2000
Jaguar-Saab of Troy	143	134
Manpower Metro Detroit	118	250
Continental Plastics	107	650
Leco	82	800
Rodgers Chevrolet GEO	75	80

Michigan Companies	2000 Revenue ($ million)	Number of Employees
Strategic Staffing Solutions	75	600
Two Men & a Truck Internat'l	75	40
Systrand Manufacturing	72	230

a. Use a calculator or computer to compute the mean and standard deviation of the year 2000 revenue for the Pennsylvania companies in the first table.

b. Use a calculator or computer to compute the mean and standard deviation of the year 2000 revenue for the Michigan companies in the second table.

c. Explain what the values you calculated in parts a and b of this exploration tell you about the data sets.

d. How would the mean and standard deviations change if the largest data value in each set were removed?

e. Find the mean and the standard deviation of the number of employees for the Pennsylvania companies in the first table.

f. Find the mean and the standard deviation of the number of employees for the Michigan companies in the second table and compare to your results in part e of this exploration.

3. Consider the distributions of IQ scores described in Example 17.5.

   a. Sketch the normal curve for the distribution of middle school student IQ scores with a mean of 100 and standard deviation of 15. Mark the measurements on the horizontal axis.

   b. Sketch the normal curve for the distribution of college student IQ scores with a mean of 115 and standard deviation of 18. Mark the measurements on the horizontal axis.

   c. Describe how the two curves you sketched are similar and how they differ.

d. On the normal curve for the middle school students, identify the area below the curve and to the right of the student's score of 134. Then, on the college curve, identify the area below the curve and to the right of her brother's score of 144. Describe how the areas show that, with respect to their age groups, the middle school student has a higher IQ than her brother.

4. Verbal SAT test scores follow approximately a normal distribution and math SAT test scores also follow approximately a normal distribution.

   a. A particular freshman student at a certain college has a standardized verbal SAT test score of $z = -1.3$ and a standardized math SAT test score of $z = 2.0$. Explain what these values tell you.

   b. Another freshman has a standardized math SAT test score of 1.5. How does the student described in part a of this exploration compare to this student?

   c. Suppose you also know that the mean math SAT test score for freshmen students at this college is 547 and the standard deviation is 74. What can you say about the math SAT test score of the student with a standardized math SAT test score of 2.0?

   d. Use the empirical rule to find an interval of values within which approximately 68 percent of the freshmen math SAT scores at this college fall.

   e. Use the empirical rule to find an interval of values within which approximately 95 percent of the freshmen math SAT scores at this college fall.

5. The following table gives the number of home runs hit by the American League home-run champions for the years 1949 to 1987, given in chronological order beginning with 1949 and proceeding left to right across the rows. (Source: *Time Almanac 2004,* page 1010.)

43	37	33	32	43	32	37	52	42	42	42	40	61
48	45	49	32	49	44	44	49	44	33	37	32	32
36	32	39	46	45	41	22	39	39	43	40	40	49

a. Find the mean and standard deviation of number of home runs hit by the American League home-run champions during the years given in the table.

b. Assuming that the data are approximately normally distributed, use the empirical rule to find an interval of values within which approximately 68 percent of the data falls. Count the number of data values that actually fall within that interval. Is it close to 68 percent?

c. Use the empirical rule to find an interval of values within which approximately 95 percent of the data falls. Count the number of data values that actually fall within that interval. Is it close to 95 percent?

d. Use the empirical rule to find an interval of values within which approximately 99.7 percent of the data falls. Count the number of data values that actually fall within that interval. Is it close to 99.7 percent?

6. The following table gives the number of home runs hit by American League home-run champions from 1988 to 2003, in chronological order. (Source: *Time Almanac 2004,* page 1010.)

42	36	51	44	43	46	40	50	52	56	56	48	47	52	57	47

a. Compute the mean and standard deviation of number of home runs for the years 1988 to 2003.

b. If you were to compute the mean number of home runs hit by the American League home-run champions during the years 1949 to 2003, do you expect the mean to increase, decrease, or stay about the same as the mean found in part a of this exploration? Explain your reasoning.

c. Compute the mean and standard deviation of number of home runs for the years 1949 to 2003 (use the data from Exploration 5) and compare with your results for part a of this exploration.

7. The following table gives typical monthly temperatures, in degrees Fahrenheit, in each of three U.S. cities. Find the mean and standard deviation of the typical monthly temperatures for each of the cities. What do these values tell you?

	Jan	Feb	Mar	Apr	May	June	July	Aug	Sept	Oct	Nov	Dec
San Diego, CA	54.9	55.9	57.6	59.9	62.2	65.1	68.7	70.2	68.5	64.8	60.1	56.1
Orlando, FL	59.5	61.2	66.6	71.1	76.8	81.0	82.2	82.4	81.0	75.2	68.0	62.1
Philadelphia, PA	30.6	32.9	42.3	52.3	62.8	71.8	76.6	75.4	68.2	56.3	46.4	35.8

8. One professor teaches a large section, section A, of a particular class, and on the first test of the term, the test scores in section A were approximately normally distributed with a mean of 78 and a standard deviation of 6. Another professor also teaches a large section, section B, of the same class, and on the first test of the term, the test scores in section B were also approximately normally distributed with a mean of 74 and a standard deviation of 10. Two students, one from each section, earned a grade of 92 on the exam. The student from section B claims that he did better because the section B test, with a mean of 74, was obviously more difficult than the section A test with a mean of 78. However, the student from section A claims that because she has a higher z-score, she actually performed "better." Calculate the z-scores for each student's test grade and settle their dispute; that is, decide who had the superior performance on this test.

9. Colleges typically use grade point average cutoffs to decide who graduates with honors and who is accepted into certain programs (such as teacher education, for example). Suppose at a particular college, a GPA of 3.0 is the cutoff for such a decision. The students in department A have a grade point average of 3.77 with a standard deviation of 0.43 (for all course grades in classes taken in the department in a particular academic year). The students in department B have a grade point average of 2.65 with a standard deviation of 1.16 for that same academic year. One student has taken most of his courses from department A; another student has taken most of her courses from department B. Both students have a GPA of 3.0.

   a. Compare the z-scores of the two students.

   b. Interpret what these z-scores mean.

TOPIC 18

Basics of Probability

OBJECTIVES

After completing this topic, you will be able to

- » Recognize basic terminology and interpret commonly used language associated with probability
- » Identify the outcomes of a sample space
- » Calculate probability as relative frequency
- » Use the technique of counting outcomes to compute probability
- » Apply basic probability rules

Shortly after the Challenger space shuttle accident in 1986, officials at NASA assessed the risk of another such disaster at approximately 1 in 225 for each shuttle mission. Another scientific organization, The National Academy of Sciences, estimated the likelihood of failure for a particular shuttle flight to be 1 in 145. Tragically, the Challenger disaster was followed by the Columbia shuttle disaster on February 1, 2003. (See www.nasa.gov for current information on shuttle programs.) Many practical issues involve an assessment of the likelihood of some event happening or the assessment of risk. Examples include statements such as those given about the shuttle disaster, as well as statements such as, "There is a 50 percent chance that it will rain tomorrow" and "Your chance of being seriously injured in a car accident is reduced if you wear a seatbelt." To understand and evaluate these and other risks and to help make informed decisions that might involve considering such risks, we will investigate the language and methods of assessing the likelihood associated with a chance event.

A **random process** is a situation that can be repeated and for which the set of *possible* outcomes is known. While there is uncertainty about what the outcome will be on any particular repetition, there is a predictable pattern over many repetitions, a regularity that appears only with many repetitions. The launching of a shuttle is a random process because we don't know what the outcome of a particular launch will be. Flipping a coin, tossing a pair of dice, or randomly selecting a set of voters for jury duty are all examples of random processes. The collection of all possible outcomes of a random process is called the **sample space** for the process. The types of outcomes in a sample space depend on what we are interested in observing for a particular random process. For example, if we choose a set of voters for jury duty, we may want to record how many females are among the set of jurors, or we may want to record how many retirees there are, each juror's political party, or some other characteristic.

Example 18.1

For each of the random processes described here, list all possible outcomes in the sample space.

a. Note people who go through the checkout line at a particular convenience store and record the gender of each.

b. Note people who go through the checkout line at a particular convenience store and record if they buy bread and/or milk or not.

c. Toss a pair of dice, one green and one white, and record the two values on the top faces of the dice.

d. Toss a pair of dice and record the sum of the two values on the top faces.

e. Select graduates of a particular college and record the GPA of each. Assume that at this particular college, the students need a GPA of 2.00 or above to graduate.

Solution

We often denote a collection or set using braces, and if there are a finite number of objects in the collection, we list the objects in the set within the braces, separated by commas. We will use S to denote the sample space for a random process.

a. There are only two possible outcomes for this process, so there are two elements in the sample space: S = {male, female}.

b. We could represent the sample space for this random process in several ways. If we record which of these two items customers buy, we have S = {bread only, milk only, both bread and milk, neither}. Alternatively, we could use a pair of yes or no replies, where the first reply answers the question, "Did the customer buy bread?" and the second reply answers the question, "Did the customer buy milk?" In this case, S = {(yes, no), (no, yes), (yes, yes), (no, no)}. Notice that the outcome (yes, no) is different from the outcome (no, yes). For each of the given ways of representing S, the number of elements in S is 4.

c. We will use pairs of numbers to indicate the elements in the sample space, with the first number representing the result on the green die and the second number representing the result on the white die. Thus,

S = {(1, 1), (1, 2), (1, 3), (1, 4), (1, 5), (1, 6), (2, 1), (2, 2), (2, 3), (2, 4), (2, 5), (2, 6),
(3, 1), (3, 2), (3, 3), (3, 4), (3, 5), (3, 6), (4, 1), (4, 2), (4, 3), (4, 4), (4, 5), (4, 6),
(5, 1), (5, 2), (5, 3), (5, 4), (5, 5), (5, 6), (6, 1), (6, 2), (6, 3), (6, 4), (6, 5), (6, 6)}

d. In this example, S = {2, 3, 4, 5, 6, 7, 8, 9, 10, 11, 12}, which represents all possible sums of the two numbers in the pairs given in the solution to part c.

e. Students need a GPA of 2.00 to graduate, so the sample space contains all numbers between 2.00 and 4.00, given to two decimal places (the accuracy with which GPA's are reported). We could denote S using an ellipsis (the symbol ...) to indicate that there are missing values not listed, but the pattern continues. Thus, S = {2.00, 2.01, 2.02, 2.03, ... , 3.95, 3.96, 3.97, 3.98, 3.99, 4.00}.

Each element in the sample space is an outcome. An **event** is any collection of outcomes from the sample space of a random process. To make it easy to refer to an event, we denote the event by a capital letter, such as A or B or some other appropriate letter chosen to signify that particular event. Because an event is a collection or set, we will again use braces to enclose the list of outcomes in the event.

Example 18.2

For each event, list the outcomes from the sample space that correspond to the given event.

a. Note people who go through the checkout line at a particular convenience store and record the gender of each. Consider the event that a male is not chosen.

b. Toss a pair of dice, one green and one white, and record the two values on the top faces of the dice. Consider the event where both numbers are greater than 4.

c. Toss a pair of dice and record the sum of the two values on the top faces. Consider the event that the sum is even.

d. We select a graduate of Moravian College and record his or her GPA. Consider the event that the graduate's GPA is 3.5 or higher.

Solution

We will choose appropriate capital letters to denote each event.

a. If a male is not chosen, then the event consists of one outcome, namely, "female," which we'll denote with F. Thus, the event "male is not chosen" = F = {female}.

b. This event consists of all outcomes where both numbers are either 5 or 6. We'll let this event be the set L (for "large"); L = {(5,5), (5,6), (6,5), (6,6)}.

c. This event is E = {2, 4, 6, 8, 10, 12}.

d. We call this event H and list all possible GPA scores between 3.5 and 4.0, assuming GPA scores are given to two decimal places. H = {3.50, 3.51, 3.52, … , 4.00}.

Many questions that require us to consider random processes involve choosing one or more items from a collection of items. If we choose more than one item from a collection, we can make our selection in one of two ways. We can replace the first item back into the original set before we choose the second item; this is called choosing **with replacement**. Alternatively, we can pick the second item (and any subsequent items) without returning the already chosen items to the set before picking additional items; this is called choosing **without replacement**. Whether we choose with or without replacement affects the outcomes we include in the sample space.

We also want to think about whether the order in which items are chosen is important to the situation at hand. For example, if we are choosing two people to receive first and second prizes, where first prize is a car and second prize is a book, order is certainly important. If we are choosing two people to form a committee, then which one is chosen first doesn't matter because both will serve on the committee. We explore these ideas in the next example.

Example 18.3

There are five students in a class: Aneesa, Benito, Carol, Donovan, and Eve. The professor, wanting to be fair, puts all names in a hat, mixes them up, and picks two to answer questions on the assigned homework.

a. If the professor picks the second name without replacing the first name in the hat before he picks the second, list the elements in the sample space.

b. If the professor picks the two names without replacement, list the outcomes in the event "Benito is chosen." How many of the outcomes in the sample space of part a are in the event "Benito is chosen"?

c. If the professor picks the second name after he has replaced the first name in the hat, list the elements in the sample space.

d. How many of the outcomes in the sample space of part c of this example are in the event "Benito is chosen"?

e. For each of the two methods of choosing names, list the outcomes in the event "no females are chosen."

Solution

We will represent the students by the first letters of their names.

a. If the second name is picked without replacing the first name, the same person cannot be picked twice. Also, the order in which the names are picked is not particularly important because both students will have similar jobs to do. This means that we don't need to list both pairs, for example, Aneesa-Benito (A-B) and Benito-Aneesa (B-A). So S = {A-B, A-C, A-D, A-E, B-C, B-D, B-E, C-D, C-E, D-E}.

b. We'll denote this event by N; N = {A-B, B-C, B-D, B-E}. N has four elements (all the elements of S with "B").

c. Now a student can be picked more than once because we are choosing with replacement, but the order in which names are chosen is still unimportant for the situation. Thus, S = {A-A, A-B, A-C, A-D, A-E, B-B, B-C, B-D, B-E, C-C, C-D, C-E, D-D, D-E, E-E}.

d. S has 15 elements; 5 of them correspond to the event "Benito is chosen."

e. Let's denote this event by M. If the two names are chosen without replacement, the event M = {B-D}, assuming that Aneesa, Carol, and Eve are female. If we choose with replacement, then M = {B-B, B-D, D-D}.

We can approach assessing the likelihood or **probability of an event** in several ways. The **probability** that an event will occur is the proportion of time the event occurs over the long run, or the **relative frequency** with which the event occurs if we repeat the random process over and over again. We can sometimes use simulation as a tool to help assess probability by creating a model of the random process; that is, we can imitate the process and repeat it a large number of times to assess what proportion of the time a particular event occurs. In this case, the probability of an event A, denoted $\mathcal{P}(A)$, is

$$\mathcal{P}(\text{A}) = \frac{\textit{number of times A occurs}}{\textit{number of times process is repeated}}$$

If the process is repeated a large number of times, the relative frequency (what we will interpret to be the probability of the event) will establish itself around a specific value. For example, we can toss a coin to simulate the playing of a game between two evenly matched teams. The coin landing heads means team A wins; the coin landing tails means team B wins. We can simulate the playing of many games, identify winning streaks for each team, and keep win-loss records over many games.

The **technique of counting outcomes** for assigning the probability of an event when the outcomes in the sample space are equally likely allows us to count the number of equally likely elements in the sample space S and in the event, call it A, whose probability we want to find. We will refer to this method as the **counting method**. Then the probability of the event A, denoted $\mathcal{P}(A)$, is the number of outcomes in the event A divided by the number of outcomes in S:

$$\mathcal{P}(A) = \frac{\textit{number of outcomes in A}}{\textit{number of outcomes in S}}$$

Another approach to probability is to use **subjective judgment**. In this situation, probability is interpreted as a measure of the strength of one's belief that a particular outcome will occur. This measure of probability is often personal and thus is subject to personal biases. Some people might be fairly good at subjective probability judgments but this method of assessing probability is not very useful because different people will make different subjective judgments. It is important to be aware, however, that sometimes given probabilities may be the result of subjective judgments. We will concentrate on the relative frequency interpretation and the counting method for assessing probabilities.

Example 18.4

Describe how one would find or estimate the following probabilities.

a. The probability of winning a particular state's lottery. (There are various ways in which state lotteries are organized, but we'll assume a state lottery in which you pick six distinct numbers, and to win, your numbers must match all six distinct winning numbers. We'll assume each number is between 1 and 70 and the order of the numbers is not significant.)

b. The probability of "snake eyes" on one toss of two dice. (Note that "snake eyes" means both dice show a 1.)

c. The risk of dying in an airplane accident.

d. The chances of getting bitten by a shark.

e. The probability of precipitation on a particular day in a particular city.

Solution

a. To assess the probability of winning the lottery, we need to know how the particular lottery system is organized. We need to figure out how many elements there are in the sample space of sets of six distinct numbers between 1 and 70. We'll call this number N. (N actually is 131,115,985.) Because there is exactly one winning set of numbers, the probability of winning is $\frac{1}{N}$, which is approximately 7.6×10^{-9}. This solution uses the counting method for finding probability.

b. We could estimate the probability of obtaining a 1 on both dice by repeatedly tossing a pair of dice a large number of times and counting the number of times (1,1) occurred. We then divide that number by the total number of tosses. This solution uses the relative frequency method. Alternatively, we could count the number of elements in the sample space of all possible outcomes of the random process of tossing two dice (which we listed in Example 18.1c). There is only one outcome in the event of "snake eyes," so we divide 1 by the number of elements in S, which is 36. Thus, we have $\mathcal{P}$(snake eyes) $= \frac{1}{36}$. This alternative solution uses the counting method.

c. We could estimate this probability by finding the total number of airplane passenger fatalities that occurred during a particular time period; we then divide that number by the total number of passengers who traveled by air during that time period. Here we used the relative frequency method for assessing a probability.

d. As in part c of this example, we could divide the number of ocean swimmers/surfers who were bitten by a shark by the total estimated number of beachgoers during the same time period. If we did this, we would be using the relative frequency method for assessing a probability. The estimates sometimes given in news articles for a particular risk such as this might in fact be "subjective probabilities" arrived at by an expert who doesn't actually count swimmers and beachgoers. A one-in-five million chance might be a way to indicate something that someone assesses is very, very unlikely to occur.

e. The relative frequency method can be used to estimate the probability of precipitation. Studying other times when weather patterns and conditions were similar and finding what proportion of them resulted in precipitation allows forecasters to model and predict current weather.

Probability Rules

There are several basic rules that assigned probabilities must satisfy:

1. The probability of an event is always a number between 0 and 1 because it represents a proportion. If an event has probability 0, then that event cannot occur. If an event has probability 1, the event is certain to occur. Thus, for any event A,

   $0 \leq \mathscr{P}(A) \leq 1$

2. If we consider all possible outcomes associated with a random process, that is, all those listed in the sample space S, the sum of their probabilities must be 1. Thus,

   $\mathscr{P}(S) = 1$

3. If a random process has n equally likely outcomes (and equally likely is extremely important!), the probability of each outcome is $\frac{1}{n}$.

4. Either an event A occurs or it does not occur. So the probability that an event A occurs is 1 minus the probability that A does not occur:

   $\mathscr{P}(A) = 1 - \mathscr{P}(\text{not } A)$

   Equivalently, if we want the probability that A does not occur, we could rewrite this rule:

   $\mathscr{P}(\text{not } A) = 1 - \mathscr{P}(A)$

5. If A and B are any two events that have *no outcomes in common*, then the probability that A or B occurs is the sum of the probability of A plus the probability of B; that is,

   $\mathscr{P}(A \text{ or } B) = \mathscr{P}(A) + \mathscr{P}(B)$

Events that have no outcomes in common are called **disjoint events** or **mutually exclusive events**. For example, let's consider students at a particular college and ask them to record the average number of hours spent exercising each week and their gender. Suppose A is the event "student spends more than five hours exercising each week," B is the event "student spends fewer than two hours exercising each week," and C is the event "student is female." Then A and B are disjoint events, but A and C are not disjoint (because the chosen student might be a female who exercises more than five hours per week). Similarly, B and C are not disjoint.

Example 18.5

Use the counting method and the basic properties of probability to find the following probabilities:

a. A professor chooses two students' names without replacement from a hat containing the names of **A**neesa, **B**enito, **C**arol, **D**onovan, and **E**ve. These two students will answer questions on the homework. Find the probability "Benito is not chosen."

b. Two dice are tossed. Find the probability that the "sum of the two dice is less than 4."

c. Two dice are tossed. Find the probability that the "sum of the two dice is less than 4 or greater than 9."

Solution

a. The sample space of this random process contains 10 equally likely outcomes, as seen in Example 18.3c. The event "Benito is chosen" contains 4 outcomes. If we let N = Benito is chosen, then $\mathcal{P}(\text{N}) = \frac{4}{10} = 0.4$. But we want the probability that Benito is not chosen. We compute $\mathcal{P}(\text{not N}) = 1 - \mathcal{P}(\text{N}) = 1 - 0.4 = 0.6$.

b. This sample space has 36 equally likely outcomes, as seen in Example 18.1c. (We need to assume these are fair, balanced dice to ensure that the 36 outcomes are equally likely.) If we denote the event A as "sum is less than 4," then A = $\{(1,1), (1,2), (2,1)\}$. Thus,

$$\mathcal{P}(A) = \frac{\text{number of outcomes in A}}{\text{number of outcomes in S}} = \frac{3}{36} = \frac{1}{12}$$

c. In part b of this example, we computed $\mathcal{P}$(A). We will let B represent the event "the sum is greater than 9." So B = $\{(4,6),(6,4),(5,5),(5,6),(6,5),(6,6)\}$ and $\mathcal{P}$(B) $= \frac{6}{36} = \frac{1}{6}$. Because A and B have no outcomes in common, that is, a dice toss cannot have a sum both less than 4 and greater than 9, $\mathcal{P}$(A or B) = $\mathcal{P}$(A) + $\mathcal{P}$(B) $= \frac{3}{36} + \frac{6}{36} = \frac{9}{36} = \frac{1}{4}$.

Tables are often used to give information about a certain population or group of individuals. If we consider the individuals to be the outcomes in the sample space, we can use the counting method to answer probability questions about this group of individuals.

Example 18.6

Use the given table to answer the questions. (The information given in the tables was taken from the website of the National Highway Traffic Safety Administration, www.nhtsa.dot.gov.)

a. The following table gives information on single-vehicle crashes in the U.S. by size of vehicle during a particular time period.

Passengers	Fewer than 5	5 to 9	10 to 15	More than 15
Crashes	1,815	77	55	10

i. If an investigator chooses a single-vehicle crash at random to carry out an additional audit of the investigation, what is the probability that the chosen crash involved a vehicle carrying more than 15 passengers?

ii. What proportion of single-vehicle crashes involved vehicles carrying 5 or more passengers?

b. The following table gives information on single-vehicle crashes in the U.S. involving rollovers by size of vehicle during a particular time period.

Passengers	Fewer than 5	5 to 9	10 to 15	More than 15
Rollovers	224	16	16	7

i. If an investigator chooses at random a single-vehicle crash involving a rollover to carry out an additional audit of the investigation, what is the probability that the crash involved a vehicle that carried fewer than 5 passengers?

ii. What proportion of single-vehicle rollover crashes involved vehicles carrying 15 or fewer passengers?

Solution

a. The total number of crashes is 1,815 + 77 + 55 + 10 = 1,957. Because we assume a crash is selected at random, each is equally likely to be chosen, and the number of equally likely elements in the sample space is 1,957.

i. There were 10 crashes involving more-than-15-passenger vehicles, so the probability that the chosen crash involved a more-than-15-passenger vehicle is $\frac{10}{1957} \approx 0.0051$. We could also say that approximately 0.5 percent of crashes during the time period involved vehicles carrying more than 15 passengers.

ii. The proportion of crashes involving vehicles with fewer than 5 passengers is $\frac{1815}{1957} \approx 0.9274$. Therefore, $1 - 0.9274 = 0.0726$ is the approximate proportion of crashes that involved vehicles carrying 5 or more passengers.

b. The total number of rollover crashes in the table is 224 + 16 + 16 + 7 = 263.

i. Of these 263 rollover crashes, 224 involved vehicles carrying fewer than 5 passengers. If the investigator chooses one rollover crash at random, the probability that the chosen crash involved a vehicle carrying fewer than 5 passengers is $\frac{224}{263} \approx 0.8517$.

ii. The proportion of rollover crashes involving vehicles with more than 15 passengers is $\frac{7}{263} \approx 0.0266$. Thus approximately $1 - 0.0266 = 0.9734$ of the rollover crashes involved vehicles carrying 15 or fewer passengers.

Summary

In this topic, we discussed basic concepts of probability: random process, sample space, outcomes and events, probability of an event. We calculated the probability of an event by computing relative frequency and by counting the number of outcomes in the sample space. We also discussed several basic probability rules.

Explorations

1. Give the sample space for each of the following random processes:

   a. You survey the customers who go to an automatic teller machine and record the amount of money withdrawn from the ATM on one visit. (You will need to make reasonable assumptions about possible increments of money that can be withdrawn and the total amount that can be withdrawn.)

   b. You place pennies, nickels, dimes, and quarters in a jar and pick one coin from the jar.

   c. You place pennies, nickels, dimes, and quarters in a jar and pick two coins from the jar. Consider whether it matters if you draw with or without replacement and whether or not the order is important to the situation.

   d. You question students at a particular college and record the last digit of their college ID number.

2. You toss two dice and compute the "difference," which you will define to be the larger top number minus the smaller top number if the numbers are different, and 0 if the numbers are the same.

   a. How many equally likely elements are there in the sample space of all tosses of two dice?

b. List all outcomes in the sample space that correspond to the event Z = the difference is 0.

c. Find $\mathcal{P}$(Z) = $\mathcal{P}$(difference is 0).

d. Find the probability that the difference is 6

e. Find the probability that the difference is greater than 0.

f. Find the probability that the difference is 5.

3. The following table contains information on the 2002 resident population of the U.S., by age. (Source: *The New York Times Almanac 2004,* page 277.)

Age	Younger than 18 years old	18 to 24 years old	25 to 44 years old	45 to 64 years old	65 years and older
Number in Thousands	1,107,108	452,196	1,270,419	1,068,243	588,542

a. If a resident of the U.S. is chosen at random, find the probability that he or she is 25 to 44 years old.

b. If a resident is chosen at random, find the probability that he or she is older than 24 years old.

c. In what age category does the median age fall?

4. One card is drawn at random from a well-shuffled standard deck of 52 cards:

a. Find the probability that the card drawn is a heart.

b. Find the probability that the card drawn is an ace.

c. Use the counting method to find the probability that the card drawn is an ace or a heart.

d. Explain why probability rule 5 given previously does not work for this exercise; that is, explain why $\mathcal{P}$(heart or ace) does not equal $\mathcal{P}$(heart) + $\mathcal{P}$(ace).

e. Explain why $\mathcal{P}$(ace or two) = $\mathcal{P}$(ace) + $\mathcal{P}$(two).

5. Two fair dice are tossed and we record the sum of the two top numbers:

   a. Find $\mathcal{P}$(sum is 7).

   b. Find $\mathcal{P}$(sum is 11).

   c. Find $\mathcal{P}$(sum is odd).

   d. Explain why $\mathcal{P}$(sum is odd) = $\mathcal{P}$(sum is 3) + $\mathcal{P}$(sum is 5) + $\mathcal{P}$(sum is 7) + $\mathcal{P}$(sum is 9) + $\mathcal{P}$(sum is 11).

6. If a state lottery involves matching five numbers between 1 and 99, where order in which the numbers are chosen is not important, there are 71,523,144 possible choices for the five winning numbers.

   a. If you buy one lottery ticket, what is the probability that you will pick the five winning numbers? What if you buy two lottery tickets?

   b. Explain in words what 1 minus the probability calculated in part a of this exploration represents.

7. Describe how the following probabilities might have been obtained:

   a. "The likelihood of another shuttle failure is 1 in 145."

   b. "A one-in-25,000 risk of being hit while crossing the street."

   c. "A one-in-a-million chance of being hit by lightning."

8. A jar contains five quarters and eight silver dollars. You pick a coin at random from the jar.

   a. Give the sample space for the random process.

   b. Find the probability that the coin picked is a silver dollar.

   c. You pick a second coin from the jar without replacing the first coin in the jar. Give the sample space for the random process of picking two coins from the jar without replacement.

   d. Are the elements listed in your sample space in part c of this exploration equally likely? Why or why not?

9. The following table contains information about the percentage of people in the U.S., 15 years old and older, who fall into each of the following marital status classes for 2002: never married, married, widowed, divorced, or separated. (Source: *The New York Times Almanac 2004*, page 281.) Assume these categories are disjoint; that is, each person is in one and only one of these categories.

Never Married	Married	Widowed	Divorced	Separated
28.5%	53.6%	6.3%		2.1%

a. Fill in the appropriate percent for "divorced."

b. If a person 15 years old and older is chosen from this population at random, what is the probability that he or she is either married or divorced?

10. Refer to Example 18.3.

a. Assume that the professor picks the second name without replacing the first name in the hat before he picks the second, what is the probability that Carol is chosen?

b. Assume that the professor picks the second name after he has replaced the first name in the hat. What is the probability that Carol is chosen?

c. Suppose that class has six students: Aneesa, Benito, Carol, Donovan, Eve, and Frank. If the professor picks the second name without replacing the first name in the hat, what is the probability that Frank is chosen?

d. Suppose the class has six students: Aneesa, Benito, Carol, Donovan, Eve, and Frank. If the professor picks the second name after replacing the first name in the hat, what is the probability that Frank is chosen?

e. In Example 18.3, you assumed that the order in which the students were chosen was not important. Describe a similar scenario where the order in which the students are chosen would be important.

TOPIC 19

Conditional Probability and Tables

OBJECTIVES

After completing this topic, you will be able to

» Obtain information from two-way tables

» Analyze data given in two-way tables

» Look for relationships using conditional probability

» Understand and calculate conditional probabilities

» Determine when two events are independent

A **two-way table** is a valuable tool for organizing information in which each individual represented in the table is characterized in two different ways. For example, if the individuals are people, they might be characterized by gender and by age, or they might be described by highest educational level and by family income level. We represent one variable using the rows and the other variable using the columns, and each cell contains the number of individuals who fall into that column and row classification. Often, looking at data organized in a two-way table allows us to explore the association between the variables and to see relationships that we may have otherwise overlooked.

In Example 18.6, we obtained information about single-vehicle crashes through two separate tables: one table gave the number of rollover accidents, while the other presented information on accidents that did not involve a rollover. We can organize that data in a single two-way table to help analyze it, as the first example shows.

Example 19.1

The following two-way table contains the crash data given in Example 18.6. In this table, each single-vehicle crash is classified in two different ways: by the number of passengers in the vehicle and by whether it involved a rollover or not.

Passengers	Fewer Than 5	5 to 9	10 to 15	More Than 15
Rollover	224	16	16	7
No Rollover	1591	61	39	3
Total	1815	77	55	10

a. Among the crashes that involved a vehicle carrying more than 15 passengers, what proportion were rollovers?

b. Among the crashes that involved a vehicle carrying 10 to 15 passengers, what proportion were rollovers?

c. Among the crashes that involved a vehicle carrying 5 to 9 passengers, what proportion were rollovers?

d. Among the crashes that involved a vehicle carrying fewer than 5 passengers, what proportion were rollovers?

e. What conclusions might be drawn from the results in parts a through d of this example?

Solution

a. To focus on crashes that involved a vehicle carrying more that 15 passengers, we look solely at the last column of the table. There were a total of 10 such crashes, and 7 of them involved rollovers. So, $\frac{7}{10} = 0.70$ or 70 percent of all crashes involved rollovers.

b. Among the 55 single-vehicle crashes involving vehicles carrying 10 to 15 passengers, 16 of them were rollovers, as seen in the third column of data in the table. This means that $\frac{16}{55} \approx 0.29$ or approximately 29 percent of the crashes involving vehicles carrying 10 to 15 passengers were rollovers.

c. For those 77 crashes involving vehicles carrying 5 to 9 passengers, 16 involved rollovers. Thus, $\frac{16}{77} \approx 0.21$ or approximately 21 percent of such crashes involved rollovers.

d. Among vehicles carrying fewer than 5 passengers, $\frac{224}{1815} \approx 0.12$ or approximately 12 percent involved rollovers.

e. It appears that vehicles that carry more than 15 passengers have a much higher risk of rollover when fully loaded than do other vehicles. The vehicles least likely to be involved in rollovers are those carrying fewer than 5 passengers.

In the previous example we looked at marginal or conditional proportions. We might also think of these as conditional probabilities if we phrased our questions in terms of probabilities. For example, we could ask the following: If we choose a single-vehicle crash at random for an audit and know that it involved a vehicle carrying 15 or more passengers, what is the probability that it was a rollover?

In general, a **conditional probability** asks us to find the probability of some event A, given that an event B has occurred. Conditional probability questions look at distributions of one of the variables in a two-way table for given values of the other variable. If the two variables of interest are gender and income, these questions might contain the following types of phrases: *assuming that* the person chosen was female, find the probability that the person earned over $100,000 last year; or find the probability that the person chosen earned over $100,000 *given that* the person was female; or *among* females, what proportion earned over $100,000?

When dealing with conditional proportions and conditional probabilities, it is important to know whether we want to look at the proportion of As that occur among the Bs or the proportion of Bs that occur among the As. Equivalently, we might ask for the probability that A occurs given that B has occurred or for the probability of B given A. For example, among crashes involving vehicles carrying more than 15 passengers, $\frac{7}{10}$ or 70 percent of them were rollovers, which is a large proportion, but only a small proportion of rollovers involved vehicles carrying 15 or more passengers ($\frac{7}{263}$ or approximately 2.7 percent).

Example 19.2

The following table gives information about winter Olympic gold medal winners (1948–2002) for seven events. (Note that there was no four-man bobsled winner in 1960.) The row and column totals are included in the table. (Source: *The World Almanac and Book of Facts 2004,* pages 874–878.)

	Men's Downhill Alpine Skiing	Men's 50K Cross-Country Skiing	Men's 500 m Speed Skating	Men's Singles Figure Skating	Women's Singles Figure Skating	Ice Hockey	Four-man Bobsled	Total
U.S.	2	0	4	6	7	2	1	22
Other Countries	13	15	11	9	8	13	13	82
Total	15	15	15	15	15	15	14	104

a. What proportion of gold medals for these seven events were earned by U.S. athletes?

b. What proportion of gold medals for these seven events were for figure skating?

c. Among U.S. gold medals for these seven events, what proportion of them won for figure skating?

d. Among gold-medal winners from other countries for these seven events, what proportion of them won for figure skating?

e. What proportion of ice hockey medals went to countries other than the U.S.?

f. Among winter Olympic gold medals in these seven events that went to other countries, what proportion were awarded for ice hockey?

g. Compare the answers to parts e and f of this example.

Solution

a. There are a total of 104 gold medals represented in the table. Of these 104, 22 were from the U.S. Thus $\frac{22}{104}$ or approximately 21.2 percent of them were awarded to U.S. participants.

b. We first observe that the set of men's figure-skating medalists and the set of women's figure-skating medalists are disjoint. So we compute the proportion of men's figure-skating medalists and add to it the proportion of women's figure-skating medalists. We have $\frac{15}{104} + \frac{15}{104} = \frac{30}{104}$ or approximately 28.8 percent of the gold medals were for figure skating.

c. Among the 22 U.S. gold medals, 13 of them were for figure skating (6 for men and 7 for women). Thus $\frac{13}{22}$ or approximately 59.1 percent of the U.S. gold medals for these events were for figure skating.

d. There were 82 gold medals awarded to other countries, of which 17 were for figure skating. So $\frac{17}{82}$ or approximately 20.7 percent of gold medals awarded to other countries were for figure skating.

e. Of the 15 ice hockey medals, 13 went to other countries. That's $\frac{13}{15}$ or approximately 86.7 percent of the ice hockey medals went to countries other than the U.S.

f. There were 82 medals that went to other countries; 13 were for ice hockey, so $\frac{13}{82}$ or approximately 15.9 percent of the medals awarded to other countries were for ice hockey.

g. Both of the fractions used to answer the questions in parts e and f of this example use the 13 ice hockey medals given to other countries as the numerator. In part e, we want to know what proportion of all ice hockey medals those 13 medals represent. In part f, we want to consider what proportion of all medals awarded to other countries those 13 medals represent.

If A and B are two specific events, then "A or B" is the event that either A occurs or B occurs or possibly both occur. "A and B" is the event that both individual events, A and B, occur. In Topic 18, we considered a probability rule for $\mathcal{P}$(A or B) when events A and B are disjoint events; that is, they do not have any outcomes in common. In that case, the rule is: $\mathcal{P}$(A or B) = $\mathcal{P}$(A) + $\mathcal{P}$(B). This rule was used in Example 19.2b where we computed the proportion of figure-skating medals as the proportion of men's figure-skating medals plus the proportion of women's figure-skating medals. (Note that if A and B are disjoint events, then $\mathcal{P}$(A and B) = 0, because A and B have no outcomes in common and so both cannot occur simultaneously.)

If A and B are not disjoint events, then we cannot compute $\mathcal{P}$(A or B) as the sum of the probability of A plus the probability of B. We must be careful not to "double count" outcomes that lie in both A and B. An additional probability rule that holds, in general, for any two events A and B is:

$$\mathcal{P}(\text{A or B}) = \mathcal{P}(\text{A}) + \mathcal{P}(\text{B}) - \mathcal{P}(\text{A and B})$$

Note that this formula is valid for any two events A and B. If A and B are disjoint, then the last term, $\mathcal{P}$(A and B), is 0 and the formula reduces to $\mathcal{P}$(A or B) = $\mathcal{P}$(A) + $\mathcal{P}$(B), which we saw earlier.

In the next example, we will consider how to figure out $\mathcal{P}$(A or B) and $\mathcal{P}$(A and B) from the information in a two-way table.

Example 19.3

A measure to authorize extension of nondiscriminatory treatment (normal trade-relations treatment) to the People's Republic of China, and to establish a framework for relations between the U.S. and the People's Republic of China passed the Senate on September 19, 2000, with the breakdown of votes as given in the following table. (Source: *The Library of Congress*, www.congress.gov.)

	Democrat	Republican	Total
Yea	46	37	83
Nay	8	7	15
Not Voting	2	0	2
Total	56	44	100

Suppose we pick a senator from the September 2000 senate at random.

a. Find the probability that the senator is a Democrat and voted Nay.

b. Find the probability that the senator is a Democrat or voted Nay.

c. Find the probability that the senator voted Nay or was Not voting.

d. Find the probability that the chosen senator voted Yea, given that he or she was a Democrat.

e. Find the probability that the chosen senator voted Yea, given that he or she was a Republican, and compare this answer with your answer in part d of this example.

Solution

a. We want to identify how many "outcomes" (that is, senators) are Democrats *and* voted Nay. These are the outcomes that fall in the "Democrat" column and in the "Nay" row. Eight senators out of the total senate count of 100 are in both the "Democrat" column and the "Nay" row. Using the counting method, $\mathcal{P}$(Democrat and Nay voter) = $\frac{8}{100}$ or 8 percent.

b. We can't use the rule for $\mathcal{P}$(A or B) for disjoint events A and B to find the probability that the chosen senator is a Democrat *or* voted Nay, because the events "is a Democrat" and "voted Nay" are not disjoint. As we saw in part a of this example, there are 8 senators who fall into both categories. So $\mathcal{P}$(Democrat or Nay voter) = $\mathcal{P}$(Democrat) + $\mathcal{P}$(Nay voter) – $\mathcal{P}$(Democrat and Nay voter) = $\frac{56}{100} + \frac{15}{100} - \frac{8}{100} = \frac{63}{100}$ or 63 percent. We could also use the counting method to count all outcomes that fall into either the Democrat column or the Nay row, taking care not to count any outcomes more than once. There are 46 + 8 + 2 = 56 Democrats and 7 Nay voters who were not yet counted in the Democrat count, for a total of 63 senators who are Democrats or Nay voters. Thus there is probability $\frac{63}{100}$ of choosing a Democrat or Nay voter.

c. Because events "voted Nay" and "Not voting" are disjoint (that is, there are no senators counted in both groups), $\mathcal{P}$(voted Nay or Not voting) = $\mathcal{P}$(voted Nay) + $\mathcal{P}$(Not voting) = $\frac{15}{100} + \frac{2}{100} = \frac{17}{100}$ or 17 percent.

d. Because we know that the chosen senator is a Democrat, we consider only the column corresponding to "Democrat." Our sample space is reduced to the 56 Democrats. Using the counting method, of these 56, 46 voted Yea, so $\mathscr{P}$(Yea voter given Democrat) = $\frac{46}{56} \approx 0.82$.

e. Because we know that the chosen senator is a Republican, we consider only the 44 senators in the Republican column. Of these, 37 voted Yea, so $\mathscr{P}$(Yea voter given Republican) = $\frac{37}{44}$ or approximately 84 percent. So the proportion of senators voting Yea among Democrats is approximately the same as the proportion of senators voting Yea among Republicans. This is roughly the same as the proportion of senators who voted Yea, which was $\frac{83}{100}$ or 83 percent.

In Example 19.3d and e, we saw that $\mathscr{P}$(voted Yea given Democrat) is approximately equal to $\mathscr{P}$(voted Yea) and $\mathscr{P}$(voted Yea given Republican) is approximately equal to $\mathscr{P}$(voted Yea). In general, when $\mathscr{P}$(A given B) = $\mathscr{P}$(A), for two events A and B, we say that A and B are **independent events**. The probability that A has occurred is the same whether we have information about whether B has occurred or not. In the context of Example 19.3, the probability that a chosen senator voted Yea on the bill is approximately the same whether we know the senator is a Democrat or not.

It is sometimes convenient to use the shorthand notation of a vertical line "|" to denote the word "given" in $\mathscr{P}$(A given B) and write it as: $\mathscr{P}$(A | B). We will look further at the notion of independent events in the next two examples.

Example 19.4

On April 2, 2001, the Senate approved a bill to amend the Federal Election Campaign Act of 1971 to provide bipartisan campaign reform. The voting tally of the senators is given in the next table. (Source: *Library of Congress,* www.congress.gov.)

	Democrat	Republican	Total
Yea	47	12	59
Nay	3	38	41
Total	50	50	100

A senator is chosen at random from those voting on this bill. Let event Y = senator voted Yea; let N = senator voted Nay; let D = senator is a Democrat; and let R = senator is a Republican.

a. Find $\mathcal{P}(\mathrm{N} \mid \mathrm{D})$ and compare with $\mathcal{P}(\mathrm{N})$. What do these probabilities tell us about the independence of the events N and D?

b. Are the events Y and R independent?

c. What meaning do the answers to parts a and b have in the context of this example?

Solution

a. $\mathcal{P}(\mathrm{N} \mid \mathrm{D})$, which we read as "the probability the chosen senator voted Nay given that he or she was a Democrat," is obtained from the 50 Democrats represented in the first column of data. Of these, 3 voted Nay so $\mathcal{P}(\mathrm{N} \mid \mathrm{D}) = \frac{3}{50}$ or 6 percent. A total of $\frac{41}{100}$ of the senators voted Nay so $\mathcal{P}(\mathrm{N}) = 0.41$ or 41 percent. Because $\mathcal{P}(\mathrm{N} \mid \mathrm{D}) \neq \mathcal{P}(\mathrm{N})$, the events N and D are not independent events. If we know the chosen senator is a Democrat, it will change the likelihood that he or she voted Nay.

b. We will compare $\mathcal{P}(\mathrm{Y} \mid \mathrm{R})$ and $\mathcal{P}(\mathrm{Y})$. $\mathcal{P}(\mathrm{Y} \mid \mathrm{R}) = \frac{12}{50}$ or 24 percent but $\mathcal{P}(\mathrm{Y}) = 0.59$ or 59 percent. Because they are not equal, Y and R are not independent.

c. We see that the probability that a Democrat voted Nay is only 6 percent, while the probability that a Republican voted Nay is 76 percent. It is clear that the majority of Democrats favored the bill while a majority of the Republicans were against it. The likelihood that a senator voted Yea on this bill depends very much on his or her political party.

There are some scenarios where we know from the setting that events are independent; that is, we know that the outcome of one event does not impact the outcome of the other event. If we conduct an experiment and then repeat it, we often want to set up the conditions so the outcome of the first experiment will not influence the outcome of the repeated experiment; that is, we want to obtain independent results. Other situations in which we can reasonably assume independence include the following: the likelihood of a baseball player getting a hit at any particular time at bat is independent of any hit obtained during his or her last time at bat; the likelihood of failure of a particular stereo system component (compact disc player, tape player, speakers) is independent of any other component's failure; the number of people in the party when making a call to a restaurant for a reservation is independent of the number of people in the party of the previous call.

Example 19.5

Explain how independence of events enters into the reasoning process in each of the following scenarios:

a. A coin is tossed repeatedly. The first 10 tosses land heads. What is the probability of heads on the 11th toss?

b. We pick a card at random from a standard deck of 52 cards and then pick a second card, also at random, with replacement (after shuffling the cards). Are the events A1, obtaining an ace on the first draw, and A2, obtaining an ace on the second draw, independent?

c. We pick a card at random from a standard deck of 52 cards and then pick a second card, also at random, without replacement. Are the events A1, obtaining an ace on the first draw, and A2, obtaining an ace on the second draw, independent?

d. A family consists of three sons. What is the likelihood that their fourth child will be a girl?

Solution

a. Assuming the coin is a fair coin, each time the coin is tossed, the probability of obtaining heads face up is $\frac{1}{2}$. The coin does not "remember" the result of any prior toss. The frequency interpretation of probability says that as we increase the number of repetitions (in this case, the number of tosses), the proportion of heads gets closer to the theoretical probability of heads. But no matter how many heads have occurred in a row, the probability of heads on the next toss is still $\frac{1}{2}$.

b. If we replace the first card in the deck before we randomly pick the second card, $\mathscr{P}(\text{A2} \mid \text{A1}) = \mathscr{P}(\text{A2}) = \frac{4}{52}$, and they are independent events. Knowing that an ace was picked on the first draw does not change the likelihood of picking an ace on the second draw.

c. If we pick without replacement, the events A1, obtain an ace on the first draw, and A2, obtain an ace on the second draw, are not independent. In this case, knowing that an ace was obtained on the first draw impacts the probability of obtaining an ace on the second draw because at the time of the second draw only 3 aces and 51 cards are left in the deck.

d. This situation is interpreted to be similar to a coin toss, although the probability of having a boy baby in the U.S. is approximately 0.51 while the probability of having a girl baby is 0.49. (This estimate is obtained from 2001 demographic information in the *Time Almanac 2004,* page 187.) The probability of having a girl on the fourth try would be 0.49, the same as the probability of having a girl on the first or second or third try.

One particularly important area in which looking at conditional probabilities can help us make decisions is in the area of testing for a condition or disease. We will construct an example within this context. This example will involve setting up a table from given conditional probabilities, which is the reverse of what we did in Examples 19.1 through 19.4, where we were given tables of data and figured out the conditional probabilities from the table.

Example 19.6

Consider a diagnostic test for a rare disease, such as hepatitis. You are being tested, and neither you nor your doctor knows whether or not you have the disease.

a. Assume a total population of 500 people tested and set up a two-way table with the columns representing categories H: "has the disease" and F: "free of the disease," and the rows representing the test results P: "test is positive" and N: "test is negative." To set up the table, assume that approximately 10 percent of all people have the disease. Also assume that, among people with the disease, the test is known to give positive results in 90 percent of the cases, and among people free of the disease, the test gives negative results in 98 percent of the cases. (These last two results are known as the test's **sensitivity** and **specificity**, respectively, and are often available from the results of clinical trials.)

b. Find the probability that if a person tests positive, he or she really has the disease.

c. What proportion of people who test negative have the disease?

d. What implications would these results have?

Solution

a. Because we have a population of 500 people tested, we first fill in the "grand total" of 500 people. We are assuming that 10 percent of them have the disease, so the total for the H column is 10 percent of 500 or 50. That leaves 450 for the F column total (the 90 percent of people who do not have the disease).

	H (has the disease)	F (free of disease)	Row Total
P (test is positive)			
N (test is negative)			
Column Total	50	450	500

Next we fill in the conditional probabilities given that among people with the disease (the H column), the test is positive for 90 percent of the cases. For the 50 people with the disease, 90 percent of 50 or 45 of them test positive; thus 5 of them test negative. Similarly, among those free of the disease (the F column) 98 percent of the 450 people, which is $0.98 \times 450 = 441$, test negative. Thus, 9 of the 450 who are free of the disease will test positive. Finally, we fill in the row totals.

True Condition	H (has the disease)	F (free of disease)	Row Total
P (test is positive)	45	9	54
N (test is negative)	5	441	446
Column Total	50	450	500

b. To find the probability that if a person tests positive, he or she really has the disease, we find a conditional probability given that a person tests positive. In symbols, we want to find $\mathscr{P}(\text{H} \mid \text{P})$. Look at the first row; we find among those 54 people who tested positive, 45 have the disease. Thus, $\mathscr{P}(\text{H} \mid \text{P}) = \frac{45}{54}$, which is approximately 83 percent. This means that approximately 17 percent of those who test positive don't have the disease.

c. This question asks us to find $\mathscr{P}(\text{H} \mid \text{N})$, so we look at the second row. Out of the 446 people who tested negative, 5 have the disease. Thus, $\mathscr{P}(\text{H} \mid \text{N}) = \frac{5}{446}$ or approximately 1 percent.

d. Approximately 17 in every 100 people who are identified as having the disease do not really have it. These people might be treated for the disease when they should not be, which might cause side effects and cost money and time. One in 100 people who test negative really do have the disease. They may go untreated. Note that these percentages depend on the proportion of all people who have this particular disease (which we assumed to be 10 percent in this example) and on the sensitivity and specificity of the particular diagnostic test.

Summary

In this topic, we used a two-way table to represent data in which each individual in the data set is characterized in two different ways. Using two-way tables, we looked at conditional proportions, which are conditional probabilities. We also discussed the concept of independent events. Finally, we explored the use of conditional probability and two-way tables to help make decisions when testing for a condition or disease.

Explorations

1. The following table gives the number of U.S. families, in thousands, with pet dogs, pet cats, and pet birds, by size of family. (Source: *Statistical Abstract of the United States 2000, 120th Edition.*)

Family Size	One Member	Two Members	Three Members	Four or More
Pet Dog	4,120	9,670	6,680	10,760
Pet Cat	4,540	8,880	5,560	8,070
Pet Bird	180	440	330	550

   a. Among one-member families with pets, what proportion have pet cats?

   b. Among families with pet cats, what proportion of them are families of size one?

   c. Given that a pet-owner family has four or more members, find the proportion with a pet bird.

   d. Given that a family has a pet bird, find the probability that the family has four or more members.

   e. If a family with one of these pets is chosen at random, find the probability that the family has two members and a dog.

   f. If a family with one of these pets is chosen at random, find the probability that the family has two members or it has a dog.

2. The following table gives participation in selected sports activities in 1998, in thousands, for persons living in the U.S. age 7 or older. (Source: *Statistical Abstract of the United States 2000, 120th Edition.*)

Sport	Walking	Swimming	Exercise Equipment	Bicycle Riding	Basketball	Golf	Total
Male	28,368	26,993	21,424	22,937	20,166	21,757	141,645
Female	49,278	31,256	24,721	20,598	9,251	5,739	140,843
Total	77,646	58,249	46,145	43,535	29,417	27,496	282,488

a. Of those who participate in walking, what proportion are female?

b. Of those who participate in walking, what proportion are male?

c. Of those who participate in golf, what proportion are female?

d. Of those who participate in golf, what proportion are male?

e. Among the males represented in the table, find the proportion who participate in walking and find the proportion who participate in golf.

f. Compare your results from parts b, d, and e of this exploration.

g. Suppose an individual age seven or older and represented in this table is chosen at random. Are the events "person is male" and "person participates in bicycle riding" independent? Explain your answer.

3. A newspaper article reported that a woman was misdiagnosed with a rare form of cancer. In fact, she was told by three different doctors that she tested positive for that type of cancer. Weeks later it was determined that she had "false positive" results. Explain what false positive results mean in this context.

4. The next table shows the final results for the vote on the Lawsuit Abuse Reduction Act in the House of Representatives on October 27, 2005. The bill, which originated in the House, is titled "To amend Rule 11 of the Federal Rules of Civil Procedure to improve attorney accountability, and for other purposes." The bill passed the House and was then sent to the Senate. (Source: *Library of Congress,* www.congress.gov.)

	Aye	Nay	Not Voting	Total
Republican	212	5	13	
Democratic	16	178	8	
Independent	0	1	0	
Totals	228	184	21	

A representative is chosen at random from those considered in the table. Let Y = representative voted Yea; let N = representative voted Nay; let D = representative is a Democrat; and let R = representative is a Republican.

a. What is the probability that the chosen representative voted Yea?

b. What is the probability that the chosen representative is a Democrat?

c. Given that the chosen representative voted Yea, what is the probability that he or she is a Republican?

d. Represent the following probability in symbols and find it: among Nay-voting representatives, what proportion are Democrats?

e. Represent the following probability in symbols and find it: if the chosen representative was Republican, what is the probability that he or she voted Yea?

5. Suppose we pick one card at random from a standard deck of 52 cards. Let K represent the event, "the card is a King"; let Q represent the event, "the card is a Queen"; let H represent the event, "the card is a Heart"; let C represent the event, "the card is a Club."

a. Find $\mathscr{P}(K)$.

b. Find $\mathscr{P}(H)$.

c. Are the events H and C disjoint? Find $\mathscr{P}(H \text{ and } C)$.

d. Are the events K and H disjoint? Find $\mathscr{P}(K \text{ and } H)$.

e. Find $\mathcal{P}$(H or C).

f. Find $\mathcal{P}$(K or H).

g. Find $\mathcal{P}$(K or Q).

6. A new test has been developed for a rare condition that occurs in approximately 1 out of 1,000 people. The test has a sensitivity of 100 percent and a specificity of 99.9 percent.

   a. Assume a population of 1,000 people and set up a two-way table, with the columns labeled *H* (*has the condition*) and *F* (*free of the condition*) and the rows labeled *P* (*test is positive*) and *N* (*test is negative*).

   b. Fill in the appropriate column totals and use the given sensitivity and specificity to fill in the column amounts. Lastly, fill in the row totals.

   c. Suppose an individual tests positive for the rare condition using this new test. What are the chances that the test results are erroneous? What should this individual do?

   d. If this were a test for a condition that could be passed between married individuals, would you argue for or against making the test mandatory for all couples prior to marriage? Why?

7. Explain how independence of events enters into the reasoning process in each scenario:

   a. A college student is chosen at random. Are the events "student has GPA over 3.5" and "student is female" independent?

   b. You are playing *Monopoly* and toss doubles three times in a row and get sent to jail. Are the events "get doubles on first toss" and "get doubles on second toss" independent?

   c. You buy a lottery ticket during the first week of June and you buy a ticket during the first week of July. Are the events "you win the first week of June" and "you win the first week of July" independent?

TOPIC 20

Sampling and Surveys

OBJECTIVES

After completing this topic, you will be able to

- » Identify the population and the sample in reported research
- » Determine the explanatory and response variables and whether a reported study is an observational study or an experiment
- » Recognize sources and types of bias in observational studies and experiments
- » Use a random number table to collect a simple random sample or a stratified random sample
- » Understand one method for collecting responses to sensitive questions

The *Newsweek* magazine cover story in the August 9, 2004, issue addressed, "The Mystery of Dreams." The article looked at the history of dream research and cited results from experiments and results from observational studies. In one of the dream-research experiments, one group of subjects was deprived of REM sleep (a phase of sleep characterized by intense brain activity and rapid eye movement), while another group was allowed REM sleep. Differences in dreaming patterns were recorded. In another study, a psychology professor collected 550 dreams from a group of twenty-four 9- to 15-year-olds and analyzed them, looking for age and gender differences in their dreams. In this topic, we will explore issues related to these types of studies.

In an **experiment**, the investigator imposes some treatment or condition on the subjects of the experiment and records how the subjects respond. In an **observational**

study, the investigator observes the subjects, without attempting to manipulate or impose any treatment, and records the responses. In both kinds of studies, researchers often will draw conclusions.

In both experiments and observational studies, researchers frequently want to find a relationship between two or more variables. A **response variable** or **outcome variable** is a variable that measures an outcome of a study. An **explanatory variable** is any variable that explains or influences the different responses measured by the response variable. For example, Harvard researchers conducted a ten-year study of more than 70,000 women and found that women getting fewer than five hours of sleep each night are more likely to develop diabetes (Source: *Prevention,* July 2003). Amount of sleep was the explanatory variable and incidence of diabetes was the response variable.

For any study, the entire group of individuals about which we want information is called the **population**. The portion of the population that we actually use to collect information is a **sample**, a subset of the population.

Example 20.1

For each of the following scenarios taken from news articles, identify the population and the sample, the explanatory variable(s) and response variables(s), and whether the scenario describes an experiment or an observational study.

a. A recent study of New Jersey drivers showed that approximately 40 percent of the 250 drivers surveyed scored highly to moderately stressed when driving. Nearly 58 percent reported that they had high to moderate levels of anger while driving. Almost half of those surveyed were prone to "take it out" on other drivers. The survey was administered at American Automobile Association offices in New Jersey.

b. Headline: "T'ai Chi Fights Shingles"

   "*T'ai Chi Chih* (TCC), a form of nonmartial art T'ai Chi, involves 20 slow movements said to help balance the flow of *chi,* or vital energy. 'TCC is a mind-body practice that combines exercise, stress reduction, and relaxation,' says Michael Irwin, MD, professor at UCLA's Neuropsychiatric Institute. Research has shown that tai chi helps improve balance, is good for your heart, and may ease arthritis pain. In a study conducted by Irwin, 18 people age 60 or older took 45-minute TCC classes three times a week for 15 weeks; 18 others were wait-listed as a control group. Both groups were given blood tests before the first class and a week after the last one to measure resistance to the varicella-zoster virus (VZV), which causes shingles, an acutely painful nerve-cell infection. VZV resistance increased by 50% in the TCC group but remained unchanged in the control group. What's more, energy and well-being increased in the TCC group. Irwin speculates that TCC could raise a person's immunity to VZV and potentially to other viruses, especially if resistance is lowered because of depression or stress." (Source: *Prevention*, April 2004.)

c. An article in *The Journal of the American Medical Association* (March 21, 2001) reported the results of a study on the benefits of walking. Researchers studied nearly 40,000 women and found that just one hour of walking per week reduced the risk of heart disease by 14 percent. Women who walked one hour to one-and-a-half hours per week cut their risk of heart disease by 51 percent. These results were seen even among women who smoked and had high cholesterol, although women with high blood pressure did not exhibit this benefit.

Solution

a. The sample consisted of New Jersey drivers who were willing to participate in the survey at American Automobile Association (AAA) offices around the state. The population would likely consist of New Jersey drivers who visit AAA offices. The explanatory variable was the levels of stress and anger people felt while behind the wheel of a car. The response variable was the number of incidents of aggressive acts while driving. This is an observational study because the researchers did not control the explanatory variable but rather "observed" by measuring people's self-perceptions of their anger levels when behind the wheel and incidents of aggressive driving.

b. The sample involved 36 individuals, probably from California, aged 60 and older. We may need more information to determine if men and women and various races were represented in the sample, but the population would likely be Californians aged 60 and older. The explanatory variable was a T'ai Chi class or no class while the response variable was resistance to the virus as measured by a blood test. This is an experiment because the researchers formed two groups, an experimental group and a control group. They controlled the explanatory variable by having one group take T'ai Chi classes, while a control group did not take the classes.

c. The sample involved nearly 40,000 women. The population would likely consist of all women. The explanatory variable in this study is amount of time spent walking per week. The response variable is occurrence of heart disease. Although the summary did not explicitly say, it appears that this is an observational study. It would be difficult for researchers to perform a controlled experiment involving 40,000 subjects, but an observational study would be possible.

Whether the study in question is an experiment or an observational study, it is very important that a sample be characteristic of the population that it represents. For example, consider the study on walking and women's heart health involving a sample of nearly 40,000 women described in Example 20.1c. We could not expect to draw conclusions about the effect of walking on heart disease risk in men based on this study because the results from the women in the sample may not be applicable to men.

We must also take care when designing an observational study or experiment that we control, as much as possible, for **confounding variables**. A confounding variable is a variable whose influence on the response variable cannot be separated from that of the explanatory variable. For example, a newspaper article titled "Study: Oscar Winners Live Longer" found that on average, Oscar winners lived to age 79.7 while nonwinners lived to be 75.8. (Source: *The Morning Call*, May 15, 2001.) The article reported on a study that included all 762 actors and actresses who had ever been nominated for an Academy Award. Each nominee was paired with an actor of the same gender and age. The researchers were investigating whether famous movie stars can benefit from a boost in self-esteem. They tried to control for the potential confounding variables of age and gender by matching each award winner with a "similar" actor. It is easier to control for confounding variables in experiments than in observational studies, but care must be taken with both types of studies.

The observational study involving Academy Award winners indicates one way to try to control for confounding variables in observational studies. When conducting an experiment, randomly assigning subjects to receive the different treatments or to be exposed to different conditions helps to control for confounding. The researcher reduces the chance of introducing any unwanted influences by randomizing. One way to choose subjects at random from a population is to put all the names in a bin, mix up the names, and pick them out one-by-one without looking at any of the names. A more sophisticated method, with a similar result, involves using a table of random digits or a random number generator on a calculator or computer. In the next example, we describe how to use a table of random digits to collect a random sample.

Example 20.2

Consider the following list of U.S. presidents and their ages at inauguration. The list is given in chronological order across the rows. We will consider this group as our population. We will then use this population to illustrate some ideas associated with sampling.

Washington, 57	J. Adams, 61	Jefferson, 57	Madison, 57
Monroe, 58	J. Q. Adams, 57	Jackson, 61	Van Buren, 54
W. H. Harrison, 68	Tyler, 51	Polk, 49	Taylor, 64
Fillmore, 50	Pierce, 48	Buchanan, 65	Lincoln, 52
A. Johnson, 56	Grant, 46	Hayes, 54	Garfield, 49
Arthur, 50	Cleveland, 47	B. Harrison, 55	Cleveland, 55
McKinley, 54	T. Roosevelt, 42	Taft, 51	Wilson, 56
Harding, 55	Coolidge, 51	Hoover, 54	F. D. Roosevelt, 51

Truman, 60	Eisenhower, 62	Kennedy, 43	L. B. Johnson, 55
Nixon, 56	Ford, 61	Carter, 52	Reagan, 69
G. H. Bush, 64	Clinton, 46	G. W. Bush, 54	

a. Choose a sample of five U.S. presidents as follows: number the presidents from 01 to 43, going across the first row for numbers 01 to 04 and proceeding left-to-right across each subsequent row. Use the following table of random digits; group the digits in the table in pairs (because we want to choose five numbers between 01 and 43). Ignore any pairs of numbers outside of the range of 01 to 43 and proceed across the first row and to the second row of digits, if necessary, continuing until you have five numbers between 01 and 43. Note that the numbers are separated into groups of five to facilitate reading them, but the grouping is not important. Record the five chosen random numbers.

12975 13258 13048 45144 72321 81940 00360 02428 96767 35964 23822
96012 94591 65194 50842 53372 72829 50232 97892 63408 77919 44575
24870 04178 88565 42628 17797 49376 61762 16953 88604 12724 62964
99612 93465 64658 27402 56319 81103 46759 14520 19807 46845 30862

Write down the name and age at inauguration of each president corresponding to your five numbers.

c. What is the average age at inauguration of the presidents in your sample and how does it compare to the average of the whole population, which is approximately 54.8?

d. Theodore Roosevelt was the first president elected after 1900. (He was elected in 1904, although he became president in 1901 after McKinley was assassinated.) How many presidents elected after 1900 are in your sample?

e. Repeat parts a, b, c, and d but use the third row in the given table of random digits to identify your sample of five. How do the samples compare?

f. How would we change this process if we had a population of 500 and wanted to pick a random sample of five from this population?

Solution

a. Grouping the digits by two's gives the following pairs of numbers: 12 97 51 32 58 13 04 84 51 44 72 32 18 19 40 and so on. We choose 12, and then ignore 97 and 51 because they are outside the range of 01 to 43. The next number chosen is 32; we ignore 58 and then select 13 and 04. Then we ignore 84, 51, 44, 72, and because we already selected 32, we ignore it also. The fifth number we pick is 18. Thus, the numbers we chose are 12, 32, 13, 04, and 18.

b. The names and ages of the chosen presidents are as follows: Taylor, 64 (he was number 12); F. D. Roosevelt, 51 (number 32); Fillmore, 50 (number 13); Madison, 57 (number 04); and Grant, 46 (number 18).

c. The average age of presidents in the sample is $\frac{64 + 51 + 50 + 57 + 46}{5} = 53.6$. It a bit less than the average age of the population.

d. Just one president in the sample was elected after 1900, F. D. Roosevelt.

e. We again group the digits in pairs and choose five numbers between 01 and 43. These pairs of digits in the third row are as follows: 24 87 00 41 78 88 56 54 26 28 17 79 74 93 76 61 76 21, and so on. The first five two-digit numbers in the range are 24, 41, 26, 28, 17. These correspond to Cleveland, 55; G. H. Bush, 64; T. Roosevelt, 42; Wilson, 56; A. Johnson, 56. Three of these presidents were elected after 1900. The average age for the presidents in this sample is $\frac{55 + 64 + 42 + 56 + 56}{5} = 54.6$. This average is close to the average of the population.

f. We could number the individuals in the population from 001 to 500 in an orderly fashion and choose three-digit numbers from our random number table. We would ignore the number 000, if it occurred, and numbers greater than 500. Using the first line of the table, with numbers grouped in threes, we would have the following: 129 751 325 813 048 451 447 232 181 940, and so on. Individuals 129, 325, 048, 451, and 447 would be in the sample of five.

Eighteen of the 43 U.S. presidents were elected after 1900, which is just over 40 percent of them. If, for some reason, we want to be sure we have a proportional representation of presidents elected before and after 1900, we might want to use **stratified random sampling** in which we would pick, in this case, three presidents from the group elected before T. Roosevelt (60 percent of our sample of five) and two from the remaining group. This would guarantee that the proportion in each of these groups or **strata** is approximately the same for both the sample and the population. We illustrate how to collect a stratified random sample in the next example.

Example 20.3

Collect a stratified random sample for the "Presidents" example using "elected before 1900" and "elected 1900 or later" as the strata or classes as follows:

a. Number the presidents elected before T. Roosevelt from 01 to 25 and use the second row (and continue to the third row if necessary) of random digits in the given table to choose three from this group, as we did in Example 20.2.

b. Number the remaining presidents from 01 to 18 and use the third row (and continue to the fourth row if necessary) of random digits to choose two more for a total sample of five. List the names and ages of the five chosen presidents and compute the average age of the five chosen presidents.

c. Why might we want to take a stratified random sample rather than a simple random sample?

Solution

a. We group the numbers from the second row of the table of random digits in pairs, and disregard any that are not in the range 01 to 25. These numbers are 96 01 29 45 91 65 19 45 08 42, and so on. We choose 01, 19, and 08, which gives us Presidents Washington, 57; Hayes, 54; and Van Buren, 54 from the first group of those elected before 1900.

b. Grouping digits from the third row in pairs, we get the following: 24 87 00 41 78 88 56 54 26 28 17 79 74 93 76 61 76 21 69 53 88 60 41 27 24 62 96 4. So far, we have only one number between 01 and 18 (and it's 17). Using the fourth row of the table, we continue our grouping, and pair the first digit in the fourth row with the 4 from the end of the third row. We get 49 96 12 93, and so on. We can stop there because we have our second sample president, number 12. Our two presidents from this group, "elected 1900 or later," are Clinton, 46 (number 17) and Nixon, 56 (number 12). The average age of this stratified random sample is $\frac{57 + 54 + 54 + 46 + 56}{5} = 53.4$.

c. In general, a stratified sample will guarantee that different groups or strata are represented in the sample. If we want to sample attitudes among college students, we might want to be sure to include students from the freshman, sophomore, junior, and senior classes in numbers proportional to their sizes. In addition, we might want to make sure we have a fair (that is, proportional) representation of both men and women.

It is not always possible to obtain results from a truly random sample for some practical reasons. If a sampling method produces results that are systematically different from the true results about the population, then the method is **biased**. **Selection bias** occurs when the sampling method either excludes entirely or includes disproportionately some particular section of the population. If there is some variable that is important to the study and on which those included and those excluded differ, the results of the study may be meaningless. In particular, phone-in surveys in which subjects self-select to respond often suffer from selection bias.

Another source of bias, **response bias**, occurs when questions on a survey or the behavior of the interviewer or the situation in which the interview is held influence the responses received. Special interest groups sometimes conduct surveys designed to promote a special cause and will word questions in such a way that the results will be biased in favor of their cause.

Nonresponse bias occurs when some of the subjects selected for the sample choose not to participate and the nonresponders are different from the responders. This might result in the received responses not being representative of the population from which the sample was taken. Many surveys, when done by telephone or mail, will have high nonresponse rates; often only those who care deeply about the results will choose to respond. Personal interview surveys usually result in a lower nonresponse rate.

Example 20.4

For each of the following scenarios, explain the type and source of any potential bias.

a. The magazine *Literary Digest* had been able to predict the results of the U.S. presidential election successfully until 1936. In that year, Republican Alf Landon and incumbent Democrat Franklin Delano Roosevelt were running for President. The *Literary Digest* sent out questionnaires to 10 million people whose names had appeared on lists of car owners, magazine subscribers, telephone directories, and registered voters. Based on 2.3 million responses, the *Literary Digest* predicted that Landon would win with 57 percent of the vote to Roosevelt's 43 percent. Roosevelt was reelected with over 60 percent of the popular vote.

b. The article, "Take Your *kids* to Work?" from *Working Woman*, April 2001, reported, "We polled the Fortune 100 to find out how many firms let boys come to Take Our Daughters to Work Day, April 26. Of 58 respondents, most allow boys to participate and a few companies have officially renamed their co-ed programs. 9% do not participate; 12% host boys-only and girls-only programs; 79% allow boys and girls to participate together."

c. A group of college students were randomly asked one of two questions: One group was asked, "Should this campus allow speeches that might incite violence?" The other group was asked, "Should this campus prohibit speeches that might incite violence?" Even though the two groups were chosen randomly, a larger proportion answered "no" to the first question than answered "yes" to the second question.

d. The article, "Most Ex-welfare Recipients Not Better Off, Survey Says" (*The Morning Call,* July 26, 2001) reported on a study that surveyed 893 former welfare recipients who were using social service facilities in ten states. The study found that their lives had not improved much even though the total number of people on welfare was lower.

Solution

a. Selection bias occurred in this survey, because by using telephone, voter, magazine, and car-owner lists, the poor tended to be excluded. The people excluded were more likely to support Roosevelt. Nonresponse bias (only 23 percent of those polled responded) also influenced the survey results because Roosevelt supporters, who were satisfied with the current situation, tended not to respond.

b. In this scenario, nonresponse bias might be present. Those responding might be more likely to include boys in their programs than the nonresponders, but we cannot know for certain unless we follow up and get information from the 42 companies polled who did not respond. (Note that 100 companies were polled. Because only 58 responded, 42 did not respond.)

c. Because the word *prohibit* in the second question carries negative connotations, people were more reluctant to answer, "yes, we want to prohibit this" than they were to answer, "no, we don't want to allow this." This is an example of response bias, because the wording of the question may have influenced the responses.

d. Some people criticized the agency for only surveying people at social welfare facilities. They claimed response bias might have been introduced because of the location chosen for the survey. Selection bias might also have occurred because, by going to a social service agency, the respondents might not have been a representative sample of former welfare recipients.

When analyzing results from surveys, researchers usually assume that the respondents are telling the truth. However, when asked sensitive questions, those surveyed might be reluctant to answer truthfully, especially in personal interviews or if they think someone might be aware of how they, personally, are responding. An interesting way to elicit candid responses to sensitive questions can be carried out using **Warner's randomized response model**, which is illustrated in the following example.

Example 20.5

We will first ask each college student respondent to toss a penny and a nickel and keep track of the result of each. We will then survey the respondents using a survey with two questions. Question 1, which we will denote by Q1 is, Have you ever cheated on a college exam? Question 2, which we will denote by Q2 is, Is the outcome on the nickel a head?

If the penny toss results in a head, the respondent answers Q1. If the penny comes up a tail, the respondent answers Q2. Each respondent tosses the coins so only he or she knows the results of the toss and then each respondent answers Q1 or Q2 and writes a single answer on his or her paper. Because no one knows which question the respondent is answering, there is no stigma attached to answering Yes. We will set up a table to find the proportion who answer Yes to Q1, the sensitive question.

a. Assume there are 100 respondents in the survey; we would expect about half of the respondents to toss a head on their penny and half to toss a tail, so half of those surveyed will answer Q1 and half will answer Q2. Fill in the row totals and the row corresponding to "Answered Q2" in the following table:

	Answered Yes	Answered No	Total
Answered Q1			
Answered Q2			
Total			100

b. Suppose that a total of 44 respondents answered Yes. Write this number in the table in the appropriate place and fill in the rest of the cells.

c. Find the probability that a respondent chosen at random answered Yes given that he or she answered question 1; that is, find $\mathcal{P}$(Answered Yes | Answered Q1). Explain what this probability represents.

Solution

a. We would expect 50 of the 100 respondents to answer Q1 and 50 to answer Q2. Of those 50 respondents who answered Q2 (Is the outcome on the nickel a head?), we expect 25 of them to answer Yes and 25 to answer No, assuming fair nickels. We can enter that information in the table as follows:

	Answered Yes	Answered No	Total
Answered Q1			50
Answered Q2	25	25	50
Total			100

b. The "Answered Yes" column total is 44. We can then fill in the rest of the cells by maintaining the row and column totals.

	Answered Yes	Answered No	Total
Answered Q1	19	31	50
Answered Q2	25	25	50
Total	44	56	100

c. This conditional probability is $\frac{19}{50} = 0.38 = 38$ percent. This gives us an estimate of the proportion of college students who have ever cheated on an exam, based on the information in the sample of 100 students.

This technique makes it possible for respondents to answer sensitive questions that have "yes" or "no" answers candidly and truthfully. Of course, how good an estimate this proportion is also depends on how representative the sample is of the population.

Summary

In this topic, we examined various issues surrounding studies and experiments. We discussed observational studies and experiments and considered the population and the sample in these studies. We discussed explanatory and response variables and looked at types of bias introduced into studies and experiments. We also investigated how to collect a simple random sample and how to collect a stratified random sample. Finally, we looked at Warner's randomized response model as a way to obtain honest responses to sensitive questions.

Explorations

1. The following article from *The Morning Call*, August 6, 2004, describes a "study." From the description, how do you think the research was carried out, was it an observational study or an experiment, and what can you conclude from the article?

Study: Lots of carbs might hike breast cancer risk

The Associated Press

High-carb diets might increase more than waistlines. New research suggests they might raise the risk of breast cancer.

Women in Mexico who ate a lot of carbohydrates were more than twice as likely to get breast cancer than those who ate less starch and sugar, scientists have found.

The study is hardly the last word on the subject, but is one of the few to examine how popular but controversial low-carb diets might affect the odds of getting cancer, as opposed to their effects on cholesterol and heart disease.

The new findings also don't mean that it is safe or healthful to eat lots of meat, cheese or fats, as many people who go on low-carb diets do, experts say.

"There are many concerns with eating diets high in animal fat," said Dr. Walter Willett, chief of nutrition at the Harvard School of Public Health. "If people do want to cut back on carbohydrates, it's really important to do it in a way that emphasizes healthy fats, like salads with salad dressings."

Willett worked on the study with doctors at Instituto Nacional de Salud Publica in Cuernavaca, Mexico. It was funded by the U.S. Centers for Disease Control and Prevention, the Ministry of Health of Mexico, and the American Institute for Cancer Research. Results were published Friday in the journal Cancer Epidemiology, Biomarkers & Prevention.

Fats, fiber, and specific foods have long been studied for their effects on various types of cancer, but few firm links have emerged. Being overweight is known to raise risk, but the new study took that into account and still found greater risk from high carbohydrate consumption.

Scientists think carbs might increase cancer risk by rapidly raising sugar in the blood, which prompts a surge of insulin to be secreted. This causes cells to divide and leads to higher levels of estrogen in the blood, both of which can encourage cancer.

A study earlier this year suggested that high-carb diets modestly raised the risk of colon cancer. Little research has been done on their effect on breast cancer, and results have been mixed. One study last year found greater risk among young women who ate a lot of sweets, especially sodas and desserts.

2. The article, "Secondhand Smoke Can Slow Circulation of Nonsmokers" (*The Morning Call,* August 7, 2001) reported on a study that examined how secondhand smoke affects the health of heart blood vessels. A research group in Japan measured the speed of coronary blood flow in 30 healthy men. Half of the subjects were smokers. The researchers took blood flow measurements, using a type of ultrasound, before and after exposing the subjects to secondhand smoke. The nonsmokers' blood flow before exposure was considerably higher than that of the smokers. After being exposed to secondhand smoke, however, their blood flow decreased to the level of the smokers' blood measurements. Blood flow of the smokers was not affected by secondhand smoke.

   a. Is this an observational study or an experiment?

   b. Identify the population and the sample described in the study.

   c. Identify the explanatory variable(s) and the response variable(s).

3. An article on cell phones and driving in *Prevention,* May 2004, reported that the risk of a traffic accident increases by 38 percent if the driver is talking on the phone while driving. Researchers estimate that using a cell phone while driving contributes to 330,000 injuries and 2,600 deaths per year.

   a. Is this an observational study or an experiment?

   b. Identify the population and the sample described in the study.

   c. Identify the explanatory variable(s) and the response variable(s).

4. The following paragraph is taken from the article, "The Evidence for Acupuncture." (Source: *Shape,* August 2004.)

   "In a study of 400 people (about 80 percent women) who suffered from severe headaches, largely migraines, for several days each month, one group received up to 12 acupuncture treatments in three months. Nine months later, the acupuncture group experienced the equivalent of 22 fewer headache days per year. They also used 15 percent less pain medication, made 25 percent fewer visits to physicians, missed 15 percent fewer days of work and had a generally improved quality of life."

a. Is this an observational study or an experiment?

b. Identify the population and the sample described in the study.

c. Identify the explanatory variable(s) and the response variable(s).

5. The following paragraphs are taken from the article, "Antibiotics' Role in Heart Attacks to Be Focus of Study." (Source: *The Morning Call,* April 5, 1999.)

"Could taking an antibiotic spare heart patients future heart attacks, bypass surgery or death?

"Researchers hope to find out with a large, federally funded study based on growing evidence a common bacterial infection that can lie dormant for years might trigger heart attacks and worsen, perhaps even cause, atherosclerosis—narrowing of the arteries from plaque deposits.

"Half of the 4,000 heart disease patients to be studied will be given a weekly antibiotic pill, Zithromax, for one year, while the comparison group gets a dummy pill. Neither will know who is taking which pill."

a. Is this an observational study or an experiment?

b. Identify the population and the sample described in the study.

c. Identify the explanatory variable(s) and the response variable(s).

d. Identify the source of any potential bias.

e. The "dummy pill" referred to in the article is also called a **placebo**. Why might that be an important consideration in a study such as this one?

6. A report in *Prevention,* March 2004, claims in its title that "Spring Veggies Fight Cancer and Stroke." The study tracked 40,000 people for 18 years and found that stroke risk was reduced by 26 percent among those who ate green or yellow vegetables almost daily.

a. Identify the explanatory variable(s) and response variable(s).

b. Give other possible explanations for the conclusion given in the quote.

7. An Internet poll posed the question, "What do you use the Internet for most?" The possible answers and the percentage of respondents who gave each answer was as follows: e-mail, 33.15 percent; surfing the web, 24.9 percent; education, 5.87 percent; online chatting 19.58 percent; shopping, 0.7 percent; online games, 6.57 percent; something else, 9.23 percent. (Source: *Internet Poll Questions,* www.jwen.com/poll/poll.html.)

   a. Identify the population and sample for this survey.

   b. Why might the results of this poll be biased?

8. Use the table of the 50 U.S. states and governors' salaries given in Example 16.3 (and repeated here) and the random digits given in Example 20.2 (also repeated here) to answer parts a through f of this exploration.

Table of States and Governors' Salaries

State	Salary	State	Salary	State	Salary
Alabama	96,361	Louisiana	95,000	Ohio	122,800
Alaska	85,776	Maine	70,000	Oklahoma	110,299
Arizona	95,000	Maryland	135,000	Oregon	93,600
Arkansas	75,296	Massachusetts	135,000	Pennsylvania	142,142
California	175,000	Michigan	177,000	Rhode Island	105,000
Colorado	90,000	Minnesota	114,506	South Carolina	106,178
Connecticut	150,000	Mississippi	122,160	South Dakota	98,250
Delaware	114,000	Missouri	120,087	Tennessee	85,000
Florida	124,575	Montana	93,089	Texas	115,345
Georgia	127,303	Nebraska	85,000	Utah	100,600
Hawaii	94,780	Nevada	117,000	Vermont	127,456
Idaho	98,500	New Hampshire	100,690	Virginia	124,855
Illinois	150,691	New Jersey	157,000	Washington	142,286

State	Salary	State	Salary	State	Salary
Indiana	95,000	New Mexico	110,000	West Virginia	90,000
Iowa	104,795	New York	179,000	Wisconsin	131,768
Kansas	98,331	North Carolina	118,430	Wyoming	105,000
Kentucky	107,130	North Dakota	87,216		

Random digits:

12975 13258 13048 45144 72321 81940 00360 02428 96767 35964 23822
96012 94591 65194 50842 53372 72829 50232 97892 63408 77919 44575
24870 04178 88565 42628 17797 49376 61762 16953 88604 12724 62964
99612 93465 64658 27402 56319 81103 46759 14520 19807 46845 30862

a. Starting in row two of the random digits, take a random sample of six states and find the average governor's salary of your sample.

b. Starting in row three of the random digits, take another random sample of six states and find the average governor's salary of this sample. How do the two samples compare?

c. What is the largest possible sample average salary that you could have obtained for a sample of size six?

d. What is the smallest possible sample average salary that you could have obtained for a sample of size six?

e. Would the largest possible sample average salary that you could have obtained increase or decrease if you took samples of size ten instead of size six?

f. Would the smallest possible sample average salary that you could have obtained increase or decrease if you took samples of size ten instead of size six?

9. Here is a list of the states that are east of the Mississippi River: Alabama, Connecticut, Delaware, Florida, Georgia, Illinois, Indiana, Kentucky, Maine, Maryland, Massachusetts, Michigan, Mississippi, New Hampshire, New Jersey, New York, North Carolina, Ohio, Pennsylvania, Rhode Island, South Carolina, Tennessee, Vermont, Virginia, West Virginia, Wisconsin.

   a. Suppose you take a sample of six states, stratified by location relative to the Mississippi River. Explain how you would get such a sample.

   b. Starting with the first line of the random digits table shown in Exploration 8, carry out the stratified sampling procedure you described in part a of this exploration.

   c. How is this sample different from the ones you collected in Exploration 8a and b?

10. Refer to Warner's randomized response model, described in Example 20.5, and find the probability that a respondent chosen at random answered Yes, given that he or she answered question 1. Assume there were 50 respondents in the survey and 23 of them answered Yes.

11. Use Warner's randomized response model to carry out a survey in class, using the question, "Have you ever used marijuana?" Set up a table to analyze the results. Find the probability that a student answered Yes given that he or she answered the sensitive question.

TOPIC 21

More on Decision Making

OBJECTIVES

After completing this topic, you will be able to

» Organize information for decisions that involve uncertainty

» Calculate the expected value (also called mean or average value) of a probability distribution

» Use the expected value to help make decisions that involve uncertainty

» Use the maximin and maximax criteria (when "payoffs" are profits) or the minimax and minimin criteria (when "payoffs" are costs) to help make a decision involving uncertain information

Suppose you have $1000 that you want to invest for the next five years; you just need to decide where to invest it. You could put your money in a standard savings account, in a mutual fund, or even buy stock with it. If you choose one kind of investment, your $1000 might grow more than if you made a different choice. You can track the earnings records for each type of investment, which will enable you to attach probabilities to your potential earnings. In Topic 11, we considered decisions that involve information that we know for certain. But many decisions involve information that contains some element of uncertainty; for some of these decisions, you might be able to estimate probabilities, but for others you might not. In this topic, we will consider some tools to help us make decisions with uncertain information.

Example 21.1

Suppose you are tracking a particular mutual fund and discover that during the previous ten-year period, the annual return was 7 percent for one year of that period, 8 percent for three of the years, 12 percent for two of the years, and 9 percent for four of the years. What was the average yearly return over the ten-year period?

Solution

We can answer the question by writing out the annual return for each of the ten years: 7, 8, 8, 8, 12, 12, 9, 9, 9, 9; then we add these ten numbers and divide by 10 to get the average: $\frac{7+8+8+8+12+12+9+9+9+9}{10} = 9.1$ percent. We could carry out the computations more efficiently by writing the average, or mean, as: $\frac{(7 \cdot 1)+(8 \cdot 3)+(12 \cdot 2)+(9 \cdot 4)}{10} = 7\frac{1}{10} + 8\frac{3}{10} + 12\frac{2}{10} + 9\frac{4}{10} = 9.1$ percent.

Note that if we were to choose a year at random from the ten-year period given in Example 21.1, the probability is $\frac{1}{10}$ that we choose a year in which the annual return was 7 percent; the probability is $\frac{4}{10}$ that we choose a year in which the annual return was 9 percent, and so on. This example suggests a definition for mean value, or average value, of a payoff if the probabilities of the various possible payoff values are known. Specifically, suppose we know possible values of a quantity and their associated probabilities, as shown in the following table:

Values V	v_1	v_2	v_3	...	v_n
Probabilities	p_1	p_2	p_3	...	p_n

The **mean value** or **expected value** is denoted by the Greek letter μ (mu) and is obtained by multiplying each possible value by its probability and then adding all these products. For the values and probabilities given in the previous table:

$$\mu = v_1 \cdot p_1 + v_2 \cdot p_2 + v_3 \cdot p_3 + \ldots + v_n \cdot p_n$$

Example 21.2

You get a special offer in the mail. A potentially winning ticket (along with special purchase price offers on a whole lot of magazines—no purchase necessary) arrives. One first prize of $1,000,000, five second prizes of $100,000, and ten third prizes of $1000 will be awarded. Your ticket contains the numbers 69-43-37-01-55-24. You read the fine print, which tells you that each individual was sent a ticket consisting of six distinct numbers between 1 and 70. To win, you must match all six numbers and the order of the numbers on the ticket is not important. The probabilities of winning first, second, and third prizes are, respectively, 7.63×10^{-9}, 3.81×10^{-8}, and 7.63×10^{-8}. To be entered into the drawing and eligible to win, you need to mail the enclosed ticket containing your numbers back to the company. Assuming you don't want to buy any magazines, is it worth the cost of a first-class postage stamp to send in your number?

Solution

We can summarize the prizes and probabilities in a table:

	1st Prize	2nd Prize	3rd Prize
Amount of Prize	$1,000,000 = 10^6	$100,000 = 10^5	$1000 = 10^3
Probability	7.63×10^{-9}	3.81×10^{-8}	7.63×10^{-8}

We'll compute the expected prize payout (in dollars) for each ticket as the sum of the amounts of the prizes times the probability: $\mu = 10^6 \times 7.63 \times 10^{-9} + 10^5 \times 3.81 \times 10^{-8} + 10^3 \times 7.63 \times 10^{-8} = 0.00763 + 0.00381 + 0.0000763 \approx .0115$ dollars. This means that, in the long run, if you sent in a large number of tickets of this type, your average payout per ticket would be 1.15¢. It probably is not worth the cost of postage to return such a ticket.

We can use the expected value to help make decisions among several possible alternatives when there is uncertainty about what may happen. Expected value is appropriate to use when an action will be done repeatedly over an extended period of time. In that case, looking at an "average" payoff or cost makes sense.

Example 21.3

You have just been offered your first full-time job and one of the benefits is health insurance. You must decide which of three healthcare plans to choose. The first plan costs $30 each month, and there is a $500 deductible annually. (This means that the employee pays all expenses until expenses for the year are $500.) After expenses have reached $500, the employee pays 20 percent of the costs, with the rest paid by the insurance company. The second plan is the same as the first, except that it costs $10 per month and the deductible amount is $1000. The third plan costs $25 per month, and there is no deductible. The employee pays 30 percent of all medical expenses with the rest paid by the insurance company. (All plans cover the same services: office visits, hospital visits, surgery, and prescriptions.)

a. How is uncertainty involved in this decision?

b. Assess yearly costs under each of the plans and for the various healthcare expense levels. Make a "decision table" containing the results and be sure to include the cost of insurance.

c. Suppose the following table gives reasonable probabilities of different healthcare expense levels for a healthy recent college graduate. (This table gives an approximation of the probabilities and a simplification of the healthcare expense levels, but it offers one way to get information to help make a decision.)

Total Medical Expenses	$200	$600	$1200	$3000	$15,000
Probability	0.50	0.30	0.10	0.07	0.03

Use these probabilities to compute the expected yearly cost of healthcare under each of the plans.

d. Recommend a health plan.

Solution

a. We are uncertain about what our healthcare costs will be for the year. We might visit the doctor's office a few times for colds and a serious case of poison ivy, or we might have considerably higher healthcare expenses.

b. We need to determine the cost for each plan and each of the possible expense levels. We'll set up a decision table with the decision choices as the rows and the possible expense levels as the columns. For example, suppose we choose the first plan and our medical expenses are \$200. The cost to us is (\$30)(12) + \$200 = \$560. If we choose the second plan and our expenses are \$200, the cost to us is (\$10)(12) + \$200 = \$320. If we choose the third plan and our expenses are \$200, the cost to us is (\$25)(12) + (0.3)(\$200) = \$360. We can verify that the costs for the other plans are as given in the following table:

Medical Expense Level	\$200	\$600	\$1200	\$3000	\$15,000
Plan 1	560	880	1000	1360	3760
Plan 2	320	720	1160	1520	3920
Plan 3	360	480	660	1200	4800

c. We want to evaluate the expected or average cost under each of the plans. For Plan 1, expected cost = (0.5)(\$560) + (0.3)(\$880) + (0.1)(\$1000) + (0.07)(\$1360) + (0.03)(\$3760) = \$852. For Plan 2, expected cost = (0.5)(\$320) + (0.3)(\$720) + (0.1)(\$1160) + (0.07)(\$1520) + (0.03)(\$3920) = \$716. Finally, for Plan 3, expected cost = (0.5)(\$360) + (0.3)(\$480) + (0.1)(\$660) + (0.07)(\$1200) + (0.03)(\$4800) = \$618.

d. If we have confidence in the probability estimates from part b, we would probably choose the plan that has the smallest expected cost: Plan 3.

It is important to realize that in this situation we must choose a health plan before we know what our healthcare expenses will be for the year. So we make the decision first and commit to the plan and then experience one of the levels of healthcare expenses. But suppose we didn't have confidence in our ability to predict the probabilities associated with each of the healthcare levels? We'll look at two additional methods, which don't require us to estimate probabilities, to help us make our decision. One could be described as an optimist's decision method and the other as a pessimist's decision method.

Suppose we have a problem such as the healthcare decision problem described in Example 21.3, where we want to choose the plan that will result in the least cost for healthcare for the year. Using the optimist's method, we want to consider each plan and record the **minimum cost** for each plan over all expense levels (since we don't know in advance what our healthcare expenses will be). That is, assuming that our medical expense levels will be either \$200, \$600, \$1200, \$3000, or \$15,000, we will be optimistic that we will incur the least cost no matter what plan we choose. The minimum cost for each plan (that is, the minimum in each row) is shown in the last column of the following table.

Medical Expense Level	\$200	\$600	\$1200	\$3000	\$15,000	MIN for each plan
Plan 1	560	880	1000	1360	3760	560
Plan 2	320	720	1160	1520	3920	320
Plan 3	360	480	660	1200	4800	360

We will choose the plan for which the minimum cost is the smallest. This is called the **minimin** decision strategy because it is the minimum of all the minimums. It is used when the decision table represents costs. For this example, if we were using the optimistic or minimin strategy, we would select Plan 2.

The pessimist's method for making a decision would consider each plan and ask, "What is the worst that can happen?" that is, what is the **maximum cost** for each plan over all expense levels (since again we don't know in advance what our healthcare expenses will be. That is, we will be pessimistic that we will incur the most cost over each plan.). The maximum cost for each plan (that is, the maximum in each row) is shown in the last column of the following table.

Medical Expense Level	$200	$600	$1200	$3000	$15,000	MAX for each plan
Plan 1	560	880	1000	1360	3760	3760
Plan 2	320	720	1160	1520	3920	3920
Plan 3	360	480	660	1200	4800	4800

We will choose the plan for which the maximum cost is the smallest. The pessimist thinks that the worst will happen but wants to make the smartest choice, given that the worst will happen. This is called the **minimax** decision strategy because it is the minimum of all the maximums. This strategy is used when the decision table represents costs. For this example, if we were using the pessimistic or minimax strategy, we would select Plan 1.

In the next example, we look at another decision involving costs and analyze it using several decision methods.

Example 21.4

We want to invest $5000 for the next year. There are three possible investments we could make, and three possible states of the economy that would affect the return on our investment. The following table shows the loss, in dollars, that we would experience for each investment and each possible state of the economy.

	State 1	State 2	State 3
Investment 1	0	-300	-200
Investment 2	2000	700	-600
Investment 3	-500	-1000	1000

a. Explain how to interpret a loss of "–500 dollars."

b. What investment would a pessimistic decision maker choose?

c. What investment would an optimistic decision maker choose?

Solution

a. A loss of –500 dollars is actually a profit of $500. A negative loss for the year represents a profit, so a loss of –500 dollars for the year means we end the year with $5500.

b. The pessimistic decision maker would identify the maximum cost for each row and then choose the decision alternative for which the maximum cost is the smallest. The row maximums are shown in the next table. The pessimist would choose the minimax alternative, that is, Investment 1.

	Maximum
Investment 1	0
Investment 2	2000
Investment 3	1000

c. The optimistic decision maker identifies the minimum cost for each row.

	Minimum
Investment 1	-300
Investment 2	-600
Investment 3	-1000

Then he or she chooses the investment alternative for which the minimum is the smallest; that is the optimist chooses the minimin alternative, which in this example is Investment 3.

In the previous example, we presented a table of the loss (or cost) associated with each investment and state of the economy pair, but we could have just as easily set up the table in terms of profits. If we had used a profit table, the sign of each table entry would change: negative entries would become positive and positive entries would become negative. In Exploration 7, you are asked to revisit this example using a profit table.

We now consider how an optimist and how a pessimist would make a decision if the decision table represents profits instead of costs. With a profit table, the strategy chosen by the optimist is the **maximax** decision strategy while the pessimist would choose the **maximin** decision strategy.

Example 21.5

a. Describe how an optimistic decision maker would make a decision if the decision table contained profits instead of costs.

b. Describe how a pessimistic decision maker would make a decision if the decision table contained profits instead of costs.

Solution

a. If the table contained profits, an optimistic decision maker would look at the largest profit for each decision alternative. This optimistic decision maker would think that the best thing that could possibly happen (that is, maximum profits) will happen. Then this decision maker would choose the decision alternative for which the maximum is the greatest. This is the maximax decision alternative.

b. If the table contained profits, a pessimistic decision maker would think the worst. This decision maker would look at the minimum profit for each decision alternative. He or she is a pessimist, but is also smart, so the smart pessimist will choose the decision alternative for which the minimum profit is the greatest. This is the maximin decision alternative.

In the next example, we consider a profit table and find what option an optimistic decision maker would choose and what option a pessimistic decision maker would choose.

Example 21.6

The senior class is planning an end-of-the-year activity. The activity must be planned now, before the planners know what the weather will be. The amount of money the class takes in will depend on the weather. The following table gives the amount of money the class will make, in dollars, for each activity and weather condition.

	Sunny and Warm	Cloudy	Rainy	Cold
Indoor Concert	50	100	140	120
Beach Trip	250	150	50	-100
Outdoor Picnic	125	200	-50	-50

a. Suppose an optimistic decision maker prevails. What activity would he or she choose using the maximax decision method?

b. Suppose a pessimistic decision maker is in charge of deciding what activity to plan. What activity would the pessimist choose using the maximin decision method?

c. Suppose we have reliable information that each of the four weather possibilities is equally likely to occur. Find the expected profit for each activity and decide what activity to plan.

Solution

a. First, we find the maximum profit for each activity. The last column of the following table shows the maximum profit for each row:

	Sunny and Warm	Cloudy	Rainy	Cold	Max ($)
Indoor Concert	50	100	140	120	140
Beach Trip	250	150	50	-100	250
Outdoor Picnic	125	200	-50	-50	200

We choose the activity for which the maximum profit is the greatest; this is the beach trip, with a maximum profit, over all weather conditions, of $250.

b. To find what activity a pessimistic decision maker would choose, we find the minimum profit for each activity and then choose the activity for which the minimum profit is the greatest.

	Sunny and Warm	Cloudy	Rainy	Cold	Min
Indoor Concert	50	100	140	120	50
Beach Trip	250	150	50	-100	-100
Outdoor Picnic	125	200	-50	-50	-50

The pessimistic decision maker chooses the indoor concert, which is the maximin choice.

c. If each weather possibility is equally likely to occur, each has a probability of $\frac{1}{4}$. So the expected profit if we choose the indoor concert is $\frac{1}{4}(50) + \frac{1}{4}(100) + \frac{1}{4}(140) + \frac{1}{4}(120) = 102.5$. Similarly, if we choose the beach trip, the expected profit is $\frac{1}{4}(250) + \frac{1}{4}(150) + \frac{1}{4}(50) + \frac{1}{4}(-100) = 87.5$, and if we choose the outdoor picnic, the expected profit is $\frac{1}{4}(125) + \frac{1}{4}(200) + \frac{1}{4}(-50) + \frac{1}{4}(-50) = 56.25$. Using this decision method, we would choose the indoor concert.

We've discussed several methods for informing decisions involving uncertain information. Ultimately, the decision still rests in the hands of the decision maker. He or she needs to decide which method, if any, to use to help make the decision.

Summary

In this topic, we looked at the concept of expected value, that is, expected profit and expected cost, in situations where uncertainty is involved. The expected value measures the average payoff or cost, in the long run if the activity is repeated over and over. We use estimates of the probability associated with uncertain events to calculate expected value. In situations where we cannot estimate probabilities or if a decision will be made just one time (rather than repeatedly), we can use the optimistic method (minimin if costs are given and maximax if profits are given) or the pessimistic method (minimax, if costs are given and maximin if profits are given). In either case, the decision still lies with the decision maker.

Explorations

1. A typical roulette wheel has evenly sized "pockets" that are numbered from 1 to 36 and two additional pockets labeled 0 and 00. The 1 through 36 numbered pockets are alternately colored red and black, with the 0 and 00 colored green. You can place many different bets; for example, you can choose a single number, even numbers, or red numbers. Suppose you bet $2 on red; this means that if the ball lands on red after the wheel is spun, you double your money. Otherwise you lose your $2.

   a. Find your expected payoff for this bet.

   b. Interpret your expected payoff from part a of this exploration.

2. Refer to Example 21.2. How might you justify returning the ticket even though the expected winnings are quite a bit less than the cost of postage?

3. You need to decide how to invest a graduation gift of $1000. The annual rate of return is given in the next table for each of three different types of investments and three different states of the economy.

	Recession	Stable Economy	Expansion
Investment A	2.5%	2.5%	2.5%
Investment B	2%	4%	5%
Investment C	-2%	4%	10%

a. Create a table that gives the amount of money for each type of account and state of the economy after one year, if interest is compounded monthly.

b. Compute the expected value of your account at the end of one year for each of the investment types, if the probability of a recession is 0.5 and the probability of a stable economy is 0.3.

c. What investment would an optimistic decision maker choose?

d. What investment would a pessimistic decision maker choose?

4. Consider again the healthcare table set up in the solution of Example 21.3.

a. How would your decision change if you had assessed the following probabilities for the healthcare expense levels:

Total Medical Expenses	\$200	\$600	\$1200	\$3000	\$15,000
Probability	0.10	0.20	0.20	0.20	0.30

b. What decision would you make if each healthcare expense level is equally likely to occur?

5. You live near a river that is prone to flooding, so you are considering purchasing flood insurance at \$150 per year. You look at storm patterns over the past ten years and estimate the yearly chance of minor flood damage (approximately \$1000 damage) to be 0.01. The chance of moderate damage (approximately \$5000 damage) is estimated to be 0.005, and the chance of major damage (\$20,000 damage) is estimated to be 0.001. The insurance will cover all damages.

a. Find the expected yearly cost of flood damage if you purchase flood insurance.

b. Find the expected yearly cost of flood damage if you do not purchase flood insurance.

c. Do you think you should purchase the flood insurance? Justify your answer.

6. A student group must order hot dogs for a concession stand at a sporting event. In the past, demand has been 7, 8, 9, or 10 dozen hot dogs for the event, depending on attendance and the weather. They need to order the hot dogs and buns the week before the event and will not be able to return any unused inventory. They purchase the hot dogs and buns for $2 per dozen and sell them for $6 per dozen.

   a. Fill in the following table to show the net profit (profit minus cost) for each of the possible order and demand levels: 7, 8, 9, or 10 dozen.

	Demand 7 Dozen	Demand 8 Dozen	Demand 9 Dozen	Demand 10 Dozen
Order 7 Dozen				
Order 8 Dozen				
Order 9 Dozen				
Order 10 Dozen				

   b. Find what size order a pessimistic decision maker would place.

   c. Find what size order an optimistic decision maker would place.

   d. What size order would you place if you knew from past experience that each demand level is equally likely?

7. Consider again Example 21.4.

   a. Rewrite the loss table given in the example as a profit table.

   b. Use your profit table to determine what investment a pessimist would choose.

   c. Use your profit table to determine what investment an optimist would choose.

   d. Explain how your results and table compare with the results of Example 21.4.

Excel Activities

ACTIVITY 1.1

World Motor Vehicle Production: Bar Graphs and Pie Charts

In this activity, you will investigate data and create several different types of graphs to help understand patterns in world motor vehicle production. You will investigate when each type of graph is appropriate for the data.

Consider the following data on leading world producers of passenger cars and trucks in 2002 (Source: *The World Almanac and Book of Facts 2004,* page 319).

Location	Cars (in thousands)	Trucks (in thousands)
United States	5,027	7,301
France	3,284	377
Germany	5,122	347
Japan	8,619	1,621
South Korea	2,651	496
Spain	2,267	577
Other	15,115	6,784

1. If you were to create a graph or chart of this data set, what would you like it to show?

2. Explain why a bar graph is appropriate for the car production data.

3. Draw axes in the space provided and create a bar graph for the car production data. Be sure to label the axes and show the scales you are using.

4. Explain what your bar graph shows that the original data table did not show and what the original table contains that your graph does not retain.

5. If you create a pie chart for the car production data, what will each "piece" of the pie show?

6. Use the Excel instructions that follow to help you fill in the third column of the next table to show the percentage of the total car production that falls into each row of the table:

Location	Cars (in thousands)	Percentage	Approximate Fraction of the Whole
United States	5,027		
France	3,284		
Germany	5,122		
Japan	8,619		
South Korea	2,651		
Spain	2,267		
Other	15,115		
Total			

Instructions to Use Excel to Calculate Percentages

a. Enter the car production data into columns A and B of an Excel worksheet. Use column A for locations and B for the number of cars produced, and use row 1 for titles. (A portion of the table follows.) Use the arrows, or press **Enter** after you type the entry in each cell, and fill in the numbers in cells B2 (column B, row 2) through B8.

b. To adjust the column width of column A to fit all the descriptions, use the mouse (or touchpad) to move the cursor to the letter **A** at the top of column A and click. This will highlight the first column. Then position the cursor at the right border of the column where the cursor changes to a black "plus sign with left and right arrows" and double-click the left mouse button. Repeat with column B, if needed.

	A	B
1	Location	Cars (in thousands)
2	U.S.	5,027
3	France	3,284
4	Germany	5,122

a. To get the total car production, position the cursor in cell B9. To instruct Excel to perform a computation, you must use the "equals" sign. Type =**sum(B2:B8)** and then press **Enter**.

d. To calculate what percentage each value in the table is of the whole, divide each number by the total, and then ask Excel to convert to a percentage. You can tell the computer to do this with the first entry in column B and then "drag down" using the mouse to get the percentages for each row. Put the cursor in cell C2 and type =**B2** / (the number you got for the total car production in cell B9).

e. To convert this decimal value to a percentage, put the cursor in cell C2, and go to **Format** (on the menu bar) and then select **Cells** and on the **Number** tab, click **Percentage**. In the **Decimal places** box, type **0**, because you just need a general idea of the percentage of the total for each country. Click **OK**.

f. Now to fill in the percentages in the rest of the column, put the cursor in cell C2 and move the cursor to the lower-right corner of the cell, until it changes to a black "plus" sign. Press and hold the left mouse button and drag down to row 8. The rest of the percentages should fill in as you drag down.

7. Use the next Excel instructions to get a rough approximate fraction of the whole represented by each row of the table and insert those values in the final column of the table shown previously. (This will help you create a pie chart.)

Instructions to Use Excel to Convert Percentages into a Fraction

a. Put the cursor in cell D2. Type =**C2** and then press **Enter**. Click cell D2 and go to **Format**, then select **Cells** and on the **Number** tab, click **Fraction**. In the **Type** box, choose **As sixteenths (8/16),** and press **OK**. This will give you the numbers as fractions with the denominator 16. You may use other forms, but for drawing the portions of the pie chart, it is easier to use denominators that are powers of 2.

b. Now fill in the rest of the column by using the "drag" function of Excel (see Step d in the previous section, "Instructions to Use Excel to Calculate Percentages").

8. Create a pie chart for the car production data using the last column of the table to help you determine approximately how big each piece of the circle should be. Be sure to label your chart.

9. Use the data given at the beginning of this activity to create a bar graph showing truck production in 2002 and explain what the graph shows.

10. For what types of data is a bar graph appropriate and for what types of data is a pie chart appropriate?

Additional Questions

1. The following table shows data on new passenger cars imported into the U.S. by country of origin in 1992 and in 2002 (Source: *The World Almanac and Book of Facts 2004,* page 320):

Country	1992	2002
Japan	1,598,919	2,046,902
Germany	205,248	574,455
Italy	1,791	3,504
United Kingdom	10,997	157,633
Sweden	76,832	87,709

Country	1992	2002
France	65	150
South Korea	130,110	627,881
Mexico	266,111	845,181
Canada	1,119,223	1,882,660
Total[1]	3,447,200	6,477,659

[1]Note that the total includes countries that are not listed separately.

a. What kinds of information, related to these data, would you want to get from a graph?

b. Would both bar graphs and pie charts be appropriate for presenting these data? Would one be preferable?

c. Create two graphs (bar and/or pie) to present these data, and explain what each graph shows specifically about these data.

2. The following data give the most popular colors for the 2002 model year for full/intermediate size cars (Source: *The World Almanac and Book of Facts 2004,* page 322):

Color	Percent
Silver	28.1
White	11.8
Light brown	11.6
Black	11.2
Medium/dark red	10.2
Medium/dark blue	9.5
Medium/dark gray	6.2
Medium/dark green	5.3
Gold	3.4
Other	2.7

a. Create a pie chart for these data or explain why it is not appropriate to do so. Explain how you determined the size of each "pie piece."

b. Create a bar graph for these data or explain why it is not appropriate to do so.

Summary

In this activity, you learned how to use Excel to find percentages. You gained experience in deciding when a bar graph or a pie chart is appropriate to represent data and in creating bar graphs and pie charts. You also practiced how to "read" and explain information from bar graphs and pie charts.

ACTIVITY 1.2

Medical Data and Class Data: Graphs with Excel

Microsoft Excel is a powerful computer program that allows you to manipulate data and create graphs. (In Excel, a graph is also called a *chart.*) In this activity, you will use Excel to create bar graphs and pie charts and more importantly, you will use these charts to help understand and interpret the data.

Consider the following data on principal reasons given by patients for emergency room visits in 2003. (Source: *The World Almanac and Book of Facts 2006,* page 190.)

	A	B
1	Reason	Number of Visits (in thousands)
2	Accidents	3,999
3	Arm/leg injuries	3,758
4	Back symptoms	2,696
5	Breathing problems	4,568
6	Chest pain	5,838

	A	B
7	Cough/throat symptoms	6,295
8	Earache/ear infection	1,867
9	Fever	5,732
10	Head/neck symptoms	4,641
11	Nausea/dizziness	5,280
12	Non-specific pain	2,234
13	Skin rashes	1,688
14	Stomach pain	7,583

1. What type of graph or chart would help you interpret the data?

2. Explain why a bar graph is appropriate for these data.

3. Use Excel to create a bar graph of these data (instructions follow).

Instructions to Enter Data and Create a Bar Graph Using Excel

a. First, enter the data and titles into the first two columns of an Excel worksheet, just as they appear in the previous table (cells A1 through B14).

b. To create a bar graph, use the mouse to point and click the **Chart wizard** icon (which looks like a bar graph) on the **Standard toolbar**. (Another way to open the **Chart wizard** is to click **Insert** and select **Chart** from the drop-down menu.) In the dialog box in the left field, select **Column** for **Chart type** and select **Clustered column** (the first type in the top row) for **Chart sub-type**. Click **Next** at the bottom of the dialog box.

c. In the next dialog box, click **Data range**; then to highlight and select your data and label cells from the worksheet, starting at cell A1, press and hold the left mouse button and drag to cell B14. Alternatively, you can type **A1: B14** in the data range box. Click **Next** at the bottom of the dialog box.

d. On the next screen, click the **Titles** tab and enter an appropriate title in the **Chart title** field. Then type **Reasons** for **Category (X) axis** and **Count** for **Value (Y) axis**. Click the **Legend** tab; place your cursor over the **Show legend** box and click to remove the check mark because you don't need a legend for your data. Then click **Next** at the bottom of the dialog box.

e. Now you can decide if you want the chart in the same worksheet as the data or in another worksheet. Select **As object in** to put your chart in the same worksheet, that is, on Sheet 1. (You also could choose another sheet in the same book.) Now click **Finish**.

f. The chart now has eight "handles" indicating that it has been "selected." If you select a handle (getting a double-headed arrow) and then press and hold the left mouse button to drag it, you can resize the chart. By clicking on the white interior of the chart and holding the mouse button down, you can drag the chart to another location. Click outside the chart to "deselect" it.

g. To sort data: Return to cell A1, and then click and drag to cell B14 to select the complete set of data. Select **Data** from the menu bar and then select **Sort**. Sort by **Number of visits**. Click **Descending**, and then click **OK**. What happens to the chart and the data?

4. Explain what your bar graph shows that the original data table did not show.

5. Now you'll use Excel to create a pie chart of these data.

Instructions to Create a Pie Chart and Change a Bar Graph to a Pie Chart in Excel

a. To create a pie chart, as you did before, highlight the data and labels. Go to the **Chart wizard** and select **Pie** and the first chart sub-type, **Pie**, from among the choices for **Chart sub-type**. Then click **Next**.

b. Make sure the data range is correct and click **Next**.

c. Click the **Titles** tab and give the pie chart an appropriate title. Using the **Legend** and **Data labels** tabs, choose whether you want to include a legend or not, and if you want to label the pie pieces and/or show values or percents. (Note that you want to be able to clearly read the chart, so including a legend and data labels will make your graph easy to read.) Click **Next**.

d. In Step 4, choose a location for your chart.

e. To change a bar graph to a pie chart, go back to the bar graph you created. Select the completed bar graph by clicking inside the border.

f. From the menu bar, select **Chart** and then click **Chart type**. In the left field, choose **Pie** for **Chart type** and **Pie** for **Chart sub-type**, the first choice in the top row. Click **OK**.

g. Now from the menu bar, choose **Chart** and then click **Chart Options**. Using the **Title**, **Legend**, and **Data Label** tabs, select a title, show the legend, and select the labels and/or percents, as described previously.

6. Is there any information in your pie chart that was not in the bar graph? Explain.

7. Now, suppose the following data on number of miles from the college to home was collected from a group of seven students. Use Excel to create a pie chart for this data set and explain what your pie chart shows. (Think about how you want your pie chart to show the data. You will need to put the data into categories first. You can use columns D and E of the same Excel worksheet.) How does your pie chart help you understand the data?

Student	Miles
John	23
Jean	45
Harry	11
Bruce	134
Fred	62
Ann	35
Sally	3

8. Now instruct Excel to convert the pie chart you created to a bar graph.
9. Which of the two charts do you think is more helpful to understand the data in #7 and why?

Some Other Useful Excel Functions

a. To save your Excel file, first select **File** from the menu bar and then select **Save as.** Choose the location in which you wish to save the Excel file. You can now give your file an appropriate name in the text box at the bottom of the dialog and click **Save**. When you want to come back and work on the file again, you can go to **File** on the menu bar, select **Open**, and retrieve the file from your saved file location. After you have accessed this file and made some modifications, if you want to save it again, just go to the **File** menu and click **Save**. Excel will save your file in the same location.

b. You may want to go to a new worksheet in your Excel file. Each Excel file document is called a *workbook*, and each workbook has multiple sheets. To go to a new sheet in the same workbook, click **Sheet 2** at the bottom of the Excel page. You will get a new worksheet to use. When you save the file, all the worksheets will be saved in the one file.

c. You will need to get data files from the CD that accompanies this textbook (or at www.keycollege.com/QRTools). On the CD, there is a file named "EA1.2 Class Data Gender.xls." Open a new sheet, and import the "EA1.2 Class Data Gender.xls" file into your worksheet by clicking **File** on the menu bar and then **Open**. In the **Open** window, you will see the words **Look in** with a small bar to the right. Choose the appropriate drive. You will see names of the files on the disk; double-click the desired file or highlight the file and then click **Open**.

Additional Questions

1. Use the "EA1.2 Class Data Gender.xls" data file from the CD or website.
   a. Create two charts that graphically illustrate this data set. You don't need to use all the data. Select appropriate data for what you want to illustrate with your graphs. (You may want to sort or group the data and create new tables, copy the data into a new portion of your worksheet, and so on.) Experiment with the copy and paste capabilities by highlighting your data and selecting Edit from the menu bar.
   b. Explain why you picked the data and the graphs you did.
   c. Explain what your graphs show.

d. What information about the variables you used is in the table but is not reflected in your graphs?

2. Look at the graphs from *The New York Times* shown in Example 1.2 of Topic 1.

a. Enter the data into an Excel worksheet and re-create the bar graph and one of the pie charts (you may choose which one) that appeared in the newspaper.

b. Describe any difficulties you encountered in creating these graphs.

Summary

In this activity, you learned to enter data in an Excel file and created bar and pie charts using Excel. You also learned how to save an Excel file and to work with more than one worksheet within an Excel workbook. You learned that sometimes it is necessary to group the data to create an appropriate pie chart.

ACTIVITY 1.3

SATs and the Super Bowl: Creating and Interpreting Histograms

A histogram is a graph of a frequency distribution and is a useful way to give a picture summary of a set of data involving a quantitative variable. In this activity, you will create and interpret histograms.

1. How is a histogram similar to a bar graph and how is it different from a bar graph?

The following table gives the percentage of high school seniors who took the SATs in the academic year ending in 2003 for a group of 14 southeastern states and Washington DC.

State	AL	AR	DE	DC	FL	GA	KY	LA	MD
Percent Taking Test	10	6	73	77	61	66	13	8	68
State	**MS**	**NC**	**SC**	**TN**	**VA**	**WV**			
Percent Taking Test	4	68	59	14	71	20			

2. What is the range of the "percent taking test" data shown in the table?

3. If you want to group the data into approximately six classes, what intervals could you use?

4. Why is it important to group the data?

5. Why do you want to choose intervals of equal width?

6. Fill in the following table, and then construct, by hand, a histogram for the percentage of high school seniors in the 15 southeastern locations who took the SATs. Be sure to label the axes of your histogram.

Interval (percent taking SATs)	Frequency (number of states in each interval)

7. Describe the overall shape of your histogram.

8. There are several ways to create a histogram using Excel. We'll illustrate the first way using the SAT data for all states. First, import the "EA1.3.1 SAT data.xls" file from the CD or website into an Excel worksheet.

Instructions to Use Excel's Chart Wizard to Create Histograms

One way to create histograms is to use the **Chart wizard** discussed previously. The **Chart wizard** is designed to use with data that have been grouped—either categorical data or quantitative data—into classes or intervals. To create a histogram using the **Chart wizard**, you will group the data, which is not too difficult if you first sort it.

a. You'll create a histogram of the "percent taking SATs" for all the states, so first sort the data by this column; however, be sure to highlight all three columns and then begin to perform the sort. From the menu bar, select **Data** and **Sort**, and then sort by the appropriate column in ascending order.

b. Next decide what the "bins" will be. That is, you need to decide what classes (of equal width) to use to group the data. For this data set, it appears that 0 < % <= 10; 10 < % <= 20; 20 < % <= 30, and so on will be convenient classes. Use your sorted data to count the number of data values that fall into each of these classes, and create a table in columns E and F of your Excel worksheet with the class labels in column E and the frequencies for each class in column F.

c. Use this table of classes and frequencies, and create a bar graph like you did previously.

d. To make this completed bar graph look like a histogram, adjacent bars should touch one another. To change the bar width, double-click on one of the bars of the finished graph to access the **Format data series** dialog box. Select the **Options** tab and change the **Gap width** from **150%** to **0%**. Then click **OK**.

e. You may add titles and experiment with some of the other features of bar graphs (like adding or taking away axis labels and so on).

9. Describe the overall shape of the SAT histogram for all states. How does this histogram compare with the one you created by hand for the southeastern states? Using the two histograms, what can you say about the percentage of seniors taking SATs in the southeastern states and in all the states.

10. Here is another method to create a histogram with Excel. For this histogram, you'll use a file containing data about the points scored by the winning team in the Super Bowl for the years 1967 to 2003. Retrieve the "EA1.3.2 Super Bowl.xls" file from the CD or website. You'll make a histogram of the variable "points scored by winner."

Instructions to Use Excel's Analysis ToolPak to Create a Histogram

You can also create a histogram using Excel's **Analysis ToolPak**, which groups the data into classes for you.

The **ToolPak** may need to be installed; to check, choose **Tools** from the menu bar and select **Add-Ins**. If the **Analysis ToolPak** box is not checked, select it by clicking and then click **OK**. If it is already selected, just click **OK**. Now when you choose **Tools** from the menu bar, you will see a **Data analysis** option.

a. From the menu bar choose **Tools** and then **Data analysis**; scroll down to the **Histogram** option and click **OK**.

b. In the dialog box, type the reference for the range of your data, **B1:B38** in the **Input range** area, or click and drag from cell B1 to cell B38. Leave the **Bin range** field blank to allow Excel to select the bins (or groupings) for your data. Check the **Labels** box because B1 (where there is a label for the column) has been included in the **Input range**. Type **C1** for **Output range** to denote the upper left cell of the output range, and check the box **Chart output** to put the histogram on the same sheet of the workbook as the data. For now, you won't use the other options (**Pareto** and **Cumulative percentage**.)

c. The output entries under **Bin** in the C column are the upper limits of the boundaries for each class, and the corresponding frequencies appear in cells in the D column. (Notice the numbers at the bottom of each bar in the histogram.)

d. Adjacent bars should touch in a histogram. To change the bar width, double-click one of the bars to bring up the **Format data series** dialog box. Select the **Options** tab and change the **Gap width** from **150%** to **0%**. Then click **OK**.

e. To create a slightly different histogram using "bin" intervals that you choose, type **My Bins** (or any other appropriate label) in cell E1. (You may need to move your histogram.) Because the data values range from 14 to 55, you will use intervals of width 5, beginning with the interval from 10 to 15. So enter the values **15, 20, 25, 30, 35, 40, 45, 50, 55** in cells E2:E10. (This creates the bins or classes of: 10 < points scored ≤ 15; 15 < points scored ≤ 20; 20 < points scored ≤ 25; and so on, to the last class, 50 < points scored ≤ 55.) Notice that the right endpoint of each interval or class is included in the interval but the left endpoint is not.

f. Use the procedure for creating a histogram described previously, but type **E1:E10** in the **Bin range** area and type **G1** for **Output range**.

Some Selected Enhancements to Your Histogram

a. One option is to get rid of the legend. To do this, click the legend (the word *frequency* on the chart). From the menu bar, select **Chart** and then **Chart Options**. Click the **Legend** tab and clear the **Show legend** box. Then click **OK**.

b. You can resize both the plot area (the gray region containing the histogram) and the larger rectangle by clicking in the boundaries and dragging the appropriate handles horizontally, vertically, or diagonally. Click outside the chart to "deselect."

c. To change the chart title, click the title word *Histogram*. A rectangular gray border with handles will surround the word. Click the word *Histogram* and type an appropriate title. When finished, click outside the graph.

d. Click the words *My Bins* at the bottom of the chart and type in an appropriate X-axis title.

e. The **More** interval with **0** counts shouldn't be there. In the workbook cell containing the word *More*, change the label to **60**. The histogram is dynamically linked to the data and the label on the X-axis changes to 60.

f. The histogram has a border around the **Plot area** and a gray shaded **Plot area**. You can change this by double-clicking on the **Plot area** to bring up the **Format plot area** dialog box. Press the **None** button for **Border** and also press it for **Area**.

11. Explain what your histogram shows about points earned by Super Bowl winning teams.

Additional Questions

1. Retrieve the data set "EA1.3.3 Governors Salaries.xls" from the CD or website.

a. Choose appropriate classes and use Excel to create a histogram for the Governors' salaries.

b. Explain why you picked the classes you did.

c. Explain what your histogram shows.

2. Create, by hand, a stemplot of the "points scored" from the "EA1.3.2 Super Bowl.xls" data file.

a. How is this graph different from the histogram for the same variable you created previously?

b. Which plot gives a better sketch of the data and why?

c. What information from the table is not reflected in your graphs?

Summary

In this activity, you grouped data to create histograms by hand and using Excel. With Excel, you created histograms in two different ways: as a bar graph using the frequencies you computed and then adjusting the width of the bars so that there were no gaps in between them, and as a histogram using the histogram function in Excel, in which frequencies are computed automatically.

ACTIVITY 2.1

Estimating Dates: Scatterplots

In this activity, you will estimate the dates of several major events and then use a scatterplot to compare your estimates with the actual dates these events occurred. You will analyze relationships between two variables that can be read from a scatterplot.

You will start with an estimation experiment. Work with one or two classmates and fill in the following table, giving your group's best estimate of the year in which each event occurred.

1. For each of the following twentieth century events, estimate the year in which the event occurred. (Record only the last *two* digits of the year; because you know all dates are in the 1900s, you don't need to record the "19.")

Event	Estimated Year Event Occurred
Oprah Winfrey born	
Martin Luther King assassinated	
Original Woodstock music festival held	

Event	Estimated Year Event Occurred
Nineteenth amendment established women's suffrage	
Long-distance telegraphic radio signal sent across the Atlantic	
Bart Simpson's character made his debut	
Aerosol can invented	
Franklin D. Roosevelt elected to a third term of office	
"Scotch" tape invented	
Hank Aaron inducted into the Baseball Hall of Fame	
Pearl Harbor bombed	
President Nixon resigned	
Schindler's List won the Academy Award for best picture	
San Francisco destroyed by earthquake and fire	
The World Wide Web developed	

2. Obtain the actual dates for those events from your instructor. Create a scatterplot of the data with x representing "year event occurred" and y representing "estimated year event occurred." (You might want to record these actual dates immediately to the left of your estimates in the previous table.)

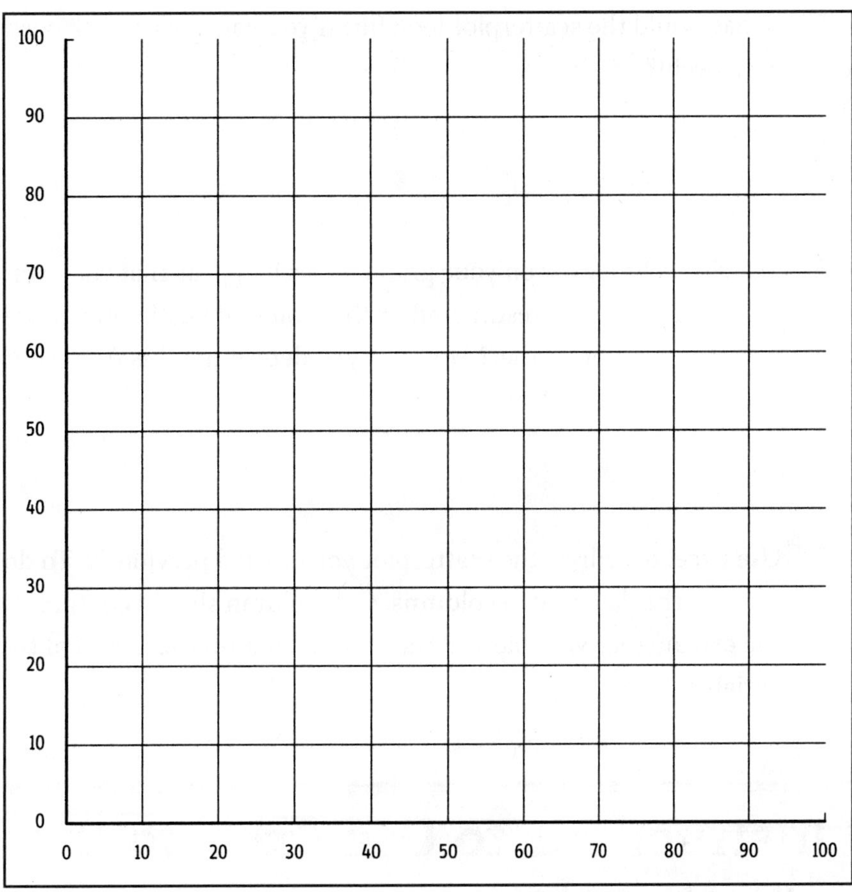

3. Why is it appropriate to use the variable "year event occurred" as the explanatory variable and "estimated year event occurred" as the response variable?

4. How close were your predictions of events that occurred in the 1980s to the actual dates?

5. What would the scatterplot look like if you had guessed the correct year for each event?

6. Sketch the line $y = x$ on your graph. Does it appear that you overestimated more than you underestimated or that you underestimated more than you overestimated or neither? How did you determine this from your graph?

7. Use Excel to redraw the scatterplot you created previously. To do so, you need to enter the data in two columns. One column should contain the values of the explanatory variable and the second column the values of the response variable.

Instructions to Use Excel to Create a Scatterplot

a. Get the file "EA2.1 Events and Dates.xls" from the CD or website. In column C of the Excel worksheet, enter your estimates of the year each of the events occurred (enter only the last two digits of the year). Add an appropriate column title in cell C1.

b. To create a scatterplot of the x-y data with x representing "year event occurred" and y representing "estimated year event occurred," select the two columns of data, including the labels. Go to the **Chart wizard** and choose **XY (scatter)** from the **Chart type** menu and the first type of graph (the "dots") for the **Chart sub-type**. Click **Next**.

c. On the next screen, click to indicate that you want the series in columns (in order to plot pairs of data) and check your data range. Then click **Next**.

d. Go to the **Legend** tab and deselect the legend by clicking the check mark to remove it. Then go to the **Titles** tab to insert the appropriate titles. Enter a relevant chart title and also labels for the x and y axes. Then click **Finish** to complete the graph. Note that you can find the coordinates of any point on your completed graph by moving the cursor to the point. If you want to change the scale on either axis, point to the axis and right-click, or double-click and select **Format axis**. Then, using the **Scale** tab, enter the appropriate minimum and maximum values for your axis.

8. Change your estimated values so the values of "estimated year event occurred" and the values of "year event occurred" are equal and look at the corresponding scatterplot. Describe the graph. Does your description agree with your answer in #5?

Additional Questions

1. The following data gives the number of strikes or lockouts involving more than 1,000 workers, and the percentage of the total labor force belonging to a union between the years 1950 and 2002. (Source: *Encyclopedia Britannica Almanac 2004,* pages 850–851.)

Year	Strikes and Lockouts	Union Membership Percentage
1950	424	31.5
1960	222	31.4
1970	381	27.4

Year	Strikes and Lockouts	Union Membership Percentage
1980	187	7.1
1990	44	5.6
1995	31	5.6
2000	39	4.0
2001	29	4.7
2002	19	5.8

a. What kinds of information, related to these data, would you want to get from a graph?

b. Create a scatterplot that relates the number of strikes and lockouts with the percentage of the total labor force with union membership.

c. Which variable did you use as the explanatory variable? Why?

d. Explain what your graph shows.

2. Create two other scatterplots using data from the previous table. Explain what each graph shows.

Summary

In this activity, you learned how to use Excel to create a scatterplot. In creating scatterplots, you decided which variable should go on the horizontal axis (that is, which variable is the explanatory variable) and which should go on the vertical axis (that is, which is the response variable). You also discovered how the line $y = x$ can help you see relationships in paired data. You will use these skills to visualize relationships between variables in bivariate data sets.

ACTIVITY 2.2

State Governors' Salaries and Per Capita Income: More on Scatterplots

In this activity, you will use scatterplots to investigate the relationship between the governor's salary and the average per capita income in the state. You will also analyze differences between scatterplots and other types of graphs.

The following table gives the average per capita income and the governor's salary in 2003 for each state in a group of nine northeastern states.

State	Per Capita Income 2003	Governor's Salary 2003
Connecticut	43,173	150,000
Maine	28,831	70,000
Massachusetts	39,815	135,000
New Hampshire	34,702	100,690
New Jersey	40,427	157,000
New York	36,574	179,000
Pennsylvania	31,998	142,142
Rhode Island	31,916	105,000
Vermont	30,740	127,456

1. Explain how you could use a bar graph to represent and interpret one or more aspects of the data. Which aspects of the data would the bar graph help interpret?

2. Would a histogram help you interpret the data, or an aspect of the data? Explain.

3. Which type of graph would help you answer the question: Is there any relationship between per capita income in a state and the governor's salary?

4. Create a scatterplot of the data using the two quantitative variables, "per capita income" and "governor's salary."

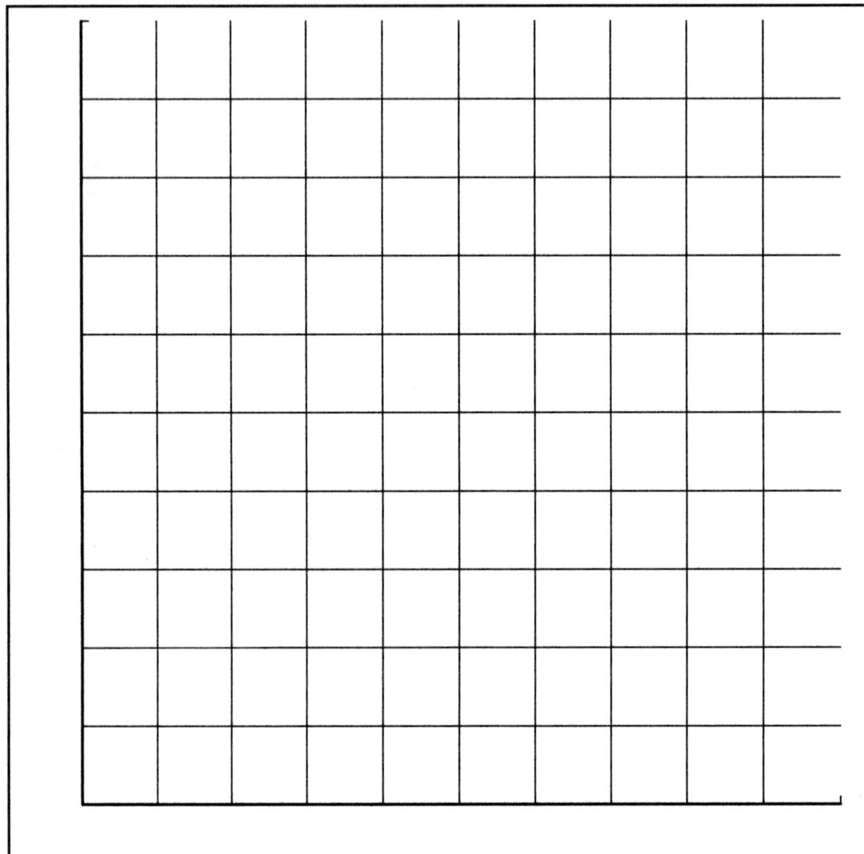

5. Which variable did you use on the x-axis? Why?

6. Which variable did you use on the y-axis? Why?

7. Explain what your graph shows about the relationship of the two variables.

8. To analyze this relationship further you will create a scatterplot using data on all 50 states. To do so, you will first create a two-column table that records the governor's salary and the average per capita income for each state.

Instructions to Use Excel to Create a Table from Other Tables

a. Retrieve the file "EA2.2.1 Per Capita Income.xls" from the CD or website. This file contains a list of all states and the per capita personal income for each for 2003. (Source: *Bureau of Economic Analysis,* www.bea.gov.)

b. Open the file "EA2.2.2 Governors Salaries.xls" from the CD or website, and then follow instructions in part c to copy and paste the list of Governors' Salaries into the worksheet "EA2.2.1 Per Capita Income.xls."

c. First, ensure that both files contain all the states listed in the same order. Then put the cursor at the top of column B, "Governors' Salaries," and click to highlight the entire column. Select **Edit** from the menu bar and then **Copy** to copy column.

d. Return to the per capita income worksheet. Place the cursor at the top of column C, and then select **Edit** and **Paste** to insert the governors' salaries into column C of the per capita worksheet. You should now have three columns in this worksheet.

9. Create a scatterplot of this data set. You should change the scale of the horizontal (or x) and vertical (or y) axes so the data are easier to read. (To change the scale on an axis, point to the axis and either double-click or right-click and select **Format axis**. Then on the **Scale** tab, enter the appropriate minimum and maximum values for the axis.) What is a reasonable minimum value of x to use? What about y?

10. Are there any trends or patterns to this data? Explain.

11. Are there any data points that appear to be "away from" the rest of the data? If so, which one(s) and what makes them stand out?

Additional Questions

1. The file "EA2.2.3 Children in Poverty.xls," which you will find on the CD or website, contains the table shown here. This table gives the percentage of children younger than 18 who were living below the poverty level in the U.S. from 1976 to 2001. Use Excel to create a scatterplot for this data table and write a paragraph describing the trends shown by your graph. Does this scatterplot represent a function? Why or why not? (Source: *Encyclopaedia Britannica Almanac 2004,* pages 828–829.)

Year	Percentage of Children	Year	Percentage of Children
1976	16.0	1989	19.6
1977	16.2	1990	20.6
1978	15.9	1991	21.8
1979	16.4	1992	22.3
1980	18.3	1993	22.7
1981	20.0	1994	21.8
1982	21.9	1995	20.8
1983	22.3	1996	20.5
1984	21.5	1997	19.9
1985	20.7	1998	18.9
1986	20.5	1999	16.9
1987	20.3	2000	16.2
1988	19.5	2001	16.3

2. The table here (also available in the "EA2.2.4 US Farms.xls" file on the CD or website) gives the total number of acres (in thousands) of land devoted to farming in the U.S. and the farm population (in thousands) at ten-year increments from 1900 to 1990. (Source: *The New York Times Almanac 2004,* page 302.)

Year	Farm Acreage (in thousands)	Farm Population (in thousands)
1900	841,202	29,835
1910	881,431	32,077
1920	958,677	31,974
1930	990,112	30,529
1940	1,065,114	30,547
1950	1,161,420	23,048
1960	1,176,946	15,635
1970	1,102,769	9,712
1980	1,039,000	6,051
1990	987,000	4,801

a. Create an appropriate scatterplot that you can use to analyze any trends about the land used for farming throughout the 20th century. Describe these trends.

b. Create a second scatterplot and use it to describe the relationship between the number of acres used for farming and the total farm population in the U.S.

Summary

In this activity, you considered the differences among a bar graph, a histogram, and a scatterplot and the information conveyed. You learned to copy and paste in Excel. You looked at trends in a sample compared to trends in a full data set and looked for unusual observations. You also investigated trends over time and relationships between variables.

ACTIVITY 3.1

Temperature Patterns: Functions and Line Graphs

In this activity, you will work with examples in which curves obtained by joining known points of the graph of a function can help you understand the data. These graphs are called **line graphs**. The temperature in a region varies over the year and is a function of the time of year. You will look at average temperature data as a function of month in various cities and investigate patterns in the data. You will also create a graph involving minimum wage data.

1. On the following axes, sketch a curve that you think shows how the average temperature in Fairbanks, Alaska, changes over the course of a year. (Think about when the temperature would be a maximum, when it would be a minimum, when it would be increasing, and so on.)

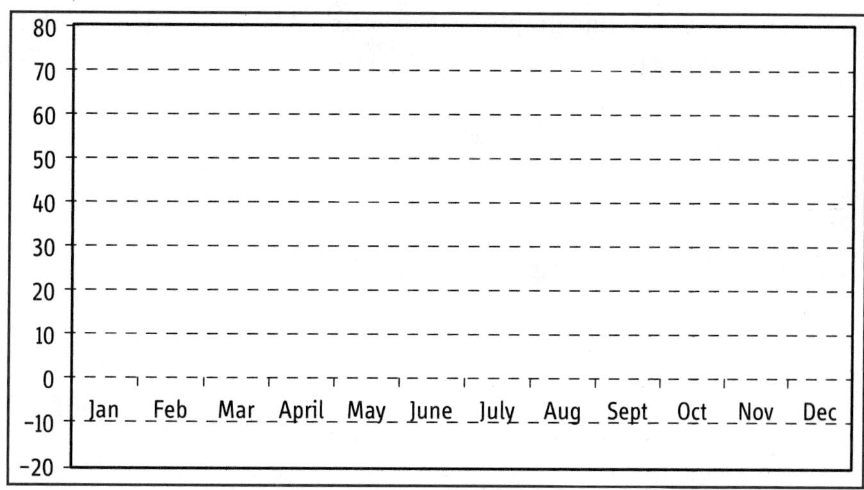

Next you'll see how your graph compares to the actual data, and how the average temperatures over the year for four cities in different parts of the world compare.

Instructions to Use Excel to Create Line Graphs

a. Retrieve the Excel file "EA3.1.1 Avg Temp Four Cities.xls" from the CD or website. (Source: *World Climate,* www.worldclimate.com.)

b. Highlight the block of data for the months Jan through Dec, including the column headings but not the Year column (column N), and access the **Chart wizard.** Select **Line** from the **Chart type** menu and select the first entry for **Chart sub-type**. Choose appropriate titles for your graph and finish the graph.

c. Note that if you click on the graph of one of the functions, the data points are highlighted. If you then point the cursor (without clicking) at any of the highlighted points, the coordinates of the point are displayed.

d. Create a second graph using the same block of data, but this time use **Column** as the **Chart type**.

2. How does the line graph for Fairbanks, Alaska, compare with the one you created in #1?

3. Explain what the graphs you created show.

4. Which of the two types of graphs (line or column) do you think is preferable for these data and why?

5. Use the line graphs you created previously to answer the following:

    a. Identify for each of the four cities when during the year the maximum average temperature occurs.

    b. Identify for each of the four cities when during the year the minimum average temperature occurs.

    c. Identify for each of the four cities when during the year the temperature is increasing.

d. Identify for each of the four cities when during the year the temperature is decreasing.

e. Over the interval Jan–May, which city's temperature increased the fastest and how does the graph show that?

f. Identify where the graph of the average temperature of Fairbanks, Alaska, is concave upward and where it is concave downward.

6. Retrieve the data set "EA3.1.2 Min Wage.xls" from the CD or website, or enter the following data on minimum wage in years in which it increased into an Excel worksheet.

Year	1974	1975	1976	1978	1979	1980	1981	1990	1991	1996	1997
Wage($)	2.00	2.10	2.30	2.65	2.90	3.10	3.35	3.80	4.25	4.75	5.15

a. Use Excel to create and label two graphs of these data: a scatterplot (make sure you select **Scatterplot** instead of **Line** in the **Chart wizard**) with just the points shown, and a scatterplot using a line to connect the points (you may choose to show the points or not, and you may choose a smooth line or data points connected by line segments).

b. Which variable should be on the horizontal axis of these graphs and why?

c. Explain why your line graph (the one with points connected) is not really appropriate for these data.

7. Retrieve the data set "EA3.1.3 Normal Avg Temp.xls" from the CD or website. Select three cities in different regions of the U.S. that appear in this file and create an appropriate graph of the data. (Source: *National Oceanic & Atmospheric Association,* www.noaa.gov/climate.html.)

Instructions to Use Excel to Copy and Paste Data

a. For each of your cities, highlight the row of data for the city, select **Edit** and then **Copy** to copy the data.

b. Then go to a new worksheet and select a new row from that sheet (leave the top row for labels, as shown in the Excel file "EA3.1.3 Normal Avg Temp.xls." Select **Edit** and then **Paste** to paste this data row into the sheet.

c. If you already have data in the first row, you can insert a blank row as follows: click on any cell in the first row. Then select **Insert** from the menu bar and then **Rows**. When you have finished copying the data from your three cities, you should have the data and labels, similar to those in the first file used, in your new worksheet.

8. Interpret your graph and explain what the graph shows about the average temperature over the year in your three cities. (Your explanation should include a lot more detail than the title of the graph conveys and should include information about when, over the course of the year, the temperature is increasing, when it is a maximum, and so on.)

Summary

In this activity, you compared bar graphs and line graphs, and used graphs to find maximum and minimum values of functions, intervals where the function is increasing, where it is decreasing, and where it is concave upward or concave downward. You learned how to use Excel to create line graphs and how to copy and paste data and insert rows in an Excel worksheet.

ACTIVITY 4.1

Blood Alcohol Levels and Credit Cards: Working with More Than Two Variables

In this activity, you will look at examples in which there is a need to graph or analyze expressions and functions that involve the interactions of more than two variables. In the first part of this activity, you will work with "blood alcohol level" as the response variable, and in the second part you will explore how some variables affect the balance on a credit card.

Blood alcohol level tables are given in Exploration 5 of Topic 4 and are contained in the file "EA4.1.1 Blood Alcohol Level.xls." You will compare the blood alcohol level of 140-pound males and 140-pound females based on the number of drinks consumed. To do so, you will create a graph that shows two scatterplots (one relates the number of drinks consumed with the blood alcohol level of a 140-pound male, the other relates the same variables for a 140-pound female) on the same axes.

1. Retrieve the Excel file "EA4.1.1 Blood Alcohol Level.xls" from the CD or website. Using these tables and the Copy and Paste functions of Excel, create a three-column table in the same worksheet. The first

column should be "Number of Drinks," and the second and third columns should be "Blood Alcohol Level of a 140-lb Male" and "Blood Alcohol Level of 140-lb Female," respectively.

2. Create and label an appropriate scatterplots (in a single graph) using the table you created. Explain what your scatterplots show about blood alcohol levels of 140-pound males and females.

3. What variables are involved in this graph and what other variables might you want to include in order to understand the issue more fully?

How fast the balance on your credit card decreases depends on the original amount charged, any additional purchases charged to the card, the monthly interest rate, and how much you pay off each month. Suppose you have charged \$1000 on a particular credit card and do not charge anything additional on the card. Suppose also that the annual interest charge on this card is 16.8 percent. Then the monthly charge, as a percent, is $16.8/12 = 1.4$. The credit card company requires that you pay off a minimum balance of \$20 each month. Then you can calculate the next month's balance using the formula:

$$\textit{balance next month} = \textit{balance this month} + 0.014 * \textit{balance this month} - 20$$

4. Explain what each part of this formula represents.

5. Set up a spreadsheet to use this formula to calculate how much you will owe after one month, after six months, and after one year, if you pay off only the minimum amount of $20 each month and don't charge anything else on your card. Here are instructions to set up the spreadsheet. Using the spreadsheet, you will be able to answer the questions that follow.

Instructions to Create a Spreadsheet and Enter a Formula Using Excel

a. Go to a new sheet in your Excel workbook. In cell A1, enter the label **Month**, and in cells 2 and 3 of column A, enter the numbers **1** and **2**, respectively. Then press the **Shift** key while highlighting cells 2 and 3 of column A (cell 2 won't look highlighted but cell 3 will, and you will see a bold line around the two boxes). Point the cursor to the lower-right corner of the highlighted cells until it changes to a bold plus sign, and then press and hold the left mouse button and drag down to cell 13 of column A. You should have the numbers 1 through 12 in cells 2 through 13 of column A.

b. Label column B with **Balance This Month** and column C **Balance Next Month**. Enter **1000** in cell B2 (without a $ sign).

c. You will enter a formula in cell C2 that will compute the balance next month for the first month. To enter a formula, use the "equals" sign. In cell C2, enter **=B2+0.014*B2-20**. Note that this is the formula given previously to calculate "balance next month" because cell B2 contains the "balance this month."

d. For month 2, transfer the amount in cell C2 (that is, the "balance next month" for month 1) to cell B3, which is the "balance this month" for month 2. To do this in an efficient way, type **=C2** in cell B3. Now, to calculate the "balance next month" for month 2, you can either retype the formula for "balance next month" in cell C3, or put the cursor on cell C2 and then move it to the lower-right corner of the cell and drag the fill handle (the bold "plus" sign) to cell C3 to auto-fill the cell.

e. To extend these same formulas to the rest of the months, highlight cells B3 and C3, move the cursor to the lower-right corner of the highlighted cells, and drag the bold plus sign down to cell C13 (that is, auto-fill down to cells B13 and C13).

f. There is a nice feature of Excel that allows you to see what formulas you used in creating the spreadsheet. Use the **Ctrl** button together with the accent key at the upper left corner of the keyboard. If you press **Ctrl** and ` together, this puts the spreadsheet in **Audit mode**. In Audit mode you can see all the formulas that were entered. Press **Ctrl** and ` together again to get out of Audit mode.

g. If you want to round the numbers to two decimal places (which is appropriate in this situation because you are dealing with money), highlight the cells you wish to round, and from the menu bar, select **Format** and then **Cells**. Then for **Category,** choose **Number** and for **Decimal places**, enter **2** and click **OK**.

6. How much progress did you make in paying off your debt in month 1?

7. What is the "balance next month" for month 2?

8. How much will you owe after six months? After one year?

9. If you continue to pay $20 each month and don't charge anything else on your card, how long will it take to pay off the whole amount? (To answer this question, highlight cells A13, B13, and C13 and auto-fill.)

10. Suppose you charged $1300 and you pay $20 per month and don't charge anything else on your card. Use Excel to set up a table to determine how much will remain to be paid on your card after one year. Explain what your table shows.

11. Suppose you want to pay off only $14 each month. If the initial balance is $1000, you don't charge anything else on your card, and the interest is 16.8 percent per year, find how much you will owe after the first month, after six months, and after one year.

12. What happens if you owe $1000 and you can pay only $13 each month?

13. What conclusions can you draw from the previous computations?

Additional Questions

1. We collected information on family size from a group of 43 students. We asked students how many children there were in the family they were raised in and how many children they would like to have when they establish a family of their own. We organized the data in a file titled "EA4.1.2 Class Data Num Child.xls"; retrieve this file from the CD or website. You'll notice that each case has three variables associated with it. What are the cases and what are the variables?

2. Sort the data by gender, highlight the portion of the data in columns B and C for female cases, and create and label a scatterplot for the female cases. Now highlight the portion of the data in columns B and C for male cases, and create and label another scatterplot for male cases. Explain what your two scatterplots show. Are the two scatterplots basically the same, or are there differences? Do your graphs show all data points? Explain.

3. What other kinds of responses might you want to gather from your respondents to understand any possible relationship between these two variables more fully?

Summary

In this activity, you learned how to enter a formula in Excel and drag down to do a series of calculations. You also learned how to use Audit mode in Excel to check formulas. You looked at explanatory variables that affect blood alcohol level. You also looked at various scenarios and explored how the original amount charged on a credit card and how much is paid off each month affect the balance remaining on the card after a period of time.

ACTIVITY 5.1

Rates of Change and Linear Functions

In this activity, you will investigate the link between the graph of a function and the rate of change and, in particular, how to identify if data describes a linear function.

1. Following are three graphs and the table of data that produced each graph. (These graphs also appear in Example 2.2 of Topic 2.) Answer the questions for each graph-table pair.

   a.

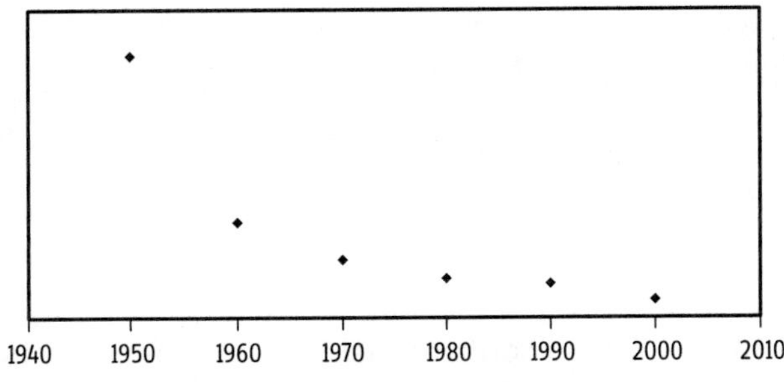

Year	Cases of Tuberculosis
1950	80.45
1960	30.83
1970	18.28
1980	12.25
1990	10.33
2000	5.82

i. Use the table to find the average rate of change in cases of tuberculosis per year.

From 1950 to 1960:

From 1960 to 1970:

From 1970 to 1980:

From 1980 to 1990:

From 1990 to 2000:

ii. Explain how these rates of change confirm that the graph represents a decreasing function.

iii. Explain how these rates of change confirm that the graph is concave upward.

b.

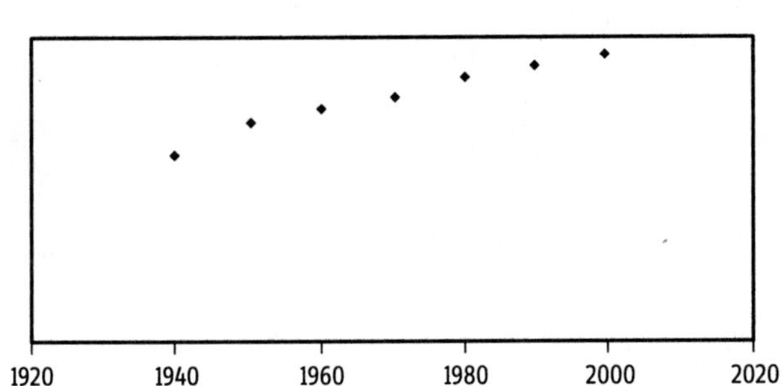

Year	Life Expectancy at Birth
1940	62.9
1950	68.2
1960	69.7
1970	70.8
1980	73.7
1990	75.4
2000	77.0

i. Use the table to find the average rate of change in life expectancy at birth per year.

From 1940 to 1950:

From 1950 to 1960:

From 1960 to 1970:

From 1970 to 1980:

From 1980 to 1990:

From 1990 to 2000:

ii. Explain how these rates of change confirm where the function represented by the graph is increasing and where it is decreasing.

iii. Explain how these rates of change confirm where the graph is concave upward and where it is concave downward.

c.

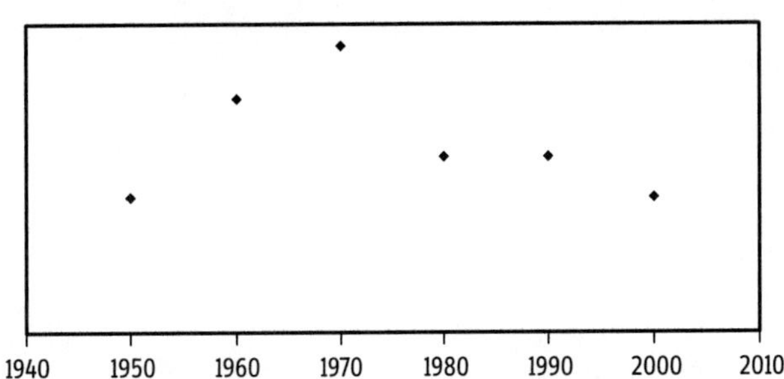

Year	Military Personnel
1950	1459462
1960	2475438
1970	3064760
1980	2050627
1990	2043765
2000	1384338

i. Use the table to find the average rate of change of number of military personnel per year.

From 1950 to 1960:

From 1960 to 1970:

From 1970 to 1980:

From 1980 to 1990:

From 1990 to 2000:

ii. Explain how these rates of change confirm where the function represented by the graph is increasing and where it is decreasing.

iii. Explain how these rates of change confirm where the graph is concave upward and where it is concave downward.

2. During the early years of the Olympics, the height of the winning pole vault increased approximately 8 inches every four years, as shown in the following table:

Year	1900	1904	1908	1912
Height in Inches	130	138	146	154

a. Explain how you can tell from the table that this is a linear function. (Hint: Look at the average rate of change in the height of the winning pole vault per year over the following intervals: from 1900 to 1904; from 1904 to 1908; from 1908 to 1912.)

b. Let t be the number of years since 1900, and let H be the height of the winning pole vault. Verify that the equation $H = 130 + 2t$ gives the height of the winning pole vault as a function of t. What does the number 2 in this equation represent? What does the number 130 in this equation represent?

c. Could you use this equation to predict the height of the winning pole vault in the next summer Olympics? Explain.

3. The following table gives calories used per hour for individuals of different weights for three different activities:

Weight	100	120	150	180	200
Hiking	225	255	300	345	375
Cross-country skiing	525	595	700	805	875
Pleasure sailing	135	153	180	207	225

For each activity, look at the rate of change of calories used per pound of weight from 100 to 120 pounds, from 120 pounds to 150 pounds, from 150 pounds to 180 pounds, from 180 pounds to 200 pounds.

a. What kind of a relationship is there between weight and calories used per hour for each activity and what will the graphs of the calories used as a function of the individual's weight for each activity look like?

b. How will the graph of calories used for hiking compare with that of the other two activities?

c. Make a graph that shows all the data on the same graph. Is your answer from part b of this question correct?

Summary

In this activity, you looked at different sets of data and the corresponding graphs. You computed rates of change and verified that the rate of change indicates whether the function is increasing or decreasing; you also explored the relationship between concavity and whether the rate of change is increasing or decreasing. You also analyzed a linear function that models the height of the winning pole vault in the Olympics and linear functions that model the calories spent in different exercise activities.

ACTIVITY 5.2

Major-League Salaries: Rates of Change and Concavity

In this activity, you will investigate major-league baseball salary data and consider connections between rates of change and the graph of a function and what these tell you about the data.

1. Retrieve the Excel file "EA.5.2 Major League Salaries.xls" from the CD or website. The table in this file gives the average yearly salary of major-league baseball players from 1980 to 2003. (Source: Haupert, Michael. "The Economic History of Major League Baseball." *EH.Net Encyclopedia*, edited by Robert Whaples. August 2003. http://eh.net/encyclopedia/article.haupert.mlb.)

2. Find the average rate of change in average yearly salary from 1980 to 1981 and from 1981 to 1982.

3. Now use the following Excel instructions to calculate the average rate of change in salary per year from one year to the next.

Instructions to Use Excel to Calculate Average Rate of Change

a. To enter the formula for rate of change of average salary, click cell C3, type **= (B3 – B2)/(A3 – A2)**, and press **Enter**. (Does the number in C3 agree with one of the two in #2?)

b. Click C3 and autofill to the end of your data.

c. Enter an appropriate title for column C in cell C1.

d. You can use **Audit mode** (**Ctrl** and ` together) to check that your formula for rate of change autofilled correctly. (Press **Ctrl** and ` together again to exit **Audit mode**.)

4. What do the numbers in column C represent?

5. Use the data now in your table (do not use Excel to graph the data yet) to answer the following:

   a. Find the interval(s) of time when the average yearly salary was increasing. Record them here.

b. Find the interval(s) of time when the average yearly salary was decreasing. Record them here.

c. Within the years 1986–1997, find the interval(s) of time when the average rate of change per year in average salary is increasing. (Think about why this question is different than the question in part a.) Give the intervals here.

d. Should the graph of the function that gives the average yearly salary of major-league players be concave downward or concave upward over the intervals of time indicated in part c? Why?

e. Within the years 1986–1997, find the interval(s) of time when the average rate of change per year in average salary is decreasing. Give the intervals here.

f. Should the graph of the function that gives the average yearly salary of major-league players be concave downward or concave upward over the intervals of time indicated in part e? Why?

6. Use the information obtained in question 5a–f to draw a rough sketch of the function from 1986 to 1997. Include an appropriate scale on the vertical axis.

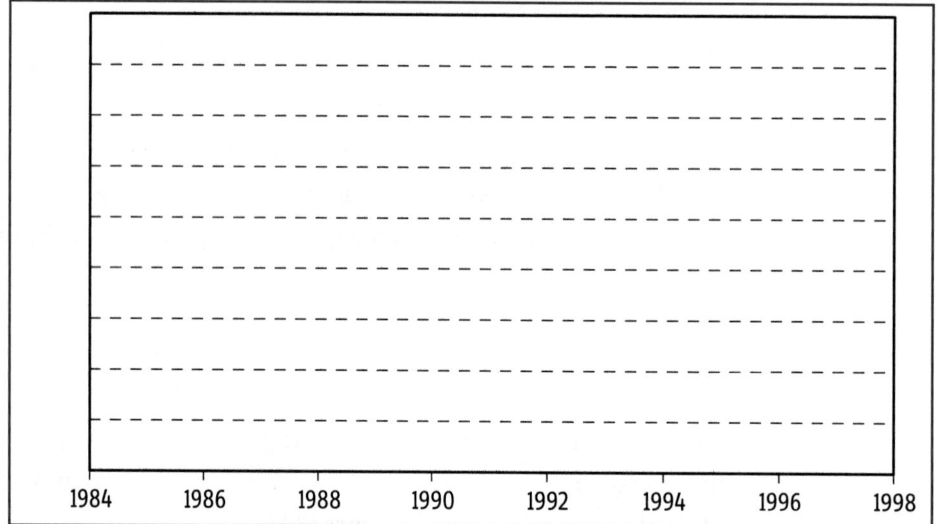

7. Use Excel to draw a graph of the average major-league salary function over time. (On your Excel graph, include all data given, starting in 1980.) How does this graph compare with your sketch?

8. What happened to the average salary of major-league players over the interval of time when the graph of the function is increasing and concave downward?

9. What happened to the average salary of major-league players over the interval of time when the graph of the function is increasing and concave upward?

10. What portion of the graph could be *approximated* by a line? Explain why you chose that portion.

11. Write an equation of a line that approximates the values of the given function from 1980 to 1986. To simplify the numbers, take the year 1980 as time $t = 0$, 1981 as time $t = 1$, and so on. Explain how you obtained your line.

Summary

In this activity, you used Excel to find average rates of change of major-league salaries for baseball players. You used these to draw a rough graph of the function that gives the players' average yearly salary and then used Excel to create the graph. You also approximated a portion of the graph using a line and then found an equation of the line.

ACTIVITY 6.1

The Genie's Offer: Exponential Growth and Linear Growth

In this activity, you will explore differences between exponential and linear growth. You will also analyze an example where value is decreasing exponentially.

1. Suppose a magic genie offers you a choice: The genie will give you \$1000 on the first day of the year and will add \$1000 to what you have on each succeeding day of January until the end of the month. Or you may choose to receive 2 cents on January 1, and each day for the rest of the month, the genie will double the amount you had on the previous day.

   a. Which deal sounds better to you and why?

   b. Set up a spreadsheet to complete the following table. Use the appropriate formulas for the second and third columns, and fill in to the end of January.

Date in January	Total $ with the First Offer	Total $ with the Second Offer
1	1000	.02
2	2000	.04
3	3000	.08
4		

(As a hint to help you fill in the spreadsheet, remember you can enter numbers **1** and **2** in the first two cells in a column and then drag down to fill in the rest of the integers. You can also enter a formula and drag it down a column and then use **Audit mode** to check the formulas entered. See Activity 4.1 if you need to refresh your memory.)

c. What does the table show?

d. At what point in the table is the amount in the third column greater than the amount in the second column and what does that mean?

e. Looking at your spreadsheet, identify a pattern and write an equation that shows how much money m you will have after d days if you take the genie's first offer of $1000 on January 1 and an additional $1000 each day after that.

f. Identify the type of function you wrote in part e of this question. How do you know it is that type of function?

g. Now you'll examine the genie's second offer to give you 2 cents on January 1 and then on each succeeding day to double the amount you had the day before. Fill in the blanks in the following exponents and identify the pattern emerging. Then analyze the pattern and develop an equation relating how much money m you will have after d days if you take the genie's second offer.

Money, in dollars, on day 1 = original amount of \$ = 0.02

Money, in dollars, on day 2 = \$ on Jan 2 = (\$ on Jan 1) * 2 = 0.02 * 2

Money, in dollars, on day 3 = \$ on Jan 3 = (\$ on Jan 2) * 2 = 0.02 * 2 * 2 = $0.02 * 2^2$

Money, in dollars, on day 4 = \$ on Jan 4 = (\$ on Jan 3) * 2 = $0.02 * 2^2 * 2$ = $0.02 * 2^3$

Money, in dollars, on day 5 = \$ on Jan 5 = (\$ on Jan 4) * 2 = $0.02 * 2^3 * 2$ = $0.02 * 2^4$

Money, in dollars, on day d = m_d = ________________. (What you fill in here should be an expression in terms of d.)

h. Fill in the blanks:

i. On January 20, you have _______ times as much money as you had on January 1.

ii. On January 31, you have ________ times as much money as you had on January 1.

i. Explain why it makes sense to call the function you wrote in part g of this question an **exponential function** and the kind of growth seen in the genie's second offer **exponential growth**.

j. On the same set of axes, graph the functions giving the amount of money you would have after *d* days with each of the genie's two offers. (You only need to graph up to day 25 to see the behavior of the two functions.) What does your graph show?

2. Appliances decrease in value as soon as you take them out of the store. A certain appliance originally cost $1200. After one year the appliance is worth $1080. Assume that the decline in value is exponential; that is, assume that

the ratio of the appliance's value in one year to the appliance's value in the previous year is constant.

a. Thus, the ratio

$$\frac{(value\ in\ year\ 1)}{(value\ in\ year\ 0)} = \_\_\_\_\_\_\_\_;\ \text{or}\ (value\ in\ year\ 1) = \_\_\_\_\_\_\_ * (value\ in\ year\ 0)$$

will be the same for all succeeding pairs of years.

b. Set up a spreadsheet starting with the following table to compute the value of the appliance for years 2 through 20. You will need to enter the appropriate formula to compute the value of the appliance in year 2 and then drag it down the column for succeeding years.

Year	Value of Appliance in $
0	1200
1	1080
2	

c. What is the appliance's value after 10 years?

d. Approximately when is the appliance's value half of its original value?

e. Approximately when is the appliance's value one-quarter of its original value?

f. If you continue these assumptions, will the appliance ever be worth $0? Explain.

g. Use the letters v and y to indicate the appliance's value in dollars and the number of years from the purchasing date, respectively. Write a formula that gives v in terms of y. (Note that your formula must be set up so that when $y = 0$, $v = 1200$, when $y = 1$, $v = 1080$, and so on.)

h. Is v an exponential function of y? Why or why not?

i. Describe the similarities and differences between the formula for v in part h of this question and the formula for m in #1g.

Summary

In this activity, you practiced creating formulas and scatterplots in Excel. You compared linear and exponential growth and explored the differences between these two fundamental types of growth. You also looked at a quantity that decreases exponentially and investigated patterns in several types of growth to find appropriate formulas.

ACTIVITY 6.2

Lines of Best Fit

It is often desirable to use a linear function to model a given set of data. In this activity, you will work with several data sets, and for each, you will look for a suitable line that approximates the given data. You will also work with the line that is generally used as the best-fitting line, the least-squares regression line. You will learn how to use Excel to find this line and its equation.

1. For the following scatterplots a, b, and c, draw an appropriate line to fit the data by "eye-balling" the graph and judging what line comes closest to all points. In the space following each graph, indicate whether the slope of the line you drew is positive, negative, or zero. For graph d, explain why a line would not be a good fit. (Source: *The Wall Street Journal Almanac 1999*.)

a.

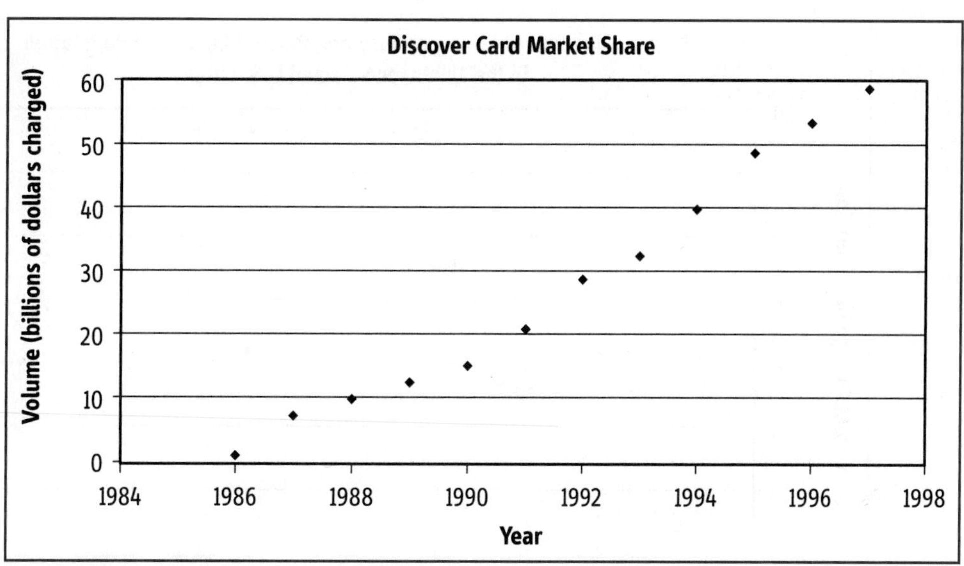

Slope (negative, positive, or 0): ________________

b.

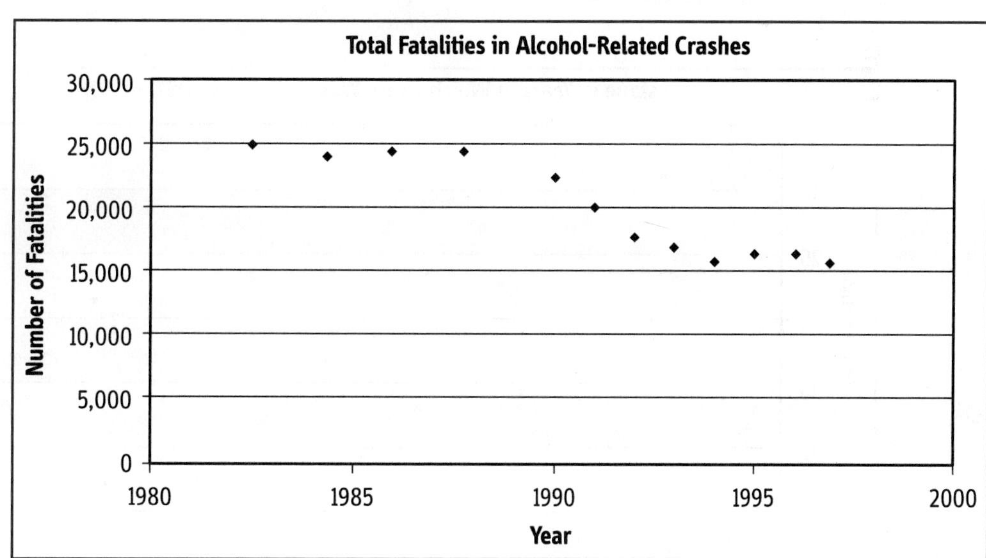

Slope (negative, positive, or 0): ________________

c.

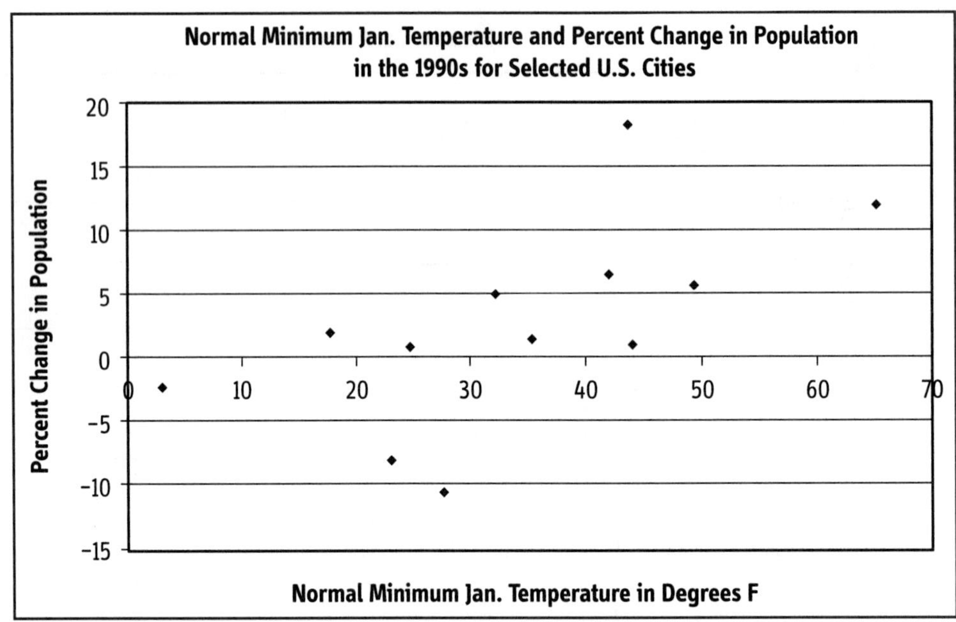

Slope (negative, positive, or 0): _______________

d.

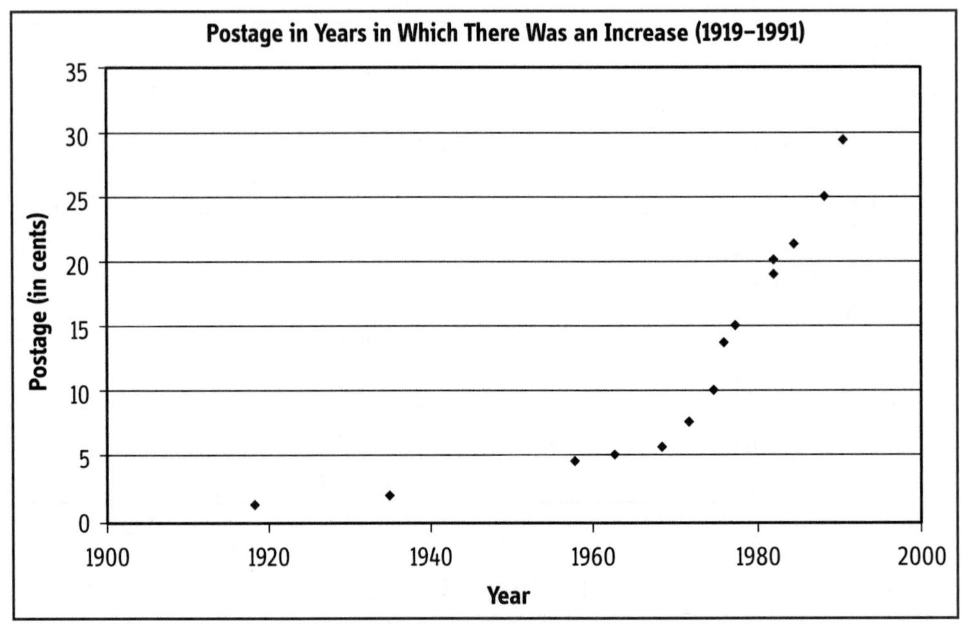

Why is a line *not* a good fit for graph d?

Because individual people might draw different lines, especially when the data is scattered as it is on graph c, we need a way to construct a line that doesn't depend on an individual's perception. The most commonly used method to construct such a line results in the **least-squares regression line** or just the **regression line**. This line, among all possible lines we could draw, is the one that makes the sum of the squares of the vertical distances of the data points from the line as small as possible. This line is easy for a computer or calculator to find because it involves calculations using straightforward (but kind of messy) formulas.

2. The regression line is used to show how a response variable changes, on average, as an explanatory variable changes. You can use such a line to predict the value of the response variable for a particular value of the explanatory variable. The least-squares regression line, for the data given in #1c, is shown on the following graph.

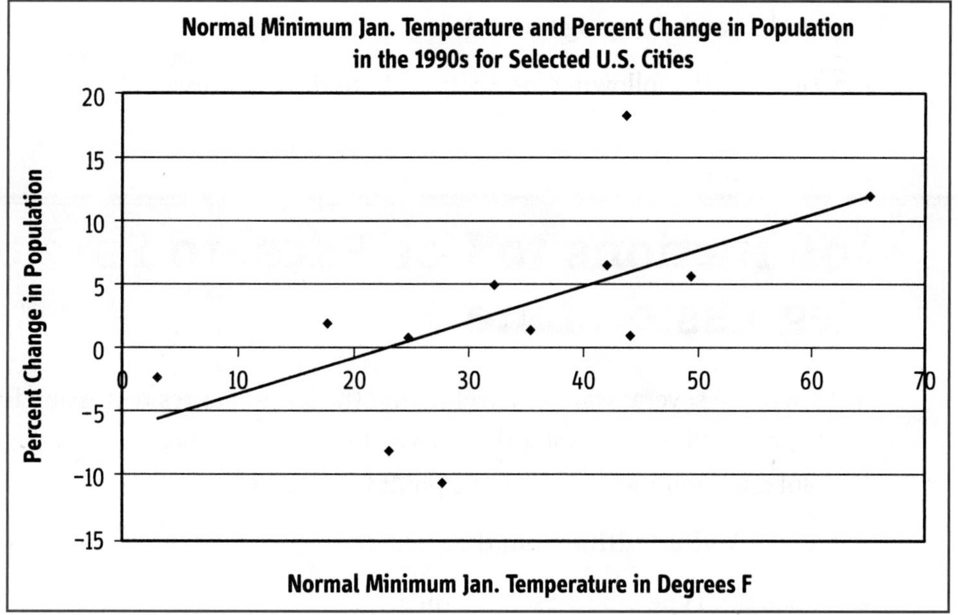

a. Draw in the vertical distances from all points in the data set to the least-squares line.

b. How many data points lie above the line? ___________

c. How many data points lie below the line? ___________

d. How can you tell from a scatterplot of the data, whether the slope of the regression line will be positive or negative?

3. Retrieve the data set "EA6.2.1 Verbal SAT Data.xls" from the CD or website and create a scatterplot of "percent taking the test" and "verbal SAT score." When creating the scatterplot, highlight only the two columns of data corresponding to the two quantitative variables; do not select the names of the states. Include appropriate titles, and change the scale on the vertical axis to go from 450 to 600. (See Activity 2.1 if you need a refresher on creating scatterplots and changing the scale.)

4. Then use the following instructions to find the regression line for these data.

Instructions to Use Excel to Find the Regression Line

a. There are several ways in Excel to find the least-squares regression line, but the easiest way is to point the cursor at one of the data points on the scatterplot and then right-click on the point to select it.

b. Select **Add trendline** from the menu bar.

c. Click the **Type** tab and select **Linear**.

d. Click the **Options** tab and select **Automatic: Linear**. Also click to place a check mark in the box **Display equation on chart**. Make sure there are no check marks in the other boxes. Then click **OK**. Your graph should display the regression line and its equation. You may need to click and drag the equation to a spot on the graph where you can read it clearly.

5. Write the equation of your line and indicate what the variables x and y represent.

6. What is the slope of the line and what does it represent? Interpret the slope in the context of the data.

7. What is the y-intercept of the line and what does it represent? Interpret the y-intercept in the context of the data.

8. Use the line you found to predict the average verbal SAT score for a state in which 60 percent of students take the exam. Where does this value appear on the graph? Mark it on a copy of the graph.

9. Retrieve the data set "EA6.2.2 Data Movies and Vid.xls" from the CD or website. This file contains data collected from a sample of college students. Create

a scatterplot of the two quantitative variables and find the regression line for these data.

a. Write the equation of the line and indicate what the variables x and y represent in the equation.

b. Is there a clear choice of explanatory variable and response variable in this data set? Why or why not?

c. Describe what your scatterplot and line show.

d. There is one clearly unusual data point in the data set—the male who estimated he saw 200 movies at a theater last year. Delete this case and look at the scatterplot for these adjusted data. Write the equation of this adjusted line and describe how the least-squares line changed when the outlier was deleted.

Summary

In this activity, you practiced creating scatterplots and learned how to find the regression line for a set of data using Excel. You interpreted the slope and y-intercept of a regression line in the context of the data set from which it was obtained, and used the regression line to predict values of the response variable. You also explored how an unusual data point can affect the regression line.

ACTIVITY 7.1

Richter Scale and Logarithms

In this activity, you will investigate earthquake data and explore the Richter scale as a measure of the intensity of an earthquake. You will consider how numbers on this scale compare with one another and study logarithms in the process.

1. How much stronger is an earthquake that measures 6.5 on the Richter scale than one that measures 6.0? How much farther away from your home is a restaurant that is 6.5 blocks away, as compared to one that is 6.0 blocks away? How are these measures similar and how are they different?

2. Retrieve the file "EA7.1.1 Deadly Earthquakes.xls" from the CD or website. (These data were obtained from the website http://earthquake.usgs.gov/.) This file contains the date, location, and magnitude on the

Richter scale of earthquakes that occurred from 1975–2003 and involved the loss of 1,000 or more lives.

3. Sort the data in ascending order by magnitude and give the date and location of the strongest and weakest of the earthquakes on the list. (See Activity 1.2 if you do not remember how to sort data.)

   Strongest:

   Weakest:

4. How many times stronger was the strongest earthquake than the weakest?

5. For each of the earthquakes listed, compute the relative energy it released using the following instructions. Recall that the relative energy E released by an earthquake of magnitude m on the Richter scale is $E = 10^m$.

Instructions to Use Excel to Calculate Energy Released

a. In cell D2, enter **=10^C2** (the ^ symbol indicates raised to a power) to calculate the relative energy of the first earthquake. Enter an appropriate title in cell D1.

b. To calculate the relative energy as a function of Richter scale magnitude for all earthquakes, use the "drag" feature of Excel to fill the column.

6. Create a scatterplot using columns C and D of your spreadsheet. (See Activity 2.1 if you do not remember how to create a scatterplot.)

   a. Explain what your graph shows.

   b. What type of function (linear, exponential, or neither linear nor exponential) does your graph show? How do you know?

   c. What variable is on the horizontal axis?

   d. What variable is on the vertical axis?

7. For each energy value E you found, compute its logarithm $\log E$, using the following instructions.

Instructions to Use Excel to Calculate Logarithms

a. To have Excel compute these logarithms and enter them in column E, place the cursor in cell E2 and enter **=LOG(D2)**. Then drag down.

b. Enter an appropriate title for column E in cell E1.

8. How are the values you just calculated in column E related to other values in your table?

9. Create another scatterplot using columns D and E.

   a. Give the name and equation of the function just graphed.

   b. What variable is on the horizontal axis?

   c. What variable is on the vertical axis?

10. Look at the two scatterplots you've created in this activity and describe how they are related. How could you obtain one from the other?

11. Here are instructions to use Excel to write in scientific notation the relative energy released by the first earthquake on the list.

Instructions to Use Excel to Write Scientific Notation

a. In cell F2, enter =**D2** to copy the number in cell D2 (which represents the relative energy released by the earthquake) into cell F2.

b. Click on cell F2, go to the **Format** menu and select **Cells**.

c. In the **Format cells** window, click the **Number** tab and choose **Scientific** from the **Category** list. For **Decimal places**, enter **2**.

12. Write the number as it appears in cell F2 and also write it in standard scientific notation (using a power of 10).

13. What is the difference between the number in cell D2 and the number in cell F2?

14. Using the "drag" feature, enter the rest of the numbers in column D into column F in scientific notation. Estimate the ratio between the largest number and the smallest number in column F. What does this ratio say about the earthquakes given in the data set?

15. Retrieve the file "EA.7.1.2 Earthquake Casualt.xls" from the CD or website. This file gives the same information as the file you used previously, except a new column has been added that shows estimated number of deaths. Create a scatterplot to show if there is any relationship between magnitude and number of deaths. Explain in detail what your graph shows. (You might want to delete one or two "unusual data values" to see what the data shows. Be sure to explain what you did.)

Summary

In this activity, you compared strengths of major earthquakes from 1975 to 2003. You explored the relationship between the earthquake's magnitude on the Richter scale and the relative energy released in the earthquake. You used Excel to draw graphs of the logarithmic and exponential functions involved and analyzed how they are related. You also used Excel to compute values of the common logarithmic function and to write numbers in scientific notation.

ACTIVITY 7.2

Estimations, Scientific Notation, and Properties of Logarithms

In this activity, you will use scientific notation to develop an estimate of a large quantity by breaking it down into small pieces. You will also use properties of logarithms to investigate and answer questions about an investment.

1. One of Ross Perot's policy recommendations in his 1992 Presidential campaign was a call for a $.50 tax on every gallon of gasoline sold in the U.S. Since that time, other lawmakers have supported additional gas taxes to help reduce consumption and raise money. This activity asks you to determine roughly how much revenue a $.50 per gallon tax on gasoline would generate in a year. Before doing any calculations, take a guess as to how much revenue this proposal would generate in a year.

2. A useful strategy for estimating a quantity such as this one is to separate the problem into its component pieces and then to estimate each piece separately. These estimates can and should be fairly rough and are useful to provide a sense of the magnitude of a quantity.

   a. How many people are in the U.S.? (You might want to use the Internet or an almanac to get a sense of this number, if you don't already have an idea.) Write this estimate in scientific notation.

   b. Based on the number of people in the U.S., estimate the number of automobiles in the U.S. Write this estimate in scientific notation.

   c. Estimate the number of miles traveled by a typical car in a year. Write this estimate in scientific notation.

   d. Perform the appropriate operation on your answers to parts b and c of this question to obtain an estimate of the number of miles driven in the U.S. in one year. Use scientific notation.

   e. Estimate the number of miles that a typical car travels per gallon of gasoline.

f. Perform the appropriate operation on your answers to parts d and e of this question to estimate the number of gallons of gasoline purchased in the U.S. in one year. Use scientific notation.

g. Use the answer to part f of this question to estimate the revenue that would be generated by Perot's $.50 per gallon tax on gasoline.

h. What other quantities might you be able to estimate using this technique?

3. Now you will investigate an equation that relates to compound interest. Suppose you deposit $1,000 in a savings account that pays 4 percent annual interest, and the interest is compounded annually. Assuming you don't withdraw any of the money or interest, how much money do you think you will have after five years?

4. After one year your savings account will have $1{,}000 + 0.04 * 1{,}000 = \$1{,}040$. Assuming you do not withdraw any money from the account, how much money will you have after two years? After three years?

5. Set up an Excel worksheet to help calculate money in the account after a period of time, using the following instructions.

Instructions to Use Excel to Calculate Interest

a. Open a new sheet in your Excel workbook and label column A as **years after initial deposit** and column B as **amount in the account**.

b. Enter numbers **0** through **20** in column A, starting in cell A2, and the corresponding amounts in column B. Use Excel to do the computations for you, by entering a recursive formula in B3 that gives the amount in B3 in terms of the amount already in B2, and then drag to fill in the rest of the column.

6. What formula did you enter in cell B3?

7. What amount did you obtain for the account balance after 20 years?

8. The account balance after n years (still assuming an initial deposit of $1,000 and that no money is withdrawn during those years) can be given by a single formula in terms of n:

   Amount after n years = $1{,}000(1.04)^n$

   In column C of your spreadsheet, enter the values obtained using this formula. (To do this, place your cursor in cell C2 and enter the appropriate formula **=1000*(1.04)^A2**. Then drag down to year 20.)

9. How do the entries in columns B and C compare and why?

10. Using the previous formula, you can find the account balance after any number of years, without using a spreadsheet. You can also calculate the number of years it would take for the account balance to reach any given amount. For example, to find out how many years it would take for the account to reach \$10,000, you need to find the value of n so that $10{,}000 = 1{,}000 * (1.04)^n$.

    Why is this equation the one we need to solve?

11. To solve for the number of years it would take to reach \$10,000 in the account, we need to get n out of the exponent. If we apply the logarithm function to both sides of the equation, we get the following:

    $\log(10{,}000) = \log(1{,}000) + n\log(1.04)$

    What properties of logarithms did we use to get this equation?

12. Because log 10,000 = 4 and log 1000 = 3, the previous equation is equivalent to $4 = 3 + n\log(1.04)$. Explain why log 10,000 = 4 and log 1000 = 3.

13. Use algebra to solve (by hand) for n in terms of log(1.04).

14. Use a calculator or the computer to find an approximate value of log(1.04), and use it to approximate n to find the number of years it would take for the account to reach $10,000.

    n = ________________

15. Use this method to find the number of years it would take the account to reach $100,000.

16. Drag down in your Excel sheet to verify your answers to #14 and #15. Do they agree with the answers Excel calculated?

Summary

In this activity, you used scientific notation to make a rough estimate of the revenue that would be generated by a $.50 tax per gallon of gasoline. You used Excel to calculate compound interest and used logarithms to solve equations to determine how many years a savings account investment should be kept in the bank to obtain a certain return.

ACTIVITY 8.1

Measurement Difficulties and Indexes

In Topic 8, you read about several indexes that are used to measure complex quantities that are not defined and measured easily. In this activity, you will work with one of those indexes, the Fog Index, which measures reading difficulty. You will analyze different reading passages to get a better sense of how the different components affect the result, and why this index gives a measurement of a text's reading difficulty.

1. Some properties are easier to understand and measure than others. For example, we all understand how to measure the height of a person or the time it takes a particular person to complete an exam. It is not as clear how to measure a person's intelligence or a customer's satisfaction with services delivered to him or her.

   a. List three additional properties of a person that are defined and measured fairly easily.

b. List three additional properties of a person that are more difficult to define and measure.

2. Read the property "reading difficulty of a passage of text." For the most part, we can probably agree that such a property exists: some passages of text are certainly easier to read than others. But how might you measure that property?

   a. List three factors that might go into measuring a passage's reading difficulty.

   b. Consider the following three passages of text:

Passage 1:

Everyone agreed to this and off they went, walking briskly and stamping their feet. Lucy proved a good leader. At first, she wondered whether she would be able to find the way, but she recognized an odd-looking tree in one place and a stump in another and brought them on to where the ground became uneven and into the little valley and at last to the very door of Mr. Tumnus' cave. But there a terrible surprise awaited them.

The door had been wrenched off its hinges and broken to bits. Inside, the cave was dark and cold and had the damp feel and smell of a place that had not been lived in for several days. Snow had drifted in from the doorway and was heaped on the floor, mixed with something black, which turned out to be the charred sticks and ashes from the fire. Someone had apparently flung it about the room and then stamped it out. The crockery lay smashed on the floor and the picture of the Faun's father had been slashed into shreds with a knife.[1]

Passage 2:

The truck drove to a part of town that George had never seen before. At last it stopped in front of a large building. It was the Museum. George did not know what a Museum was. He was curious. While the guard was busy reading his paper, George slipped inside.

He walked up the steps and into a room full of all sorts of animals. At first George was scared, but then he noticed that they did not move. They were not alive, they were stuffed animals, put into the Museum so that everybody could get a look at them.[2]

Passage 3:

To engage in a serious discussion of race in America, we must begin not with the problems of black people but with the flaws of American society, flaws rooted in historic inequalities and longstanding cultural stereotypes. How we set up the terms for discussing racial issues shapes our perception and response to these issues. As long as black people are viewed as a "them," the burden falls on blacks to do all the "cultural" and "moral" work necessary for healthy race relations. The implication is that only certain Americans can define what it means to be American, and the rest must simply "fit in."

The emergence of strong black-nationalist sentiments among blacks, especially among young people, is a revolt against this sense of having to "fit in."[3]

c. Based on your impression from one reading, rank the three passages from least difficult to most difficult to read.

   i. Least difficult

   ii. Next in difficulty

   iii. Most difficult

3. Because longer sentences tend to be more difficult to read than shorter ones, the number of words per sentence in a passage of text would seem to be a reasonable measure of reading difficulty. Set up a table on an Excel spreadsheet, like the one shown here, and record the number of words and the number of sentences for each of the three passages. Then tell Excel to use the appropriate

formula to compute the number of words per sentence. Record your answers in this table.

Passage	Number of Words	Number of Sentences	Words per Sentence
1			
2			
3			

4. Based on "words per sentence," rank the three passages from least difficult to most difficult.

   a. Least difficult

   2

   b. Next in difficulty

   1

   c. Most difficult

   3

5. Another measure of reading difficulty might be the number of "big" words in a passage of text. Count the number of "big" words (words having three or more syllables) in each of the previous passages. Note that compound words, such as "everything," and words in which the third syllable is formed from a suffix such as "ed" or "ing" should not be counted as "big" words. Add another

column to your spreadsheet, to the right of the **Words per Sentence** column, and label it **Big Words**. Circle all the "big" words in each passage, count them, and record the number in the appropriate column of your spreadsheet.

6. Add a sixth column to your spreadsheet, and using the appropriate formula, compute the percentage of "big" words in each passage. Write your answers here.

7. Based on "percentage of 'big' words" rank the three passages from least difficult to most difficult.

   a. Least difficult

   b. Next in difficulty

   c. Most difficult

8. The Fog Index is a measure of reading difficulty used by newspaper and magazine editors. This measure takes into account both sentence length and word size. Set up another column on your spreadsheet to calculate the Fog Index.

Use the following formula, also given in Topic 8, and record the Fog Index for each passage:

Fog Index = 0.4 * (*words per sentence* + *percent of "big" words*)

a. Passage 1 Fog Index =

b. Passage 2 Fog Index =

c. Passage 3 Fog Index =

9. A Fog Index value of 9 purportedly indicates ninth-grade reading level, a value of 12 indicates twelfth-grade reading level, and a value of 14 indicates college-sophomore reading level. Based on your calculations, does this seem reasonable? Why or why not?

10. Use the spreadsheet you set up to help answer the following questions:

    a. How would the Fog Index for each of the passages change if the number of "big" words is doubled?

    b. How would the Fog Index for each of the passages change if the number of words and the number of "big" words stayed the same, but the number of sentences is doubled?

Summary

In this activity, you examined the Fog Index for several passages of text and looked at whether the Fog Index is a reasonable measure of reading difficulty. You also considered how the Fog Index would change if there were more "big" words or more sentences in the passages.

[1] Lewis, C.S. *The Lion, the Witch and the Wardrobe.* New York: Scholastic Inc., 1981, pp. 53–54.

[2] Rey, H. A. *Curious George Gets a Medal.* Boston: Houghton Mifflin Company, 1957, pp. 30–31.

[3] West, Cornel. *Race Matters.* Boston: Beacon Press, 1993, p.3.

ACTIVITY 8.2

Consumer Indexes

In this activity, you will look at how the Consumer Price Index can be used to understand trends involving changes in costs and salaries over time.

Several "Consumer Indexes" are used to measure items related to consumers. In particular, the Consumer Price Index (CPI) is used as a measure of inflation by considering how the price of commonly purchased commodities changes; it measures the price of a market basket of a large number of goods and services purchased by consumers.

1. The following table gives the median weekly earnings, in dollars, of full-time wage and salary workers, 25 years and older, by educational attainment from the years 1980 to 2000 along with the CPI for the same years. (Source: *U.S. Bureau of Labor Statistics.*)

Year	CPI	High School, 4 Years Only, Median Weekly Earnings ($)	College, 1 to 3 Years, Median Weekly Earnings ($)	College, 4 or More Years, Median Weekly Earnings ($)
1980	82.4	266	304	376
1982	96.5	302	351	438
1984	103.9	323	382	486

Year	CPI	High School, 4 Years Only, Median Weekly Earnings ($)	College, 1 to 3 Years, Median Weekly Earnings ($)	College, 4 or More Years, Median Weekly Earnings ($)
1986	109.6	344	409	525
1988	118.3	368	430	585
1990	130.7	386	476	639
1992	140.3	404	485	697
1994	148.2	421	499	733
1995	152.5	432	508	747
1996	157.6	443	518	758
1997	160.5	461	535	779
1998	163.0	479	558	821
1999	166.6	490	580	860
2000	172.2	506	598	896

a. By what percentage did the CPI increase from 1980 to 2000?

b. By what percentage did the median weekly salary for workers with four years of high school only increase from 1980 to 2000?

c. By what percentage did the median weekly salary for workers with one to three years of college increase from 1980 to 2000?

d. By what percentage did the median weekly salary for workers with four or more years of college increase from 1980 to 2000?

e. Explain what these percentages show about salaries for these groups of workers from 1980 to 2000.

2. Retrieve the file "EA8.2 Median Salaries and CPI.xls" from the CD or website. This file contains the information given in the previous table. You will first create one graph showing median weekly earnings of the three groups of workers. To do this *without* graphing the CPI column, follow these instructions.

Instructions to Create a Scatterplot from Noncontiguous Columns in an Excel Spreadsheet

Highlight the **Year** column (that is, column A) from cell 1 to cell 15. While pressing the **Ctrl** key, highlight the three columns of salary data. When you've highlighted the **Year** column and the three salary columns, release the **Ctrl** key and proceed to create a graph, selecting **Scatterplot** along with one of the connected-line options.

3. Explain what your graph shows about the salaries of the three groups of workers.

4. You will now add columns that will contain these salary figures converted into constant 2000 dollars. To the right of *each* of the salary columns, create *one* additional column.

Instructions to Insert a New Column in an Excel Spreadsheet

a. Place the cursor anywhere in the column immediately to the right of where you want to insert the new column. From the menu bar, select **Insert** and **Columns**.

b. After you insert three new columns, one to the right of each salary column, enter appropriate headings in cell 1 of each of these three new columns.

5. Next you'll convert each of the salary values to dollars in year 2000, starting with cell 2 of the "High School, 4 Years Only" column. To convert this value to 2000 dollars (interpreted to be dollars at time O), use the ratio:

$$\frac{\textit{Dollars at time A}}{\textit{CPI at time A}} = \frac{\textit{Dollars at time O}}{\textit{CPI at time O}}$$

Because you want to convert values to *time O* dollars, when solving for *dollars at time O*, you get:

$$Dollars\ at\ time\ O = Dollars\ at\ time\ A * \frac{CPI\ at\ time\ O}{CPI\ at\ time\ A}$$

With *time A* as 1980 and *time O* as 2000 (because you're converting to 2000 dollars), you will calculate for the first year in the "High School, 4 Years Only" column of the table,

$$\text{1980 salary in 2000 dollars} = 266 * \frac{\text{CPI in 2000}}{\text{CPI in 1980}}$$

Compute this value and write it here: _________

Similarly, for the next year in the "High School, 4 Years Only" column, you'll have

$$\text{1982 salary in 2000 dollars} = 302 * \frac{\text{CPI in 2000}}{\text{CPI in 1982}}$$

Compute this value and write it here: _________

6. In column D (labeled something along the lines of "High School, 4 Years Only in 2000 $") of your spreadsheet, use the following Excel instructions to compute the "High School" salaries from 1980 to 2000 in 2000 constant dollars.

Instructions to Name a Cell and Use an Excel Formula with a Value Kept Constant When the Formula Is Dragged

a. Because you are using CPI in 2000 in all your computations, you will "name" the cell containing this value. Then anytime you enter that name in a formula, it will use that constant value. First, put the cursor in cell B15. You will then see "B15" appear in a white box immediately above column A. This is the **Name box**. Place your cursor in the **Name box** and click the mouse button to highlight **B15**. Type an appropriate name for the value in the cell, for example, **CPI2000**. Then press **Enter**.

b. Now you are ready to use this named constant in your formula. Place the cursor in cell D2 and enter the formula: **=C2*CPI2000/B2**. Note that the 1980 dollar value that you want to convert is in cell C2 and the CPI values for the various years are in column B. Because you need to use the CPI in 2000, which is in cell B15, for all computations in that column, use the name of the constant **CPI2000** so when you drag down, the cell address doesn't change. (Alternately, you can use an absolute cell reference, **B15**, instead of naming the constant in cell B15, so the value won't change when dragged vertically or horizontally.)

c. Now highlight cell D2 and drag down to fill in the other salaries in column D in 2000 dollars. Check that the two values you computed in #5 appear in cells 2 and 3 of column D.

7. Enter the appropriate formulas and convert the "College, 1 to 3 Years" and "College, 4 or More Years" salary values to 2000 dollars in the columns you added for this purpose.

8. Now create a scatterplot using the "Year" column and the three columns containing the salaries converted to 2000 dollars. (Remember to use the **Ctrl** key to highlight noncontiguous columns, as you did when making the graph in #2.) Explain what this graph shows about the salaries of the three groups of workers: "High School, 4 Years Only," "College, 1 to 3 Years"; and "College, 4 or More Years."

Summary

By investigating the Consumer Price Index over a 20-year period and using it to compare salaries, in constant dollars, during that period of time, you experienced the effects of looking at values over time in constant dollars. You also learned how to construct a scatterplot using noncontiguous columns, how to add a column, and how to name a cell in Excel.

ACTIVITY 9.1

Mortgages

In this activity, you will investigate some considerations that arise when looking for a home mortgage loan. You will determine basic characteristics of several options to compare the consequences of the various choices.

Purchasing a home is a big decision. Generally, a down payment of at least 10 percent of the purchase price of the home is required. A lender will loan you the money for the rest of the purchase price but with additional closing fees. These closing costs include direct fees for things such as a home appraisal and title insurance and also fees for **points** charged; each point is 1 percent of the loan amount. For a **fixed rate mortgage**, you are guaranteed that the interest rate stated when you sign the loan papers will not change for the term of the loan. With an **adjustable rate mortgage** (ARM), which is more attractive to the lender, your interest rate changes over time. ARMs are generally given as a fixed rate period and then an interval, such as 3/1, which means the rate is fixed for three years and then will change every year based on a rate index. In considering mortgage loan options, you will need to take into account the following:

Interest rate

Amount of the loan

Closing costs, including points

Fixed rate or adjustable rate mortgage

Any prepayment penalties

You have decided to purchase a home and have enough money for a decent down payment as well as the money to cover direct fees for closing costs (which you'll assume do not vary much from one bank to another). You are considering several mortgage options and want to evaluate them. Not counting points, you will need a loan for $130,000. If a mortgage option requires you to pay points, you will need to add that money to your loan amount. Assume the mortgage options you are considering have no prepayment penalties.

1. Lender I offers you a fixed rate 15-year mortgage at 5.25 percent APR, compounded monthly, with no points.

   a. Find your monthly payments under this option.

   b. Find the total amount of interest you will pay over the life of the loan.

2. Lender II offers you a fixed rate 30-year mortgage at 5.75 percent APR, compounded monthly, with one point.

   a. Find your monthly payments under this option. (Don't forget to include the "point" in the loan amount.)

   b. Find the total amount of interest you will pay over the life of the loan.

3. Excel has a special function that computes the payment required on a loan. Here are the instructions to use this function to check the previous computations.

Instructions to Use Excel to Calculate a Mortgage Payment

a. Open a new sheet in your Excel workbook. In cell A1, enter the label **Loan Amount**; in cell A2, enter the label **Annual Interest Rate**; in cell A3, enter the label **Length of Loan in Yrs**; and in cell A4, enter the label **Monthly Payment**. To check your work from #1, enter **$130,000** in cell B1, **5.25%** (do not forget to type the symbol %) in cell B2, and **15** in cell B3.

b. Now go to cell B4 on your worksheet. Move your cursor to the symbol *fx*. (This symbol might be located to the left of the formula bar or it might be near the **Chart wizard** icon on the toolbar.) This is the "insert function" symbol; click on it. In the **Insert function** window that appears, select **Financial** for category and **PMT** for function, and then click **OK**.

c. Now you want to fill in the requested values. For **Rate**, the interest rate is in cell B2, so enter **B2/12** (because the mortgage will be paid monthly and this value gives the monthly rate).

d. In **Nper**, enter the total number of payments for the life of the loan. The length of the loan in years is found in cell B3, so in this box enter **B3*12**.

e. **Pv** is the amount of the loan; that amount is found in cell B1, so enter **B1** in this box.

f. You can leave the remaining two boxes blank because the values that are used if omitted are what you want to use. If you want to enter values in these fields, you can enter **0** in the **fv** box and **0** in the **Type** box. Press **Enter**. Does this value agree with your answer to #1?

g. Change the interest rate to **5.75%** in cell B2 and change the amount and length of the loan to the values used in your calculations for #2. Does this value agree with your answer for #2?

---

4. Suppose you think that the payments required for the 15-year mortgage are too high for you to afford, but that you'd be able to afford higher payments than the 30-year mortgage requires. You are considering obtaining a 30-year mortgage at the offered rate but paying it off sooner by making larger payments than are required.

   a. Use Excel to find the monthly payments needed to pay the loan off in 20 years and record the monthly payment here.

b. What monthly payments would you need to make in order to pay it off in 25 years?

5. You want to compare how much interest is paid during the first two years under each lender's plan. Excel's function to calculate interest can help.

Instructions to Use Excel to Calculate Interest

a. On the same Excel worksheet, in cell D1, enter the label **Month**; in cell E1, enter the label **Interest**; in cell F1, enter the label **Interest, Lender I**, and in cell G1, enter **Interest Lender II**. (You'll use column E as a "calculation column" to calculate interest, and then you'll save the values you calculate in columns F and G.) Enter the numbers **1** and **2** in cells D2 and D3, respectively, and drag down to fill in the rest of the months to cover the two-year period.

b. Now you want to find the interest under Lender I's option, so start by entering the appropriate values for "Loan Amount," "Annual Interest Rate," and "Length of Loan" in cells B1, B2, and B3, respectively. Then select cell E2, move your cursor to the symbol *fx*, and click it. For **category**, select **Financial**, and for **function**, select **IPMT** (for interest payment) and then click **OK**.

c. Now you want to fill in the requested values. For **Rate**, the interest rate is found in cell B2, so enter **B$2/12** (since the mortgage will be paid monthly and this value gives the monthly rate). Because you want to drag down to fill in the interest for the remaining months, make this an absolute reference that won't change when you drag down by adding a **$** sign between the **B** and the **2**.

d. For **Per**, enter **D2**, because the period is the month.

e. For **Nper**, because the length of the loan in years is found in cell B3, enter **B$3*12**.

f. For **Pv**, which is the amount of the loan found in cell B1, enter **B$1**.

g. You can leave the remaining box blank, or you can enter **0** in the **fv** box. Then click **OK**.

h. Drag down to fill in the interest for the first two years. Then go to cell E27 and find the sum of all the interest payments under Lender I's option. To do this, type =**SUM (E2:E25)**. Record the total interest here:

Total interest paid under Lender I's plan: ______________________

i. You want to save these values so that they won't change when you change the interest rate and length of the loan (which you will do to evaluate Lender II's option). To put these interest values in column F, highlight cells E1 through E27 and select **Edit** and then **Copy**. Then go to cell F2 and select **Edit** and **Paste special**. Choose **Values** by clicking on the white circle to the left of that word and click **OK**. This command pastes the numbers in column F so they are no longer linked to cells B1, B2, and B3.

j. Change the numbers in cells B1, B2, and B3 to reflect Lender II's option. (Remember to include the "points" in the amount borrowed.) The numbers in column E reflect the costs under Lender II's option. Copy and paste (as you did in part i of this set of Excel instructions) the values in column E to column G. Record the total interest here:

Total interest paid under Lender II's plan: ______________________

6. You have another mortgage option—Lender III offers you an adjustable rate 30-year mortgage, with no points, at 4.95 percent interest for the first year. After that time, the mortgage rate will likely change, but you don't know what the new rate will be. Find your monthly payments for the first year under this option and record that value here.

7. Let's suppose that under Lender III's adjustable rate mortgage option, the interest rate increases by 1 percent for the second year. Calculating interest paid for the first two years under Lender III's plan is more complicated. Here's one way to do it.

   a. First, use a method similar to what you did for Lenders I and II to find the interest paid for each month of the *first year* under Lender III's plan and record that total here:

      Total interest paid during the first year under Lender III's plan: ________

b. Now use your work in #6 to find the total payments made under this plan during the *first year*; then find the amount of principal paid during the first year and the total principal remaining. Record those values here:

Total principal paid during the first year: ______________________________

Total principal remaining: ____

c. Finally, use the IPMT function with the new principal remaining as the amount of the loan, the new interest rate of 5.95 percent, and 29 years remaining on the loan, to calculate the interest paid during each month of the second year of Lender III's plan. Find the total interest paid during the two years under Lender III's plan and record that total here:

Total interest paid for the first two years under Lender III's plan: _______

8. Consider the monthly payments, the total interest, and any other factors that you think are important and discuss the pros and cons of each of the lender's plans. Which lender would you choose and why?

Summary

In this activity, you learned how to use Excel's loan payment and interest calculation functions. You learned how to copy and paste values so they are no longer linked to cells that may change. Finally, you examined and compared several realistic options for home mortgage loans.

ACTIVITY 10.1

Savings and Loans: Problem Solving and Using Scroll Bars

In the first part of this activity, you will identify problem-solving techniques. In the second part of this activity, you will learn how to create a scroll bar using Excel and then use it to solve problems about savings and loans.

1. Identify all problem-solving techniques used in #1 of Activity 6.1 (where you decided which of the magic genie's two offers to choose). For each technique you identify, describe where in the activity it was used.

2. Describe a problem (you can use one solved previously, or you can construct one) that can be solved using the stated problem-solving technique (by itself or in combination with other techniques):

   a. Problem-solving technique *2b. Draw a picture, graph, or diagram.*

   b. Problem-solving technique *3. Examine a simple case or try several special cases.*

   c. Problem-solving technique *5. Work backward.*

3. In the next part of this activity, you will create an Excel scroll bar to see how interest grows in a savings account.

Scroll Bars

A scroll bar is an exciting addition to a spreadsheet that can allow you to vary a value easily. You can vary, for example, the interest rate on an investment in an account over an interval from 1 percent to 20 percent to see how the total amount in the account is changing dynamically.

Instructions to Use Excel to Create a Scroll Bar

a. You'll first set up a spreadsheet to calculate the interest and total money in the account with an investment of $1000 over a 15-year period. In cell C1, enter the label **Interest Rate as a Percent** and in cell C2 enter **=1**. In cell B1, enter the label **Interest Rate as a Decimal** and in cell B2 enter **=C2/100**. (The value in cell C2, which is the interest rate as a percent, will be linked to the scroll bar. The value in cell B2 is the interest rate as a decimal.)

b. In cells A4, B4, and C4, enter the labels **Year, Interest, Total,** respectively.

c. In cell A5, enter **0**; in A6 enter **1**. Highlight both of these cells and drag down to cell A20 to represent 15 years.

d. In cell C5, enter **$1000.00**, which will be the starting amount in the account. In cell B6, enter **=B2*C5**. (Note that you can type **=B2** and then press the **F4** function key to get the result B2.) In cell C6, enter the formula, **=C5 + B6**.

e. Highlight cells B6 and C6 and drag down to show the interest and total for each of the 15 years.

f. Use the **Chart wizard** to sketch a scatterplot of x = **Year** and y = **Total**. You will need to highlight the label and values in the **Year** column, press **Ctrl**, release the left mouse button, and then go to the **Total** column and highlight the label and values in this column. Now using the **Chart wizard**, create the scatterplot, with the points connected by a line. Choose appropriate titles for the axes and chart.

g. You will now set up a scroll bar to allow you to vary the interest rate conveniently. Go to **View** on the menu bar, select **Toolbars** and then **Control toolbox**. Find the **Scroll bar** icon on the **Control toolbox** toolbar. (Look for the small picture with the up and down arrows and check that the words **Scroll bar** appear when you place the cursor over it. Click on this icon. The **Scroll bar** icon will light up and the cursor becomes a thin "plus sign." Notice that the top-left icon on the **Control toolbox** toolbar lights up, too, to indicate you are in "design mode."

h. Move the cursor (the "plus" sign) to the spot where you want the scroll bar. Press and hold the left mouse button. Drag the mouse horizontally to the right to create a horizontal scroll bar, or drag it down to create a vertical scroll bar. When the scroll bar is the size you want, release the mouse button. You should see small open circles, called handles, around the outside of the scroll bar.

i. With the handles still showing, move the cursor to the **Control toolbox** toolbar and click the **Properties** icon, which should be immediately to the right of the **Design mode** icon.

j. This brings up the **Properties** box containing a list of properties in alphabetical order. You will need to fill in three values: **LinkedCell**, **Max**, and **Min**. Linked cell refers to the cell where you have placed the value you want to change when you scroll. Point to the words **LinkedCell**, click, and type **C2** here. Point to **Min**, click, and enter **1,** and then point to **Max**, click, and enter **20**. Move the cursor to a cell outside the **Properties** box and click. To close the **Properties** box, click the **x** in the upper-right corner of the box.

k. Go to the **Control toolbox** toolbar, which should still be showing on your screen and click the icon in the top-left corner that is still lit up to exit **Design mode**. Also click the **x** in the top-right corner of the **Control Toolbox** toolbar to close it. You are almost ready to use the scroll bar you designed.

l. You need to change the scale on the *y*-axis of the scatterplot so the scaling is not automatic, but is fixed. On your finished graph, point to the *y*-axis and double-click to access the **Format axis** menu; select the **Scale** tab.

Click the check mark in the box next to **Minimum**; then enter **1000** in the **Minimum** box. Also click the check mark in the box next to **Maximum** and enter **15000** in the **Maximum** box. Click **OK**. (Fixing the scale will allow you to see how the graph changes as the scroll bar changes the value of the interest rate.)

m. You are now ready to move the little box on the scroll bar to increase or decrease the value of the interest rate. This will in turn alter the Total column of the spreadsheet and change the graph.

4. Describe how the graph changes as the scroll bar moves.

You can also use a scroll bar to solve Josh's problem (see Example 10.5 in Topic 10). To do this, you will first create a graph of the monthly balance remaining, and then you can change the monthly payment using the scroll bar. Here is how:

5. Go to a new sheet in your Excel workbook. Set up the new sheet as follows:

   a. In cell A1, enter the label **APR as a Decimal**, and in cell A2, enter **0.06**, the loan's APR (as a decimal) for the loan from the credit union.

   b. In cell B1, enter the label **Monthly Interest Rate**, and in cell B2 enter = **A2/12**, the monthly interest rate. In cell C1, enter the label **Tentative Payment** and in cell C2, enter **$100.00,** a tentative payment amount (this will be "attached" to the scroll bar and moved until you find the payment needed to pay off the loan in 60 months).

c. In cells A4, B4, and C4, enter **Year**, **Balance Last Month**, and **Balance This Month**, respectively.

d. In column A beginning at cell A5, fill in numbers **1** through **60** (for the 60 months in 5 years).

e. In cell B5, enter **$13000.00** (this is the starting loan principal or balance at the end the month before you start payments).

f. In cell C5 enter **=B5 + B5*B2 – C2** to calculate the loan balance after one month, that is, at the end of the first month.

g. In cell B6, enter **=C5**.

h. Drag to fill in columns B and C up to month 60.

i. Create a graph that shows "balance this month" as a function of "year." Because you are setting up a scroll bar, you will want to format the y-axis so scaling is fixed.

j. Create a scroll bar with a **LinkedCell** of **C2** (so the payment can be changed by scrolling), with a **Min** of **0**, and a **Max** of **1000**.

6. How would you know from the graph that the choice of monthly payment in C5 is the correct one (or a good approximation of it) to pay off the debt in five years? Write the correct amount here.

7. Use the same spreadsheet, changing the interest rate, principal, and compounded period as appropriate, to find all the information needed to solve Josh's problem. Explain how you found the information.

8. Identify which problem-solving techniques (see the list in Topic 10) you used to solve Josh's problem in this activity. Compare the techniques used here with the techniques used to solve the same problem in the examples in Topic 10. Which techniques were used in both solutions? Which were used in one and not the other?

Summary

In this activity, you worked with several problem-solving techniques and looked for situations where specific techniques could be used. You learned how to create a scroll bar in Excel to see how the interest paid and the accumulated total in a savings account change for different annual interest rates. Finally, you used a scroll bar to solve the loan decision problem stated in Example 10.5 and solved in Topic 10.

ACTIVITY 11.1

Ranking Cities: Ratings and Decisions

You will practice using several methods for making decisions that involve analyzing characteristics of possible alternative choices. These methods can be used for a variety of decisions and ratings, from consumer purchases to rating cities or films.

The following table gives information about twelve U.S. cities. Suppose you are considering job offers in each of these cities and want to decide, on the basis of the characteristics presented in the table (population, average January daily temperature, serious crimes per 100,000 population, percent unemployed, per capita income, and geographic region), which city you would most prefer. Assume that the jobs are fairly similar, so the only characteristics of the cities will influence your decision. (Source: *Time Almanac 2004*, pages 236–259, 390.)

City	Population	Average Daily Temperature, Jan	Crimes per 100K	Percent Unemployed	Per Capita Income	Region
Atlanta	416,474	41.0	13,318.5	5.3	33,769	Southeast
Boston	589,141	28.6	6,088.5	4.8	39,873	Northeast
Charlotte	540,828	39.3	7,904.0	6.3	31,526	Southeast
Dallas	1,188,580	44.6	8,838.3	7.1	34,697	South
Honolulu	371,657	71.4	5,324.5	3.9	31,115	South Pacific
Minneapolis	382,618	11.8	7,184.5	4.3	38,131	North central
New York City	8,008,278	31.5	3,600.2	7.3	40,450	East
Omaha	390,007	21.1	6,876.5	3.8	33,249	Central
San Francisco	776,733	51.1	5,429.4	5.9	57,714	West
Seattle	563,374	40.1	8,040.8	6.8	41,229	Northwest
Tucson	486,699	51.3	9,148.4	4.9	24,767	Southwest
Washington DC	572,059	34.6	7,272.7	3.7	41,754	East

1. Without examining the data in the table closely, based on your impression of the cities, which one do you think you'd choose and why?

2. Suppose you decide on the following cutoffs for the characteristics: population, no more than 600,000; average January temperature, at least 25 degrees; crimes per 100,000, less than 8000; percent unemployed, 4.5 or less. Use the

cutoff screening method with these cutoffs, and additional characteristics and cutoffs if you need any, to determine which city you would choose.

3. Retrieve the data set containing the previous table, "EA11.1 Cities.xls," from the CD or website. In the next several steps, you'll look at the given characteristics one-by-one, and set up a ranking of the characteristics based on your preferences.

4. For the characteristic of population, set up a ranking based on a system in which *your* most preferred ranking receives a 10 and the least preferred gets a 1. (Your rankings are based on your personal preferences, which may be different from your neighbor's or instructor's preferences.) Record these rankings for each city in the corresponding row of column H of the spreadsheet, and enter an appropriate heading for column H. Explain the rationale for your ranking system.

5. For the characteristic of average January temperature, set up a ranking based on a system in which your most preferred ranking receives a 10 and the least preferred gets a 1. Record these rankings in column I of the spreadsheet and enter an appropriate heading for column I. Explain the rationale for your ranking system.

6. Choose three of the four remaining characteristics that appear in the table and that you feel could influence your decision of which city to choose. List these characteristics here.

7. For each of the characteristics you just listed in #6, set up a ranking system in which the most preferred ranking receives a 10 and the least preferred gets a 1. Record these rankings in columns J, K, and L of the spreadsheet. For each of these columns, include an appropriate column heading and explain the rationale for your ranking system.

8. Now look at the characteristics on which you will base your decision. These characteristics include population, average January temperature, and the three additional characteristics you chose. Assign weights to these characteristics, using a scale of 1 to 10, with 10 being the most important characteristic for you and a weight smaller than 10 chosen to indicate that a characteristic is less important to you. Record your five characteristics and corresponding weights here, and explain why you picked the weights you did.

(1)

(2)

(3)

(4)

(5)

9. You are now ready to calculate the weighted sum of the rankings for each city. For each city, you will calculate *weight* ∗ *assigned rank* for each characteristic and then add these.

Instructions to Use Excel to Calculate the Weighted Sum

a. In your spreadsheet, enter the weights you picked for each characteristic in #8 in row 15 of the corresponding columns that contain the rankings, that is, columns H, I, J, K, and L. In cell G15, enter the label **weights=** .

b. Name each of the cells containing the weights using an appropriate name. (Activity 8.2 described how to do this by using the **Name box** above column A on the spreadsheet.) For example, you could name the value in cell H15 **popwt**, with no spaces in the name.

c. In cell M2, enter a formula such as the following, depending on the names you chose for the named cells, for the weighted sum ranking for Atlanta:

=popwt*H2 + tempwt*I2 + crimewt*J2 + unempwt*K2 + incomewt*L2

Write your result for Atlanta's weighted sum: ___________

d. Drag the formula down to find the total weighted ranking for each city in the list. (Note that the named cells for the weights do not change when you drag down the formula.)

10. List the city with the highest weighted ranking and the city with the lowest weighted ranking. (Before you answer, you may want to sort the data by "weighted sum." To do this, highlight all columns, then click **Data** on the menu bar, and choose **Sort**. In the **Sort by** box, choose **column M**.) Are you surprised by these rankings?

11. List at least three additional characteristics you might want to include in a ranking of desirable places to live.

12. For what other types of decisions might you use an analysis such as this?

13. Discuss the advantages and disadvantages of this method for helping you make decisions.

Summary

You investigated two methods for making decisions and also explored how to rate cities using various criteria and how to rank each city on those criteria. You also learned how to decide on weights for the criteria and used Excel to compute the weighted sum. Finally, you practiced naming a cell and entering a formula in Excel.

ACTIVITY 12.1

Analyzing Studies: Inductive Reasoning

In this activity, you will investigate possible flaws when using inductive reasoning. First, you will read newspaper articles that summarize situations in which inductive reasoning is used. You will then decide if the study described is an observational study or an experiment. Finally, you will analyze each situation and decide which inductive reasoning flaws might be present.

Inductive reasoning often involves a general conclusion or relation that results from specific examples or experiences. Four types of inductive reasoning are: prediction, generalization, analogy, and causal inference. Such reasoning is sometimes done by sampling a population, obtaining results from the sample, and inferring that these results will hold for the whole population. One way to do this is through an **observational study** where individuals are observed and some variable or variables of interest are measured, but the researcher does not attempt to influence the responses. On the other hand, in an **experiment**, the researcher deliberately imposes some treatment on individuals in order to observe their responses.

Because of the nature of inductive reasoning, the conclusions resulting from an inductive argument are not guaranteed to hold. Studies and experiments must be carefully planned and carried out so errors are minimized and the results are likely to hold (even though still not guaranteed). The following are some of the most commonly occurring inductive reasoning errors in quantitative research.

Correlation and Causation

If a group of people or objects is measured with respect to two variables, and if neither of the variables is experimentally manipulated, then the simple finding of a relationship between the two variables does not necessarily mean that one variable has a causal influence on the other. It is quite possible that both of the variables are linked to some unmeasured third variable.

Experimenter Effect

If there are two or more conditions of a manipulated treatment variable, and if the treatments are administered to the subjects by someone familiar with the researcher's hoped-for results, then it is possible that the comparison groups will be treated differently, biasing the results. This unconscious biasing of the results is sometimes called **expectancy**.

Sampling Bias

If subjects are sampled in such a way as to make them unrepresentative of the total group from which they are drawn, conclusions about the total group are not valid. Particular care must be taken when interpreting the results of questionnaires returned anonymously. If the return rate is 30 percent, for example, the researcher cannot assume the 30 percent responding is representative of the total population.

Subject Selection

When two or more groups of subjects receive different treatments, usually with one of the groups serving as a control group, only random selection of subjects to the groups will protect against the possibility that some extraneous variable (age, sex, education level, for example) is having an unwanted effect on the results.

Subject Effect

The subjects who receive a particular treatment might figure out, even if not told directly, what treatment they have been given, what treatment the other group(s) in the study received, and the intent of the research. This information may cause the groups to perform differently on the response variable(s) simply because of the way the subjects expect their treatments to affect their behavior, attitudes, or knowledge.

Valid Data/Self-Report

When subjects self-report data, they might be consciously or unconsciously motivated to withhold their honest thoughts. Or the data collected might constitute quantitative information on the wrong response variable. Additionally, sometimes subjects simply make mistakes when responding to a questionnaire.

On the following pages, you will find newspaper articles containing inductive reasoning. For each article, answer these questions:

a. Write out, in question form, a question that the research and reasoning was designed to answer. (Note that the article's headline might guide you but might also be misleading.)

b. Determine if the research was an observational study or an experiment. (Be careful about assuming it is a study just because the word *study* is used in the headline or article.) Explain how you arrived at your answer.

c. Identify any inductive reasoning flaws you think might be present in the study or experiment and what the researchers might have done to avoid the flaws.

d. Think about the researcher's conclusion. If possible, suggest a plausible explanation for the outcome of the study that is different from the explanation that the researcher has offered.

Activity 12.1 Response Sheet

1. "Small, Frequent Caffeine Doses May Work Best," *The Morning Call,* June 1, 2004.

Small, frequent caffeine doses may work best

By Jane E. Allen
Special to *The Morning Call*

Most people tank up on caffeine in the morning, often ingesting up to 500 milligrams of the stimulant (the amount in a Starbucks grande coffee).

But if you need to stay awake for extended periods, a better way to stay sharp—without suffering from sleepless nights—might be to consume caffeine slowly and steadily, beginning halfway through the workday. This intake pattern can counteract the internal system that pushes the body toward sleep as the day wears on, new research has found. Morning consumption spikes caffeine levels in the blood and brain, which then drop off.

James K. Wyatt, now lab director of the sleep disorders center at Rush University Medical Center in Chicago, led a study of 16 healthy Boston men whose days were lengthened to 42.85 hours, including 25.57 waking hours. The schedule was meant to simulate the long shifts of emergency workers, military personnel, and truckers.

Half received a pill every waking hour containing the amount of caffeine in two ounces of coffee; the others received dummy pills.

Jane E. Allen is a writer for the Los Angeles Times, a Tribune Publishing newspaper.

a. Research question:

b. Observational study or experiment and how you determined this:

c. Possible inductive reasoning flaws and remedies:

d. Possible alternative explanation for conclusion:

2. "Tanning Salons Raise Skin Cancer Risk, Study Indicates," *The Morning Call,* October 15, 2003.

Tanning salons raise skin cancer risk, study indicates

By PAUL RECER
Of The *Associated Press*

WASHINGTON —Regularly baking to a golden tan under sun lamps can increase the risk of malignant melanoma, a sometimes fatal skin cancer, and the younger a woman starts the greater the risk, a study says.

The study, which analyzed the lifestyles and melanoma risks for women between the ages of 30 and 50, found what the researchers said was the strongest evidence yet that artificial sun tanning can be dangerous to healthy skin.

Melanoma risk is highest among fair-skinned people in Australia, New Zealand, Europe and North America. Since the 1950s, the rate of the skin cancer has tripled in Norway and Sweden, where light skin is common. About 50,000 cases of melanoma are diagnosed annually in the United States and about 7,500 people die of the disease each year, according to American Academy of Dermatology officials.

In the study, appearing this week in the Journal of the National Cancer Institute, an international group of researchers analyzed data from the Women's Lifestyle and Health Cohort Study in Norway and Sweden. In 1991 and 1992, 106,379 women completed extensive questionnaires about their exposure to sunlight and to artificial tanning. In 1999, the researchers rechecked the women's cancer status using the national health registries in Norway and Sweden.

The researchers found 187 cases of malignant melanoma diagnosed among the study group during the eight-year follow-up period.

They found that women of any age or hair color who regularly visited tanning salons once or more per month increased their chance of developing melanoma by 55 percent.

The risk was highest for young adults. Compared with women who never used a solarium, women who reported using artificial tanning systems once or more per month when they were between the ages of 20 and 29 increased their risk of melanoma by about 150 percent.

"Our results provide stronger evidence than those of other studies that solarium use is associated with an increased risk of melanoma," the authors of the study wrote.

"This is just one of many papers that have suggested a link between indoor tanning and the development of melanoma skin cancer," said Dr. James M. Spencer, vice chairman of dermatology at the Mount Sinai School of Medicine in New York. He said studies have also linked artificial tanning to basal cell and squamous cell carcinoma, two common types of skin cancer.

Spencer said it is well known that ultraviolet light causes skin cancer.

"Whether you get it at the indoor tanning parlor or at the beach, (UV light) is a carcinogen," he said.

People may have good reasons to work and play in natural sunlight, where they can protect themselves with sun block, Spencer acknowledged. However, "There is no compelling reason to go to a tanning salon," he said. "It is just for a cosmetic tan that fades in a couple of weeks and can cause you a lifetime of trouble."

a. Research question:

b. Observational study or experiment and how you determined this:

c. Possible inductive reasoning flaws and remedies:

d. Possible alternative explanation for conclusion:

3. "Study: Lots of Television Time Can Hurt Children's Reading Ability," *The Morning Call,* October 29, 2003.

Study: Lots of television time can hurt children's reading ability

PARENTS MUST UNDERSTAND PITFALLS OF MEDIA, RESEARCHER SAYS.

By Siobhan McDonough
Of The *Associated Press*

WASHINGTON — Children who live in homes where the television is on most of the time might have more trouble learning to read than other children, a study says.

Tuesday's report, based on a survey of parents, also found that children 6 months to 6 years spend about two hours a day watching television, playing video games or using computers. That's roughly the same amount of time they spend playing outdoors and three times as long as they spend reading or being read to.

The study, by the Kaiser Family Foundation and Children's Digital Media Centers, found about one-third of children 6 and younger live in homes where a television is on most or all the time. In those "heavy TV households," 34 percent of children ages 4 to 6 can read, compared with 56 percent in homes where the TV is on less often.

CHILDREN AND THE MEDIA

Children's exposure to electronic media, according to a study:

» Children 6 and younger spend about two hours a day with a TV, video games or a computer.

» 48 percent of children 6 and under have used a computer; 30 percent have played video games.

» 90 percent of parents set rules about what their children can watch and 69 percent control how much time their children can watch TV.

» Children with time-related rules spend about a half-hour less per day in front of the TV than other children.

» 68 percent of children under 2 will be in front of a screen for an average of just over two hours a day.

» 36 percent of children 6 and younger have TVs in their bedrooms.

» 27 percent of 4- to 6-year olds use a computer each day, and those who do spend an average of about an hour at the keyboard.

» Those who live in households where the TV is on always or most of the time are less likely to read every day.

» 72 percent of parents say computers mostly help with children's learning.

—*Associated Press*

a. Research question:

b. Observational study or experiment and how you determined this:

c. Possible inductive reasoning flaws and remedies:

d. Possible alternative explanation for conclusion:

4. "Hypnosis a Help in Surgery, Study Says," *The Morning Call,* April 28, 2000.

Hypnosis a help in surgery, study says

Benefits reported are less need for pain medicine, time saving, better vital signs.

LONDON (AP) — People who were hypnotized while undergoing surgery without a general anesthetic needed less pain medication, left the operating room sooner and had more stable vital signs than those who were not, according to a study in this week's issue of The Lancet medical journal.

Trance states have been used for hundreds of years by both witch doctors and modern surgeons to help sick people.

But there had been little scientific evidence that hypnosis really works to reduce pain during surgery.

"Despite how long hypnosis has been around and the dramatic effects it has shown, there are very few properly designed clinical studies that demonstrate that it is more than a placebo," said David R. Patterson, professor of rehabilitation medicine, surgery and psychology at the University of Washing ton in Seattle, who was not connected with the study.

"This really solidifies the evidence. For acute pain ... it's not arguable anymore."

The study, led by Dr. Elivira Lang of Beth Israel Deaconess Medical Center in Boston, involved 241 people of similar health and age who had operations to open clogged arteries and veins, relieve blockages in the kidney drainage system or block blood vessels feeding tumors.

The patients were divided into three groups—one that experienced normal interactions with doctors and nurses, another that received extra attention from an additional person in the operating room who made sure nobody said anything negative, and a third who were helped to hypnotize themselves.

All the patients were able to give themselves as much pain medication as they wanted through an intravenous tube.

The hypnosis group—who were guided through visualizations of scenarios they found pleasant —fared best, but the patients receiving extra attention also benefited.

About half the patients in those two groups needed no drugs, while the rest gave themselves only half the amount of medication as those undergoing the operation with no special attention.

The hypnotized patients were the only ones who said the pain did not get worse as the surgery progressed. They also had fewer problems with blood pressure and heart rate during the operation.

Their operations also finished 17 minutes earlier than those conducted without the special attention.

Lang suggested that time was saved in those operations because the surgeon's attention was diverted less often by events such as the patient's being oversedated, unstable blood pressure or vomiting.

The study was funded by the *National Institutes of Health.*

a. Research question:

b. Observational study or experiment and how you determined this:

c. Possible inductive reasoning flaws and remedies:

d. Possible alternative explanation for conclusion:

Summary

In this activity, you investigated possible flaws in inductive reasoning by analyzing newspaper reports of studies.

ACTIVITY 13.1

Code-Breaking and Deductive Reasoning

In the first part of this activity, you will exercise your deductive reasoning skills in the same way you do when you play some board games. Work with another student so you can discuss your reasoning. In the second part of this activity, you will work with truth tables and the contrapositive and converse of a conditional statement.

Breaking the Code

Many code-breaking schemes use mathematics and deductive reasoning to decipher a code. The following setup is based on the game *Mastermind* by Invictus.

The object is to decipher a secret three-letter code word consisting of the letters A, E, I, O, and U. Letters may be repeated one, two, or three times in the word. For example, each of the following could be possible secret code words: EEE, OEO, IOU.

The code-breaker makes a series of three-letter guesses, and each one is followed by a reply containing information about whether any of the letters is correct and in the correct position or if any letter is correct but not in the correct position. Each reply of **x** means that one letter is correct and is in the correct position, although there is no information about which letter is the correct one. Each reply of **o** means that one letter is correct but is not in the correct position. Again there is no information about which letter is the correct one.

For example, suppose the code-breaker guesses I O U and the reply is **x o**. This might mean that the I is correctly placed in the first position and U is in the code but is not in the third position. If that's the case, then the U would have to be in the second position, and you also know that there is no O in the secret code word. (How do you know this?) However, there are other possibilities: the O could be the correct letter in the correct position; the I might be in the third position in the secret code word; and there might not be a U in the word.

1. Here is a series of guesses and replies to help decipher a secret code word.

Guess Number	Code-Word Guess	Reply
1	A I I	x
2	O I U	x o
3	E I O	o

Answer the following questions about the secret code word that you are trying to find.

a. If there is an I in the secret code word, where is it? How do you know this?

b. If there is an I in the secret code word, can there be an A? Why or why not?

c. If there is an I in the secret code word, can there be an O? Why or why not?

d. If there is an I in the secret code word, can there be an E? Why or why not?

e. Can there be two I's? Why or why not?

f. If there is an I, can there be a U? Why or why not?

g. What do these answers lead you to conclude about the existence of an I in the secret code word?

h. If there is an A, where is it?

i. If there is an A, what are the other two letters?

j. If there is an A, what is the code?

k. Is this the only possible code that fits the previous clues?

2. Here is another series of guesses and replies for a new secret code word.

Guess Number	Code-Word Guess	Reply
1	A E O	o o
2	E I A	x
3	U O A	x

Fill in a "useful" consequent in each of the blank spaces, and answer the questions.

a. If U is the correct letter in guess #3, then ______________________________.

b. Can U be the correct letter in guess #3?

c. If O is the correct letter in guess #3, then ______________________________.

d. Can O be the correct letter in guess #3?

e. If A is the correct letter in guess #3, then ______________________________.

f. What is the secret code word? Is this the only possible code word?

If-then Statements

3. Write the converse and the contrapositive of each of the following statements.

   a. "If there is an I in the code word, then there is not an A in the code word."

      i. Converse:

      ii. Contrapositive:

   b. "If teenagers are caught in a car with alcohol, then they will lose their license."

      i. Converse:

      ii. Contrapositive:

4. Look at the examples of if-then statements and their converses and contrapositives in #3. Using these examples, explain why the contrapositive says the same thing as the original statement, but the converse does not.

5. In the following exercises you will compare the truth values of the conditional statement "if P, then Q" with the truth values of its contrapositive and converse.

   a. Fill in the following truth table:

P	Q	not P	not Q	if P, then Q	if (not Q), then (not P)
T	T				
T	F				
F	T				
F	F				

b. Explain why the truth table you completed in part a shows that the contrapositive of the statement "if P, then Q" is equivalent to the original conditional statement.

c. Fill in the following truth table:

P	Q	if P, then Q	if Q, then P
T	T		
T	F		
F	T		
F	F		

d. Explain why the truth table you completed in part c shows that the converse of the statement "if P, then Q" is not equivalent to the original conditional statement.

Summary

In this activity, you practiced using logic by playing a code-breaking game and formulating conditional statements to help you break the code. You also composed truth tables for the contrapositive and converse of the conditional statement "if P, then Q" and constructed the contrapositive and converse of several conditional statements.

ACTIVITY 13.2

Compound Statements Used in Reasoning

In this activity, you will analyze compound statements found in news articles and quotes from famous people. You will formulate and analyze the negation of these statements. You will also investigate the equivalence of statements using truth tables.

A compound statement is formed from two other statements using logical connectors. You'll work with three different connectors: *and*, *or*, and *if-then*.

1. For each of the following compound statements, identify the two statements P and Q that are joined to make the compound statement. Then give the form of the compound statement ("P and Q," "P or Q," "if P, then Q"), and identify the type of compound statement (conjunction, disjunction, conditional).

   a. "[I believe that] If you do the kind of alliance-building that is available to us, it is appropriate to have a goal of reducing our troops over that period of time." (Source: "Kerry Says His Vote on Iraq Would Be the Same Today," *The New York Times*, August 10, 2004.)

      i. P:

ii. Q:

iii. Form:

iv. Type:

b. "If no one gets more than half the vote on Nov. 2, a runoff will be held in December." (Source: "Party Switch in Louisiana Leaves Democrats Fuming," *The New York Times*, August 8, 2004.)

i. P:

ii. Q:

iii. Form:

iv. Type:

c. "If the freedom of speech is taken away then dumb and silent we may be led, like sheep to the slaughter." (Source: George Washington.)

i. P:

ii. Q:

iii. Form:

iv. Type:

d. "You can keep Johnson in Washington D.C. or you can send him back to his Texas cotton patch." (Source: Malcolm X "The Ballot or the Bullet," speech given on April 12, 1964.)

i. P:

ii. Q:

iii. Form:

iv. Type:

e. This is a world of compensations, and he who would *be* no slave, must consent to *have* no slave." (Source: Abraham Lincoln, Letter to Henry Pierce, April 6, 1859.)

i. P:

ii. Q:

iii. Form:

iv. Type:

f. "[Senator Kerry said Monday that] he would have voted to give the president the authority to invade Iraq even if he had known all he does now about the apparent dearth of unconventional weapons or a close connection with Al Qaeda." (Source: "Kerry Says His Vote on Iraq Would Be the Same Today," *The New York Times,* August 10, 2004.)

i. P:

ii. Q:

iii. Form:

iv. Type:

g. "The Cubans have lost some prominent players to defection and retirement, but still have more than enough to win another gold." (Source: "Olympic Baseball Pallid without U.S-Cuba Clash," *The Providence Sunday Journal,* August 15, 2004.)

i. P:

ii. Q:

iii. Form:

iv. Type:

h. "They [the different mouse strains] are deficient in or cannot respond to growth hormone, the substance that prompts the liver to churn out another hormone called IGF-1, for insulin-like growth factor." (Source: "In Aging Small May Have Its Advantages," *The New York Times,* August 17, 2004.)

i. P:

ii. Q:

iii. Form:

iv. Type:

The negation of a statement is a statement that is false if the original statement is true and is true if the original statement is false. One way to negate a statement P is to say, "It is not the case that P." However, there are generally more useful ways to negate a statement, especially a compound statement.

2. For a statement of the type "P and Q" to be true, both statement P and statement Q must be true. So the negation of the statement "P and Q" is that at least one of the statements P, Q is not true; that is, "(not P) or (not Q)." Complete the following truth table that summarizes the relationship between the compound statements "P and Q" and "(not P) or (not Q)."

P	Q	P and Q	not P	not Q	(not P) or (not Q)
T	T				
T	F				
F	T				
F	F				

3. Explain how the truth table shows that the negation of the statement "P and Q" is equivalent to the statement "(not P) or (not Q)."

4. Complete the following truth table and explain how it shows that the negation of the statement "P or Q" is equivalent to the statement "(not P) and (not Q)."

P	Q	P or Q	not P	not Q	(not P) and (not Q)
T	T				
T	F				
F	T				
F	F				

5. Complete the following truth table and explain how it shows that the negation of the statement "if P, then Q" is the statement "P and (not Q)."

P	Q	if P, then Q	not Q	P and (not Q)
T	T			
T	F			
F	T			
F	F			

6. Give a useful negation of the following statements relating to the code-breaking activity (Activity 13.1):

   a. A is in the code or I is in the code.

   b. O is in the code and U is not in the code.

   c. If E is in the code, then it is in the first position.

   d. If U is in the code, then A is not in the code.

7. Give a useful negation of the following statements:

   a. "Children caught with cigarettes or other tobacco products on public school property in Pennsylvania often pay larger fines than stores that are cited for selling tobacco to minors."

b. "If no one gets more than half the vote on Nov. 2, a runoff will be held in December."

c. "If teenagers are caught in a car with alcohol, then they will lose their license."

d. "You can keep Johnson in Washington D.C. or you can send him back to his Texas cotton patch."

e. "... the findings will help scientists better understand the biology of nicotine addiction and lend more plausibility to the idea that some people may be more genetically susceptible to it than others."

Summary

In this activity, you practiced identifying P and Q in the three types of compound statements. You used truth tables to confirm the relationship between the negation of a compound statement and an equivalent compound statement. You also formulated the negation of a variety of compound statements.

ACTIVITY 13.3

Quantified Statements and Deductive Reasoning: Direct and Indirect Reasoning

In the first part of this activity, you will identify the type of quantified statement and will formulate negations of quantified statements. In the second part of this activity, you will use the two forms of deductive reasoning to draw conclusions from given statements; you will also identify the type of deductive reasoning used to draw a conclusion.

Quantified statements can be expressed using one of the two quantifiers: the universal quantifier *all*, or the existential quantifier *there exists*.

1. For each of the following statements,

   i. Identify the type of quantifier it contains, and write an equivalent statement using the quantifier *all*, or *there exists*.

   ii. Write the negation of the statement.

   a. "Some economists said the pullback in their forecast reflects indications that household spending has come under pressure as oil prices have soared to new highs." (Source: "Outlook for U.S. Economic Growth Is Trimmed," *The Wall Street Journal*, August 13, 2004.)

i.

ii.

b. "Each period of our national history has had its special challenges." (Source: Harry S. Truman, Inaugural Address, January 20, 1949.)

i.

ii.

c. "None of the veterans featured in the advertisements served on the river patrol boats Kerry commanded during Vietnam." (Source: "GOP Donor's Influence Is Far More Visible Than He Is," *The Providence Sunday Journal*, August 15, 2004.)

i.

ii.

d. "Several other Republicans and a chorus of Democrats have also questioned the change, with some proposing legislation to prohibit it." (Source: "Senators Criticize Decision to Allow Scissors on Planes," *The New York Times,* December 13, 2005.)

i.

ii.

A deductive argument consists of premises (or hypotheses or assumptions) and conclusions that follow logically from those premises. There are two key elements for good deductive reasoning: (1) the premises are true and (2) the reasoning is valid.

The two forms of valid deductive reasoning are summarized here:

Modus Ponens (Direct Reasoning)

If the statement "if P, then Q" is true, and the statement P is true, then the statement Q is true.

Modus Tollens (Indirect Reasoning)

If the statement "if P, then Q" is true, and the statement Q is false, then the statement P is false.

2. In each of the following cases, use Modus Ponens or Modus Tollens (say which one you are using) to draw a conclusion from the statements given.
   a. If I drink coffee between 7 and 8 P.M., I cannot fall asleep until midnight.

      This evening I drank coffee at 7:15 P.M.

      i. Conclusion:

      ii. Type of reasoning used:

   b. If you are younger than 21, you cannot drink alcohol legally.

      You can legally drink alcohol.

i. Conclusion:

ii. Type of reasoning used:

3. In each of the following situations, if the hypotheses allow a conclusion through valid reasoning, state the conclusion and describe why it is valid. If a conclusion is not possible, write "no valid conclusion" and explain why there is no valid conclusion.

   a. If a salesperson is rude with the customers, then the salesperson will be fired.

      Shanon, a salesperson, is rude with the customers.

b. If a salesperson is rude with the customers, then the salesperson will be fired.

John, a salesperson, was fired.

c. If a salesperson is rude with the customers, then the salesperson will be fired.

Bob, a salesperson, is not rude with the customers.

d. If you liked the play, then you'd like the movie.

You did not like the movie.

e. If you liked the play, then you'd like the movie.

   You liked the movie.

f. If you liked the play, then you'd like the movie.

   You liked the play.

g. All American presidents have been men. (Note this is equivalent to stating: If a person has been an American president, then the person has been a man.)
   John Kennedy was a man.

h. All American presidents have been men.

   Dana Brown has not been an American president.

i. All American presidents have been men.

   T. Roosevelt was president.

Summary

In this activity, you practiced working with quantified statements and their negations. You also used the two basic forms of deductive reasoning: direct reasoning (Modus Ponens) and indirect reasoning (Modus Tollens), and identified invalid reasoning.

ACTIVITY 14.1

Methods of Apportionment: Quota Methods

In this activity, you will work with the two quota methods of apportionment: Hamilton's method and Lowndes' method. You will use each of these methods to determine how many representatives in the U.S. House would correspond to each state if the method were currently used.

All apportionment methods assign to each state a number of representatives based on the state's population. The Excel file "EA14.1 State Population.xls" contains the population of each state, as given by the December 28, 2000, census figures. (Source: *Census Bureau webpage,* www.census.gov.)

1. Open the file "EA14.1 State Population.xls," and then find the total U.S. population. (Instructions to remind you how to do this follow.)

   Total Population: _____________

Instructions to Use Excel to Add Consecutive Elements of a Column

To find the total population, you want to add all the state's populations, which in the file "EA14.1 State Population.xls," are the numbers in cells B2 through B51. To do this, enter = **SUM(B2:B51)** in cell B53. Name this cell using the **Name box** above column A on the spreadsheet. (You could use the name **totalpopulation**, or any other appropriate name.) Also add the label **Total Population** = in cell A53.

2. In cell A54, enter the label **Total Seats** = and in cell B54, enter the number **435**; this is the total number of seats in the House of Representatives. Name this cell. You might want to call it **totalseats**, or another appropriate name. Recall that the standard divisor is given by

$$\textit{standard divisor} = \frac{\textit{total population}}{\textit{total number of seats}}$$

Ask Excel to find the standard divisor in cell B55 and name this cell. Add the label **Standard Divisor** = in cell A55.

Write the standard divisor here: _____________

3. Recall that a state's standard quota is defined by the equation,

$$\textit{standard quota} = \frac{\textit{state's population}}{\textit{standard divisor}}$$

In column C, enter each state's standard quota. First, enter the column title **Standard Quota** in cell C1. Then ask Excel to do the computations for you by entering the appropriate formula in cell C2. Drag-and-fill to enter the rest of the values.

4. Can each state's standard quota be the apportioned number of representatives for that state? Why or why not?

In both quota methods, apportionment is done by rounding up or down each state's standard quota.

In Hamilton's method, each state is initially given the number of representatives equal to the integer value of the state's standard quota, unless this number is 0. For this case the initial and final apportioned number of seats for that state is 1 (so that every state is represented).

5. Suppose that three states, State A, State B, and State C, have standard quotas as given here. Record in the table the initial number of representatives apportioned to those states when using Hamilton's method.

	Standard Quota	Initial Number of Representatives
State A	3.56	
State B	2.1	
State C	0.38	

After the initial apportionment is done using the integer value (or 1 when the integer value is 0), the remaining seats are assigned to the states with a standard quota larger than 1. The states with the largest **fractional part** of the standard quota are each given an extra seat.

6. Suppose the total number of representatives that correspond to the three states is 7. Record in the table the number of representatives (seats) that correspond to each of the three states using Hamilton's method.

	Number of Seats by Hamilton's Method
State A	
State B	
State C	

You will now use Excel to determine the number of representatives apportioned to each of the 50 states under Hamilton's method.

7. In column D of your spreadsheet, you'll ask Excel to enter the initial number of seats for each state; that is, enter the integer part of the state's standard quota, unless this is 0. If the integer part of the standard quota is 0, then you'll enter **1**. To do this, enter the integer value of the first state's standard quota in cell D2 by typing = **INT(C2)**; then drag to autofill. If column D contains any 0s, go through and replace each 0 with **1**. Label column D with an appropriate label.

8. Use an appropriate formula to enter in column E the difference of each entry in column C minus the corresponding entry in column D. Label column E **Fractional Part**. List here the states for which the entries in column E are negative. Explain why these are negative.

9. Add the numbers in column D (use Excel) to decide how many seats have already been apportioned. Write that number here: __________

   Determine how many seats remain after the initial apportionment. Write the number of remaining seats here: __________

   This is the number of seats you will apportion using the fractional part of the standard quota.

10. To apportion the remaining seats using the fractional part of the standard quota, sort the data in columns A, B, C, D, and E by "Fractional Part" in descending order.

    In column F, enter the final number of seats apportioned by Hamilton's method to each state. (Note: you do not need to enter these numbers one by one. Enter = **D2 + 1** in cell F2 and drag until you have apportioned all the remaining seats; then have Excel copy the numbers in column D for the remaining states.) Label column F **Seats Using Hamilton's Method**. Then sort the data (remember to highlight all columns) by "Seats Using Hamilton's Method," to help answer parts a and b of this question.

    a. List the four states that get the most number of seats and say how many seats each of them gets.

    b. What is the smallest number of seats assigned? Which states get the smallest number of seats?

Lowndes' method also uses the standard quota to apportion representatives. However, Lowndes' method uses the relative fractional part of the standard quota instead of the fractional part. The relative fractional part is used to determine which states' standard quotas will be rounded up to fill out the remaining seats.

The relative fractional part is the following quotient:

$$\textit{relative fractional part} = \frac{\textit{fractional part}}{\textit{integer part}}$$

11. Assuming that State A and State D have the following standard quotas, record their relative fractional parts in the table.

	Standard Quota	Relative Fractional Part
State A	3.56	
State D	1.56	

To apportion the number of seats for each of the 50 states by Lowndes' method, you will first compute the relative fractional part of each state's standard quota.

12. Insert two columns to the left of column F in your spreadsheet. In the new column F, recompute the integer part of the standard quota for each state. In cell G2, compute the relative fractional part of the standard quota for the state in row 2. Drag-and-fill to compute the relative fractional part for each state, and label column G **Relative Fractional Part**.

13. Notice that you see #DIV/0! in several places in column G. Explain why this error message occurred and what it means.

14. Label column I, **Seats Using Lowndes' Method** and sort the data by "Relative Fractional Part," in descending order. Give one seat to each state showing the error message and explain why that's needed.

15. Then apportion the rest of the seats, using the relative fractional part.

  a. List the four states that get the most number of seats and say how many seats each of them gets.

  b. What is the smallest number of seats assigned? Which states get the smallest number of seats?

16. Name the states that would prefer Hamilton's method and those that would prefer Lowndes' method. Explain your answer.

17. Which states, those with a large population or those with a small population, are favored by Lowndes' method? Explain.

Summary

In this activity, you used Excel to investigate two quota methods of apportionment and compared the results of using them to apportion the seats of the House of Representatives. You learned the Excel command that gives the integer part of a real number and used it to find fractional parts.

ACTIVITY 14.2

Apportionment: Divisor Methods

In this activity, you will work with two divisor methods of apportionment, Adams' method and the Huntington-Hill method. With the help of Excel, you will determine how many representatives in the House of Representatives would correspond to each state if these methods were used.

Each divisor uses a modified divisor, that is, a divisor other than the standard divisor, to determine each state's modified quota. If the divisor D is used, then a state's modified quota is

$$\textit{modified quota} = \frac{\textit{state's population}}{D}$$

Then depending on the method used, all modified quotas are rounded up or down. In Adams' method, they are all rounded up; in Jefferson's method, they are rounded down. In Webster's method each modified quota is rounded up or down to its nearest integer. The method currently used in the U.S. to apportion representatives is the Huntington-Hill method. This method uses the geometric mean to round off quotas.

Adams' Method

1. In this method modified quotas are rounded up; thus the modified divisor needs to be greater than the standard divisor. Why? Explain what would happen if the modified divisor were less than the standard divisor.

2. Open the "EA14.1 State Population.xls" file. Compute the total population in cell B53 and record the total population here: __________________________. To label the "total" row, enter **total** = in cell A53.

3. Compute the standard divisor in cell B55 using the relationship,

$$\textit{standard divisor} = \frac{\textit{total population}}{\textit{total number of seats}}$$

   Record it here: ______________________. Enter the label **Standard divisor** = in cell A55.

4. In column C, enter each state's standard quota using the relationship,

$$\textit{standard quota} = \frac{\textit{state's population}}{\textit{standard divisor}}$$

   Use cell C53 to find the sum of all states' standard quotas and write the sum here: ____________. Label column C by entering **Standard Quota** in cell C1.

5. In cell A56, enter **Divisor** = and in cell D56 enter a number greater than the standard divisor. (Pick any number greater than the standard divisor—you'll change it later to find the "best" one.)

   Record the number you chose here: __________________

6. Label column D by entering **Modified Quota** in D1. In cell D2, enter an appropriate formula and then drag it down to compute the modified quota for each state obtained by using the modified divisor you entered in cell D56. (Make sure you enter a formula that will keep the location "D56" fixed when you drag the formula. Recall that you can do this by either naming cell D56 or by using **$** signs.)

7. In each cell of column E, enter the integer number obtained by rounding up the modified quota. (To accomplish this enter = **INT(D2)** + **1** in cell E2 and drag to autofill the column.) Label the column **Number of Seats Using Adams' Method**.

8. In cell E53, enter the sum of the numbers in cells E2 through cell E51. Record the sum here: ____________________, and explain what this number represents.

9. If the previous sum is exactly 435, you have succeeded in using Adams' method to apportion House seats. Explain why this is so.

10. If the sum is not 435, then you need to try other numbers for the modified divisor until 435 is reached. Note that when you change the number in cell D56 (the modified divisor), the numbers in column E, including the sum in E53, will also change automatically.

Using this table, record the divisors you've tried and the total number of House seats resulting from each divisor. Indicate which one is the modified divisor that actually works.

Divisor							
Total # of Seats							

11. List the four states that would get the largest number of seats if you used Adams' method and the number of seats each would get.

12. List the four states that would get the smallest number of seats if you used Adams' method and the number of seats each would get.

Huntington-Hill Method

In this method, each modified quota is rounded up or down according to the **geometric mean**. The geometric mean of two numbers x and y is the square root of the product of the two numbers. That is,

Geometric mean of x and $y = \sqrt{xy}$

13. Find the geometric mean of 30 and 17, and compare it with the average of the same two numbers.

14. In the Huntington-Hill method, the geometric mean of the integer part of the modified quota and the next integer is calculated. If the modified quota is greater than the geometric mean, then the modified quota is rounded up. Otherwise, it is rounded down. For example, if the modified quota is 12.742, the geometric mean of 12 and 13 is calculated: $\sqrt{12 * 13} = 12.489$. Because the modified quota 12.742 is greater than 12.489 (the geometric mean of 12 and 13), the modified quota is rounded up to 13.

    Suppose the modified quota is 7.49. Would the modified quota be rounded up or down in this method? Show your computations and explain your conclusion.

15. In cell F56, enter **660,000** (this will be the initial trial divisor). In column F of the spreadsheet, enter the new modified quotas using the number in F56 as the divisor. Label the column **New Modified Quotas** and record what you entered in F2 here: ________________.

16. Label column G by entering **Geometric Mean** in cell G1. Use the following Excel instructions to compute the geometric mean of the integer part of the modified quota and the next integer for each state.

Instructions to Use Excel to Compute the Geometric Mean

In cell G2, enter = **SQRT(INT(F2)*(INT(F2)+1))**. (This is the square root of the product obtained by multiplying the integer part of the number in cell F2 times the next largest integer.) Then drag the formula down to compute the appropriate geometric mean for each state.

17. Record the number you obtained in G2: ____________. Using a calculator, verify that it is the geometric mean used by the Huntington-Hill method to round off the modified quota in F2.

    Modified Quota in F2: _________ Integer Part: ______________

    √____________ * ______________ = _________________

18. Label column H **Number of Seats Using Huntington-Hill Method** and tell Excel to round each modified quota up or down, according to the geometric mean, using the following instructions.

Instructions to Use Excel to Round Numbers According to the Geometric Mean by Using an "IF Statement"

In cell H2, enter = **IF(F2 < G2, INT(F2),INT(F2)+1)**. (This formula tells Excel to see if the number in F2 is less than the number in G2; if it is, enter the integer part of F2; otherwise, enter the integer part plus 1, which is the next largest integer.) Then drag the formula down to obtain the number of seats for all 50 states.

19. Record the number obtained in H2 here: _______________. Explain how this is consistent with your computations in #17.

20. In cell H53, enter the sum of the numbers now in H2 through H51. If this number is 435 you have apportioned the House seats using the Huntington-Hill method. If the number is not 435, change the modified divisor used (which is in cell F56) until the sum of apportioned seats is 435.

Record the divisors you've tried and the total number of House seats resulting from each divisor. Indicate which one is the modified divisor that actually works.

Divisor							
Total # of Seats							

21. List the four states that have the largest number of seats and the number of seats each has.

22. List all states that have just one seat.

23. Name the states that would prefer Adams' method and those that would prefer the Huntington-Hill method. Explain your answer.

Summary

In this activity, you used Excel to investigate two divisor methods of apportionment: Adam's method and the Huntington-Hill method, which is the method currently used to apportion seats in the House of Representatives. Using each of these methods, you found the number of representatives that corresponds to each state. You learned how to use Excel to compute the geometric mean and to round numbers according to the geometric mean.

ACTIVITY 15.1

Making a Purchase Decision

In this activity, you will use a variety of problem-solving techniques along with the decision-making process you learned in Topic 11 to make a purchase decision of your choice.

You are planning (either for real or hypothetically) a major purchase in the near future. (You may want to look into purchasing a car, a stereo, a computer, a television set, a grill, or other item of your choice.) You need to do the analysis to assure the purchase you are contemplating would be carried out in optimal fashion.

Follow the guidelines described here, which divide the problem into sub-problems, and address the issues raised in a coherent narrative. You can use the spaces provided to note your ideas, but you should write a single essay that incorporates thoughtful responses to all the questions.

1. Describe the product upon which you have decided to focus and explain how/why you made that particular choice.

2. Collect information about products in your category of interest. You may need to consult a variety of sources. Include at least two websites and give the URL of each site; you might also use consumer periodicals

relating to your product, newspapers, almanacs, and possibly personal visits to "test use" a product. Make certain your sources are up-to-date and relevant.

3. List all the criteria (at least six) that might affect your decision and assign relative priority to those criteria. Justify why you chose the criteria you did. Determine if the criteria are readily measurable and decide how you will measure or rank your choices relative to them. You should use at least five choices of the item you are planning to purchase to compare against each other according to your selected criteria.

4. Decide how you will organize your information and make a table or organize it using an Excel spreadsheet.

5. Use the two processes (cutoff screening and weighted sum methods) discussed in Topic 11 for making decisions about which product to purchase. Explain these processes fully.

6. Reflect on your entire process. What problem-solving techniques did you use? How does this approach compare with the way you have made important purchase decisions in the past or with how you have rated things?

Summary

In this activity, you used problem-solving techniques to help you make an important purchase decision.

ACTIVITY 16.1

Visualizing Football Scores: Measures of Center and Spread

In this activity, you will examine some data from the National Football League and investigate several numerical measures of center and spread and a graphical method particularly useful for comparing data sets.

The following table gives the points scored by the Philadelphia Eagles in their sixteen 2004 regular season games. (Source: *National Football League,* www.nfl.com/teams/stats.)

Date	Points Scored—Eagles
9/12	31
9/20	27
9/26	30
10/3	19
10/17	30
10/24	34
10/31	15
11/7	3
11/15	49

Date	Points Scored—Eagles
11/21	28
11/28	27
12/5	47
12/12	17
12/19	12
12/27	7
1/2	10

1. Without doing any computations, look at the "points scored" values and estimate the "center" of the data set. Why did you choose this value?

2. Enter the dates and points scored and the column titles, as they appear in the table, into columns A and B of an Excel worksheet.

3. Sort the data by "points scored" (in ascending order) and find the median number of points scored; write it here: __________.

4. Use the following Excel instructions to find the mean and median points scored by the Eagles in 2004.

Instructions to Use Excel to Find the Mean and Median of a List of Data

a. In cell A19, type the word **mean**, and in cell A20 type the word **median**.

b. In cell B19, instruct Excel to calculate the average (that is, mean) points scored by entering the command **=average(B2:B17)**; in cell B20, calculate the median points scored by entering the command **=median(B2:B17)**.

5. Record the average and median points scored per game by the Eagles in 2004 here:

   average = ______________

   median = ______________

   Describe how these values compare with your estimate of the "center" of this data set in #1.

6. Looking at the center of a distribution doesn't give the whole picture; you might also want to measure the spread of the data. One way to do this is to give the range of the data; that is, *maximum–minimum*. Give the range of the Eagles football data here: ____________________. (This shows the spread of the data, but may be misleading if one or both of these values are outliers.)

7. To get a better sense of the spread, you'll now look at the **quartiles**. The first quartile, denoted Q_1, is the median of the lower half of the data values.[1] The third quartile, denoted Q_3, is the median of the upper half of the data values. The following list of numbers is the **five-number summary** of a data set: *minimum, Q_1, median, Q_3, maximum.* Use Excel to find Q_1 and Q_3 for the data set of Eagles' points scored by using the median command and the appropriate halves of the data for each. Record the five-number summary of this data set:

   ____

8. Sketch a **boxplot** for this data set in the space provided. First, set up a horizontal axis. Next, mark a scale on your axis, then locate each of the five numbers in the five-number summary on the axis, and complete the boxplot.

9. The following table gives the points scored in the 2004 regular season games by the Washington Redskins and the N.Y. Giants teams. Look at the "points scored" by each team and estimate the "center" of the data for each data set.

[1]Note that some texts and software give slightly different descriptions of how to find the first and third quartiles. Excel contains a built-in function **QUARTILE** that could also be used to find the quartiles. In this activity you are asked to find quartiles using the description of quartiles as medians of subsets of the data, because doing so helps enhance understanding of what quartiles represent.

Date	Points Scored—Giants	Date	Points Scored—Redskins
9/12	17	9/12	16
9/19	20	9/19	14
9/26	27	9/27	18
10/3	14	10/3	13
10/10	26	10/10	10
10/24	13	10/17	13
10/31	34	10/31	14
11/7	21	11/7	17
11/14	14	11/14	10
11/21	10	11/21	6
11/28	6	11/28	7
12/5	7	12/5	31
12/12	14	12/12	14
12/18	30	12/18	26
12/26	22	12/26	10
1/2	28	1/2	21

Your estimate of "center" for the Giants: ______________________________

Your estimate of "center" for the Redskins: ____________________________

10. Retrieve the file "EA16.1 Points Scored 2004.xls" (which contains the data shown in the previous tables) from the CD or website, or enter the data into Excel for the New York Giants and Washington Redskins NFL teams.

11. Find the five-number summary for points scored by each of these teams and record them here.

    Giants: ____

    Redskins: ____

12. Go to the website www.nfl.com/teams and click the "Stats" link below the team you want, to get the most recently completed season points–scored information for one additional NFL team. Find the five-number summary for your team and sketch comparative boxplots showing the five-number summaries for points scored by the Eagles and the three additional football teams on the same scale.

13. Write a paragraph describing what these four comparative boxplots show about the scores of the four NFL teams. Who do you think would win a game between the Eagles and Giants? Explain your reasoning. Who do you think would win the other five games that could be played between pairs of these four teams?

Summary

In this activity, you used Excel to calculate mean and median. You found the five-number summary for several data sets of football scores. You also drew boxplots and used them to compare the performance of four different football teams.

ACTIVITY 17.1

Coins, Presidents, and Justices: Normal Distributions and z-Scores

In the first part of this activity, you will generate some data that should have an approximately normal (or bell-shaped) distribution. In the second part, you will use the definition of standard deviation and compare the standard deviations for two different data sets.

1. Work with a partner to generate the following data.

   a. Toss 10 coins and record the number of heads you obtained.

b. Repeat this 24 more times until you have a list of 25 numbers, each of them between 0 and 10.

c. Retrieve the file "EA17.1 Coins and Presidents.xls" from the CD or website, and you will find the results of 35 tosses of 10 coins that someone else carried out. When you first retrieve the file, column B contains the number of times 0 heads was obtained in the 35 tosses of 10 coins, the number of times 1 head was obtained in the 35 tosses, and so on, up to the number of times 10 heads was obtained. Add your results to the list so you have a total of 60 in column B.

d. Create a scatterplot of these data, using one of the versions of the scatterplot with the dots connected. Describe what your curve looks like, including where it is "centered" and what its "spread" is.

e. Change your graph to a bar graph (instructions follow).

Instructions to Use Excel to Change a Scatterplot to a Column Graph

a. Click on the plot area and select **Chart** from the menu bar.

b. From the drop-down menu, select **Chart type**. Now go to **Column** to change your graph to a column (that is, a bar) graph and click **OK**.

c. Click one of the bars, go to **Format** on the menu bar, and choose **Selected data series.** Click the **Options** tab. Change the **Gap width** to **0**, so adjacent bars will touch. (This will make it look like a histogram, and then you can sketch a curve along the tops of the bars.)

f. Print your bar graph, with appropriate titles on the axes, and by hand draw in a bell-shaped curve that "fits" this data. How does your hand-drawn curve compare with the curve you described in part d of this question?

2. In the second part of this activity, you will examine one measure of the spread of a data set, the standard deviation. The standard deviation plays an important role in understanding the spread of a distribution, especially a bell-shaped or **normal distribution**.

   a. You'll use the data set of ages of U.S. presidents at their inauguration, which can be found on sheet 2 of the file "EA17.1 Coins and Presidents .xls," and calculate the standard deviation of this data set. To do this, first find the mean (average) of the ages and store that value in cell B46. (Source: *The World Almanac and Book of Facts 2004,* page 563.)

      Record the mean age here: _____________

b. In cell C1, enter the label **Deviation from the mean**. In cell C2, enter an appropriate formula to subtract the mean age from the value in cell B2 and that will allow you to drag down to compute each data value minus the mean. (Remember to use **B46** to keep the value of the mean fixed when you drag.) Drag down to compute each data value minus the mean.

How many of the values in this column are negative? _______________

What property makes these values negative?

c. Add the values in column C and record their sum here: ________________

Write this number without using scientific notation: ___________________

This number should be 0 or very, very close to 0. Is it? _________________

d. Now you want to compute the square of the deviation from the mean of each data value. In cell D1, enter the label **Squared deviation** and use the instructions that follow to enter the squares of the deviation values in column D. Then add all the values in column D, store that number in cell D46, and record the sum here: _____________.

Instructions to Use Excel to Compute Squares

In cell D2, enter the formula =**C2^2**. Then drag this formula down to cell D44.

e. Next you need to divide the sum of the squared deviations (the value in cell D46) by one less than the number of data points. There are 43 data points, so divide the value in cell D46 by 42; store this number in cell E46.

f. Finally, compute the standard deviation by taking the square root of the number in cell E46 (see the following instructions for a reminder of how to compute the non-negative square root of a number). Store this value in cell F46 and record it here:

standard deviation = ___________________________

Instructions to Use Excel to Compute a Square Root

To compute the non-negative square root of the value in E46, use the command **=sqrt(E46)**.

g. Check the computations by using the following Excel command to compute the standard deviation of the ages in column B. Enter the value obtained in cell F47, and write this value here: _______________.

Instructions to Use Excel to Compute the Standard Deviation

To compute the standard deviation of the values in cells B2 through B44, use the command **=stdev(B2:B44)**.

h. The standard deviation of a set of data is a measure of the spread of the data values. It incorporates the sum of the squared deviations from the mean, of all data values. Suppose Clinton had been 86 years old instead of 46 at his inauguration. What would change in the computations you just performed?

i. Change Clinton's age on your spreadsheet from 46 to 86. Notice that all of your computations change automatically. Record the new mean and the new standard deviation.

mean = ___________; standard deviation = ____________

j. Change Clinton's age back to its correct value of 46.

3. Go to sheet 3 of the file "EA17.1 Coins and Presidents.xls," where you will find another data set. This data set contains the names and ages at time of appointment as chief justice for all the chief justices of the U.S. Supreme Court. (Source: *The World Almanac and Book of Facts 2006,* page 53.)

   a. Compute the mean and standard deviation of these ages, and record these values here:

   mean = ___________; standard deviation = ____________

   b. Describe how the means and standard deviations of the two data sets, "Presidents' Ages" and "Supreme Court Chief Justices' Ages," compare.

c. Pick the maximum data value in the "Presidents' Ages" data set. Call it x, and compute its z-score by computing: $z = \frac{x - mean}{standard\ deviation}$, using the mean and the standard deviation of the presidents' ages.

Record this z-score here: ____________________

d. Find the z-score for the largest data value in the "Supreme Court Chief Justices' Ages" data set, using the mean and standard deviation of the chief justices' ages.

Record this z-score here: ____________________

e. What do the two z-scores tell you?

Summary

In the first part of this activity, you generated data and created a graph to see that the data has an approximately normal distribution. In the second part of the activity, you worked with the standard deviation to see the impact of a large data value on this measure of spread. You also compared data from two data sets by looking at z-scores.

ACTIVITY 18.1

Simulations

In this activity, you will explore some ideas of probability by using Excel to simulate tossing a coin and throwing a free throw in basketball.

1. Toss a coin 10 times and after each toss, record in the following table the result of the toss and the proportion of heads so far. For example, suppose you obtain the following sequence of heads and tails for the first five tosses: H T T T H. After the first toss, the proportion of heads so far is one out of one: $\frac{1}{1}$ or 1. After the second toss, the proportion of heads so far is one out of two: $\frac{1}{2}$. After the third toss, the proportion of heads is one out of three: $\frac{1}{3}$. After the fourth toss, the proportion of heads is one out of four: $\frac{1}{4}$. After the fifth toss, the proportion of heads is two out of five: $\frac{2}{5}$.

Toss #	1	2	3	4	5	6	7	8	9	10
H or T?										
Prop of H So Far										

2. On the following axes, plot the proportion of heads so far, for each toss from your table. What does the graph show?

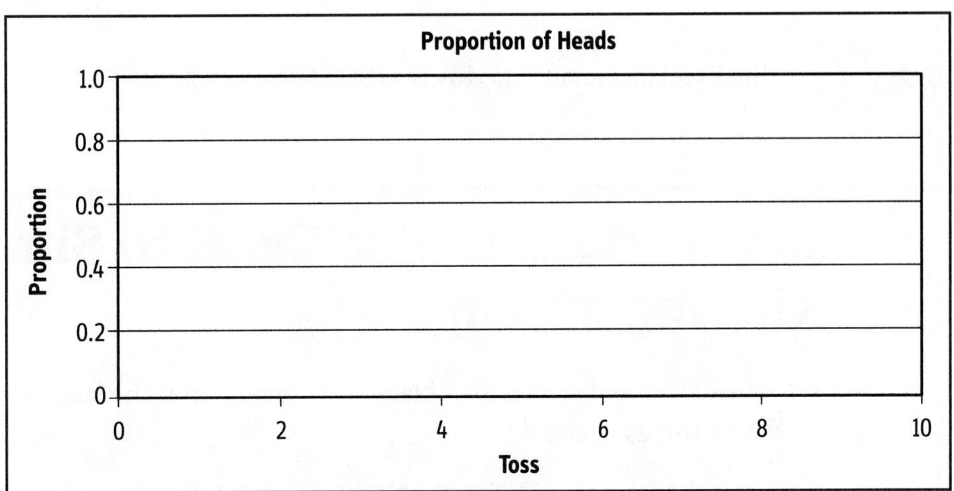

Now, you will use Excel to simulate 1000 independent tosses of a fair coin and plot on a graph the proportion of heads so far after each toss using the instructions that follow #3.

3. In Excel, the function **RAND()** (that is, RAND followed by two parentheses) produces a decimal number between 0 and 1, in such a way that every decimal number between 0 and 1 is equally likely to be produced. You will use the **RAND()** function to generate integers 0 or 1 with equal probability. The integer 1 will signify "heads" and the integer 0 will signify "tails." To get a 0 or 1 with equal probability, you'll multiply the random number by 2 and then take the integer part of it; that is, you will drop all digits after the decimal point.

   Suppose the decimal number produced is 0.13061. What value do you get if you multiply that number by 2 and then take the integer part of it?

   ____

Suppose the decimal number produced is 0.78934. What value do you get if you multiply that number by 2 and then take the integer part of it?

____

The Excel formula to do this is =**INT(2*RAND())**.

Instructions to Use Excel to Simulate Tossing a Coin

a. Open Excel and start with a blank worksheet. Enter the label **Results from 1000 Tosses** in cell A1.

b. Enter the formula =**INT(2*RAND())** in cell A3. Drag this formula down to cell A1002 to generate a column of 1000 0s and 1s, representing 1000 tails and heads.

c. Enter the label **Heads, So Far** in cell B1. In cell B2, enter the value **0**, and in cell B3, enter the formula =**A3+B2**. (This formula will keep a running count of the number of heads so far.) Drag this formula down to cell B1002.

d. Because you want to keep a running count of the proportion of heads, you'll start by recording the number of tosses so far. In cell C1, enter the label **Tosses, So Far**. In cell C3, enter **1**, and in cell C4, enter **2**. Highlight cells C3 and C4; then drag down to cell C1002. You should have a column of integers 1 through 1000.

e. Now fill the cells D3 to D1002 with the value **0.5**, so the graph you construct will have a horizontal line at the height 0.5. (Do this efficiently using autofill.)

f. In cell E1, enter the title **Proportion of Heads, So Far**, and in cell E3 enter =**B3/C3**. Drag this formula down to cell E1002.

g. To display the results of the coin toss simulation on a graph, click the **Chart wizard** button. Select **Line** for **Chart type** and **Line** for **Chart sub-type** (this will be the first sub-type choice). Click **Next**.

h. Now (in step 2 on your Excel screen) for the **Data range** enter **D3:E1002**, or highlight these cells and select the **Series in: Columns** button. Click **Next**.

i. For this next step, on the **Titles** tab, enter a graph title and labels for the axes. On the **Axes** tab, select **Automatic** for **Category (X) axis** and place a check mark in the **Value (Y) axis** box. On the **Gridlines** tab, turn off all gridlines, and on the **Legend** tab, clear the legend.

j. Click **Finish** to complete the graph and format the X and Y axes as you choose.

4. Write a paragraph explaining what your graph shows.

5. Put the cursor in any blank cell near your graph. Press **Ctrl=** to change the random numbers and your graph. Do this several times and describe how the graph changes.

6. Shaquille O'Neal is not particularly good at free throws. He makes about 50 percent of his free throws over an entire season.

   a. Go to sheet 2 and set up a new worksheet to simulate 100 free throws shot independently by a player who has probability 0.5 of making each shot. In column A, keep a record of the attempt number by generating integers 1, 2, . . . , 100 in cells A2 through A101. (Enter an appropriate title in cell A1.)

b. Label cell B1 **Hit or Miss** and generate a random sequence of 1s (each 1 represents a hit) and 0s (each 0 represents a miss) in cells B2 through B101.

c. Compute the overall proportion of hits by calculating the mean of the B column values. Also look at your data and identify the length of the longest streak of hits and the length of the longest streak of misses. Write a paragraph commenting on the proportion of hits and the "streaks."

7. A certain college's men's basketball team is quite accomplished at making free throws. According to the coach, in their most recent season the team made approximately 75 percent of free-throw attempts, and ranked 10th in the country among similar schools in successful free-throw attempts.

   a. Set up another new worksheet to simulate 100 free throws shot by team members who have probability 0.75 of making each shot.

   b. Again, use column A to keep track of the attempt number in cells A2 through A101.

   c. Label cell B1 **Hit or Miss**. Now you will generate a sequence of equally likely occurrences of the numbers 0, 1, 2, or 3 with 0 representing a miss and any of the other three numbers representing a hit. By entering the formula **=INT(4*RAND())** into cell B2, you can set the sequence up so each of the numbers occurs with equal probability. Then autofill to cell B101.

   d. In cell C2, enter the formula **=IF(B2=0,0,1)**. This will give you a "1" in cell C2 if cell B2 recorded a hit, and a "0" if cell B2 recorded a miss. Autofill the formula down to cell C101.

e. Find the overall proportion of hits, and identify the length of the longest streak of hits and the length of the longest streak of misses. Write a paragraph commenting on your proportion of hits and your "streaks."

f. Describe how the "streaks" compare for the 50-percent and 75-percent scenarios.

Summary

In this activity, you learned how to use Excel to simulate random processes. You simulated the random process of tossing a coin and graphed the frequency of heads to visualize the probability that the toss comes up heads. You also used Excel to simulate basketball free throws and analyzed the proportion of hits and streaks of hits and misses.

ACTIVITY 19.1

Diagnostic Testing and Conditional Probability

In this activity, you will investigate how to interpret the results of diagnostic tests and other tests that have some possibility of erroneous results associated with them. You will also explore ways to represent graphically the information given in a two-way table.

1. Suppose you are a varsity athlete at a college that has decided to institute drug testing to screen out those athletes taking drugs. Here are some important questions to consider: How concerned should you be that the test will detect drug use when, in fact, you are not using drugs? What are the chances that the test will report that you are not using drugs if, in fact, you are? How sensitive is the test in picking up real drug usage?

   Here is a review of some terminology associated with such a test.

A test's **sensitivity** refers to the proportion of drug users that the test detects accurately. These are also referred to as **true positives**. (In general, a test's sensitivity to diagnose a certain condition refers to the proportion of cases in which the test actually detects the condition.)

A test's **specificity** refers to the proportion of non-drug users (or, in general, cases that do not have the condition) that the test accurately identifies. These are also referred to as **true negatives**.

Sensitivity and specificity for a particular diagnostic test can be determined from tests run on people known to be using or not using drugs (or known to have a certain condition or not). Of course, the administrators who are testing the student athletes for drug use don't know whether the students are using drugs or not. And if a clinician is testing you for a particular disease, he or she doesn't know whether or not you have the disease. (That's why the test is done!)

Screening tests for drugs, as well as diagnostic tests for diseases, can give false positive or false negative results. A **false positive** occurs when someone is identified by the test as using drugs when, in fact, he or she is not using drugs. A **false negative** occurs when someone is identified by the test as not using drugs when, in fact, he or she is using drugs.

As an athlete, your primary concern would probably be that you might be a false positive; that is, you would be identified as using drugs when you, in fact, do not. The College should be concerned both with false positives (it doesn't want to make false accusations) and false negatives (it wants to identify anyone who is using drugs).

To figure out what proportion of student athletes fall into these categories, it is helpful to set up a two-way table like this:

	Student Uses Drugs	Student Is Drug-Free	ROW TOTAL
Test Is Positive			
Test Is Negative			
COLUMN TOTAL			

Now, suppose the total population of student athletes taking the test is N, and values a, b, c, and d have been entered into the cells of the two-way table as follows:

	Student Uses Drugs	Student Is Drug-Free	ROW TOTAL
Test Is Positive	a	b	
Test Is Negative	c	d	
COLUMN TOTAL			N

a. Explain why the sensitivity is given by the proportion $\frac{a}{a+c}$.

b. Explain why the specificity is given by the proportion $\frac{d}{b+d}$.

c. Explain why the proportion of false positives is given by $\frac{b}{a+b}$.

d. Explain why the proportion of false negatives is given by $\frac{c}{c+d}$.

e. What fraction gives the proportion of student athletes who use drugs? Explain why this is so.

2. Now you'll enter some values in the table, using the following information:

Assume a population of 1000 student athletes takes the test.

It is estimated that 9 percent of all student athletes use drugs.

The test's sensitivity is known to be 95 percent.

The test's specificity is known to be 96 percent.

	Student Uses Drugs	Student Is Drug-Free	ROW TOTAL
Test Is Positive			
Test Is Negative			
COLUMN TOTAL			

a. First, enter the total number of student athletes being tested into the appropriate cell of the table.

b. Given that 9 percent of athletes take drugs, fill in the appropriate total number of students who use drugs; then fill in the number who are drug-free.

c. Use the test's sensitivity to fill in the appropriate cell of the table.

d. Use the test's specificity to fill in the appropriate cell of the table.

e. Fill in the rest of the table.

f. Now use the filled-in table to find the probability that if a student athlete tests positive, he or she is not taking drugs.

g. Find the probability that if a student athlete tests negative, he or she is using drugs.

h. What are the implications (for the athletes, for the coaches, for the school's administration) of your answers to parts f and g of this question?

3. Now suppose that 4 percent of all athletes take drugs, the test's sensitivity is 98 percent, and the test's specificity is 99 percent. Recompute the table for a population of 1,000 athletes using the changed values and find the probabilities asked for in #2f and g. What are the implications?

	Student Uses Drugs	Student Is Drug-Free	ROW TOTAL
Test Is Positive			
Test Is Negative			
COLUMN TOTAL			

4. Think of additional situations in which the results of tests such as the one described here might be used to make important recommendations and/or decisions. How should the information from such tests be used?

You will now look at how to represent conditional probabilities pictorially. Consider the following table of data, which gives the number of physicians in selected specialties in the U.S. in 2001. (Source: *The World Almanac and Book of Facts 2004,* page 82.)

	Female Physicians	Male Physicians
Family Practice	22,391	52,672
Dermatology	3,351	6,476
Ob/Gyn	15,032	26,010
Pediatrics	32,265	32,735
Psychiatry	12,223	27,793

5. Because the two variables (gender and medical specialty) are categorical, you can display these data using a column chart. Retrieve the Excel file "EA19.1 Physicians.xls" from the CD or website and use these instructions to create two column charts.

Instructions to Use Excel to Construct a Stacked Column Chart

a. Highlight the rectangle containing the data and labels and go to the **Chart wizard**. For **Chart type**, choose **Column**, and for **Chart subtype**, choose the second type, **Stacked column**. Click **Next.**

b. In Step 2, construct the graph with series in columns; click **Next**. Add an appropriate chart title at Step 3; click **Finish.**

c. Repeat the steps to construct a stacked-column chart, but this time construct the graph with series in rows. Be sure to title this chart as well.

6. Use your table to answer the following conditional probability questions, and for each question, indicate whether one of your charts would help visualize the probability. If one would help, describe how it helps visualize the probability. Suppose a doctor from the set of doctors represented by this table is chosen at random.

   a. Among all the doctors in pediatrics, what proportion are female? That is, find the probability that the chosen doctor is female, given that he or she is in pediatrics.

b. Among doctors in psychiatry, what proportion are male? That is, find the probability that the chosen doctor is male, given that he or she is in psychiatry.

c. Among all male doctors, what proportion are in psychiatry? That is, find the probability that the chosen doctor is in psychiatry, given that he is male.

d. Among all female doctors, what proportion are in family practice? That is, find the probability that the chosen doctor is in family practice, given that she is female.

Summary

In this activity, you explored the concepts of sensitivity and specificity of a diagnostic test. You used conditional probability and two-way tables to analyze the accuracy of a diagnostic test. As an aid to visualize conditional probabilities, you used Excel to draw stacked-column charts.

ACTIVITY 20.1

Sampling and Surveys

In this activity, you will investigate why random sampling is important and also consider some issues involved in the design of surveys.

1. Here is a list of the 50 United States.

Alabama	Hawaii	Massachusetts	New Mexico	South Dakota
Alaska	Idaho	Michigan	New York	Tennessee
Arizona	Illinois	Minnesota	North Carolina	Texas
Arkansas	Indiana	Mississippi	North Dakota	Utah
California	Iowa	Missouri	Ohio	Vermont
Colorado	Kansas	Montana	Oklahoma	Virginia
Connecticut	Kentucky	Nebraska	Oregon	Washington
Delaware	Louisiana	Nevada	Pennsylvania	West Virginia
Florida	Maine	New Hampshire	Rhode Island	Wisconsin
Georgia	Maryland	New Jersey	South Carolina	Wyoming

a. Using your sense of the land area of each state, choose what *you think* is a representative (that is, your *subjective*) sample of six states. List them in the following table. (It might help to try to visualize a map of the U.S. to get your sample of six "representative" states.)

State	
1.	
2.	
3.	
4.	
5.	
6.	

b. Refer to the table at the end of this activity and record the land area for each of your chosen states, in square miles given to the nearest integer, in the second column of the table. (Source: *Department of Commerce, Bureau of the Census,* www.infoplease.com/ipa/A018355.html.)

c. Compute the sample mean land area for your *subjective* sample of six states and record it here: _________________.

d. Now use Excel's random number generator to generate six random integers between 00 and 49. (Recall how you generated random integers 0, 1, 2, and 3 in Activity 18.1 and adapt that technique to generate integers between 00 and 49. It is possible that you might get repeated numbers when you generate the random integers. If you do, discard the second occurrence of the repeated number and generate another integer so you have six distinct integers.) Describe how you generated the six integers and list them here.

e. Number the states in a systematic way, and use the numbers you generated in part d of this question to pick a *random* sample of six states. Explain your systematic method for numbering the states.

f. Enter the states you chose in the following table and record the land area for each of your states.

State	
1.	
2.	
3.	
4.	
5.	
6,	

g. Compute the sample mean land area for your *random* sample of six states and record it here: ___________.

h. Recall that a *population* refers to the whole group about which you want to draw a conclusion and a *sample* refers to a subgroup of the population. The population mean land area for the population of all 50 states is approximately 70,748 square miles. Count how many of your classmates found *subjective* sample means (in part c of this question) greater than the population mean and how many found *subjective* sample means less than the population mean. Record those values here:

Number of subjective sample means greater than the population mean:

____

Number of subjective sample means less than the population mean:

____

i. Now count how many of your classmates found *random* sample means (in part g of this question) greater than the population mean and how many found *random* sample means less than the population mean. Record those values here:

Number of random sample means greater than the population mean:

____

Number of random sample means less than the population mean:

____

j. Explain how the class results using the two sampling methods (subjective and random) were different, if they were, or how they were similar. Were the class results unexpected?

2. Use Excel to generate a stratified random sample of six states, stratified by location relative to the Mississippi River. (Use the list of states east of the Mississippi given in Topic 20, Exploration 9.) Describe how you generated your stratified random sample of states.

a. List your six states in the following table; refer to the last table of this activity and record the *water area* of each state in your sample.

State	
1.	
2.	
3.	
4.	
5.	
6,	

b. Compute the sample mean water area for your *stratified random* sample of six states and record it here: ___________.

c. Describe how your sample mean water area compares to the population mean water area of approximately 5,133 square miles.

d. Explain why you might want to stratify as you did in part b of this question.

3. Work with one or two classmates on this question. Choose two of the following topics, and for each, discuss how you would design and implement a survey to obtain information about how college students feel about the issue. Be sure to consider what kinds of questions you would ask, how you would obtain truthful answers to sensitive questions, how you would get information from a representative sample, whether you would want to use a stratified sampling technique, how you would avoid bias, and any other factors surrounding sampling that you would need to bear in mind.

a. Students' attitudes about changing to an academic system in which no grades are given for college course work. (In this system, professors provide a written evaluation about each student's work in individual courses but no grades are given.)

b. Students' attitudes about the topic of health insurance coverage for abortions.

c. Students' attitudes about laws that establish a voting age of 18 and a drinking age of 21.

d. Students' attitudes about whether municipalities should contribute to the cost of professional sports stadiums.

e. Students' attitudes about restricting Internet access at public libraries to approved sites.

Summary

In this activity, you investigated why you would want to collect random samples rather than "subjective" samples. You also collected a stratified random sample. To collect these random samples, you used Excel's random number generator. Finally, you considered various aspects of how to design a survey to collect information about students' attitudes on a specific topic.

State	Land Area	Water Area	Total Area
Alabama	50744	1675	52419
Alaska	571951	91316	663267
Arizona	113635	364	113998
Arkansas	52068	1110	53179
California	155959	7736	163696
Colorado	103718	376	104094
Connecticut	4845	699	5543
Delaware	1954	536	2489
Florida	53927	11828	65755
Georgia	57906	1519	59425
Hawaii	6423	4508	10931
Idaho	82747	823	83570
Illinois	55584	2331	57914
Indiana	35867	551	36418
Iowa	55869	402	56272
Kansas	81815	462	82277
Kentucky	39728	681	40409
Louisiana	43562	8278	51840
Maine	30862	4523	35385
Maryland	9774	2633	12407
Massachusetts	7840	2715	10555
Michigan	56804	39912	96716
Minnesota	79610	7329	86939
Mississippi	46907	1523	48430
Missouri	68886	818	69704
Montana	145552	1490	147042

State	Land Area	Water Area	Total Area
Nebraska	76872	481	77354
Nevada	109826	735	110561
New Hampshire	8968	382	9350
New Jersey	7417	1304	8721
New Mexico	121356	234	121589
New York	47214	7342	54556
North Carolina	48711	5108	53819
North Dakota	68976	1724	70700
Ohio	40948	3877	44825
Oklahoma	68667	1231	69898
Oregon	95997	2384	98381
Pennsylvania	44817	1239	46055
Rhode Island	1045	500	1545
South Carolina	30109	1911	32020
South Dakota	75885	1232	77116
Tennessee	41217	926	42143
Texas	261797	6784	268581
Utah	82144	2755	84899
Vermont	9250	365	9614
Virginia	39594	3180	42774
Washington	66544	4756	71300
West Virginia	24078	152	24230
Wisconsin	54310	11188	65498
Wyoming	97100	713	97814
Total	**3,537,379**	**256,641**	

ACTIVITY 21.1

To Purchase a Warranty or Not: Making a Decision

In this activity, you will create and evaluate a decision table that will help you decide whether to purchase a limited warranty, an extended warranty, or no warranty at all for a major electronics purchase. You will consider various decision strategies to help you make your decision.

1. Suppose you just purchased a major piece of electronic equipment and you are considering whether you want to purchase a two-year limited warranty, a full two-year warranty, or no warranty at all.

   a. What additional information do you need to collect to construct a decision table to help you make your decision?

   b. What are your three decision alternatives? (These alternatives will form the rows of the table.)

c. Suppose that a limited warranty costs $100, has a $50 deductible (you pay the first $50 of repairs), and covers repairs up to $500. (So insurance pays a maximum of $450. If repairs cost more than $500, you are responsible for the first $50 and any amount greater than $500.) A full warranty costs $200 and covers all repair costs. Repair records have shown that the two-year repair costs can be grouped into three levels of repairs: minor repairs cost around $150; medium-level repairs cost approximately $500; major repairs cost $800. Fill in the following table, giving your cost for the warranty plus repairs for each decision alternative and each repair level.

	No Repairs Needed	Minor Repairs	Medium-level Repairs	Major Repairs

d. Suppose you want to use the pessimistic (or minimax) decision strategy. What would you decide to do about purchasing a warranty? Explain how you made this decision.

e. Suppose you want to use the optimistic (or minimin) decision strategy. What would you decide to do about purchasing a warranty? Explain how you made this decision.

f. You have a friend who has accessed the repair records for this type of electronic equipment. The following table shows the probability of each level of repairs. Use these probabilities to evaluate the expected cost for each decision alternative. What would you do about purchasing a warranty using the expected cost to decide?

Repair Level	No Repairs	Minor	Medium-level	Major
Probability	0.75	0.1	0.1	0.05

g. Suppose the probability of each level of repairs is as given in the following table. Use these probabilities to evaluate the expected cost for each decision alternative. What would you do about purchasing a warranty using these probabilities and the expected cost to decide?

Repair Level	No Repairs	Minor	Medium-level	Major
Probability	0.5	0.2	0.15	0.15

h. What would you do if you assumed that the four repair levels were equally likely to occur and you used expected cost to make a decision?

i. After considering several decision strategies, if you had to decide to purchase the warranty, what would you do? Explain how you made your decision.

j. In analyzing this decision, you made some simplifying assumptions. Explain how you could enhance your analysis.

Summary

In this activity, you looked at a decision that involved uncertain information. You analyzed it using a variety of decision strategies and obtained different recommendations about what to do. You then had to make and defend a decision, based on the analysis and your personal preferences about decision-making methods.

Excel Commands by Activity

Topic 1

Activity 1.1: World Motor Vehicle Production: Bar Graphs and Pie Charts

Activity 1.2: Medical Data and Class Data: Graphs with Excel

Activity 1.3: SATs and the Super Bowl: Creating and Interpreting Histograms

Topic 2

Activity 2.1: Estimating Dates: Scatterplots

Activity 2.2: State Governors' Salaries and Per Capita Income: More on Scatterplots

Topic 3

Activity 3.1: Temperature Patterns: Functions and Line Graphs

Topic 4

Activity 4.1: Blood Alcohol Levels and Credit Cards: Working with More Than Two Variables

Topic 5

Activity 5.2: Major League Salaries: Rates of Change and Concavity

Topic 6

Activity 6.1: The Genie's Offer: Exponential Growth and Linear Growth

Activity 6.2: Lines of Best Fit

Topic 7

Activity 7.1: Richter Scale and Logarithms

Activity 7.2: Estimations, Scientific Notation, and Properties of Logarithms

Topic 8

Activity 8.1: Measurement Difficulties and Indexes

Activity 8.2: Consumer Indexes

Topic 9

Activity 9.1: Mortgages

Topic 10

Activity 10.1: Savings and Loans: Problem Solving and Using Scroll Bars

Topic 11

Activity 11.1: Ranking Cities: Ratings and Decisions

Topic 14

Activity 14.1: Methods of Apportionment: Quota Methods

Activity 14.2: Apportionment: Divisor Methods

Topic 16

Activity 16.1: Visualizing Football Scores: Measures of Center and Spread

Topic 17

Activity 17.1: Coins, Presidents, and Justices: Normal Distributions and z-Scores

Topic 18

Activity 18.1: Simulations

Topic 19

Activity 19.1: Diagnostic Testing and Conditional Probability

Topic 20

Activity 20.1: Sampling and Surveys

Index

This is the general subjects index. *See also* the Excel Commands by Activity index.

A

E

J

L

M

N

O

P

Q

R

S

T

U

V

W

Z